The Aviation & Aerospace Almanac

1993 Edition

Published by
Aviation Week Group Newsletters
McGraw-Hill
1200 G Street, N.W.
Suite 200
Washington, D.C. 20005

PREFACE

The Aviation/Aerospace Almanac is a reference book whose time is long past due. Its pages were compiled with help of the writers and editors of Aviation Daily and Aerospace Daily, who somehow found time to lend assistance to the project despite an already incredibly busy work day. Dozens of other individuals within the McGraw-Hill family and aviation groups willingly joined with us to collect the reams of data that makes this book such a useful tool for the aviation professional. That's what makes aviation/aerospace so great an industry, the willingness of all of us work together to help one another.

A full list of contributors follows, but special note should be made of some individual efforts, without which this project could not have been completed.

Richard Lampl, copy editor for Aviaton Daily, was the steady hand that held the controls and guided the Almanac from takeoff to landing. Without his special brand of patience and talent for organization we would still be waiting for finished pages to bring to the printer. Working closely with Richard was desktop printing expert, Doug Hyde of HyTech Design, who created a design for the 400-plus pages that brought order to thousands of pieces of data, text, and telephone listings. He somehow tamed the increasingly technologically complex world of the computer to translate into a single usuable format information that was provided from a variety of wordprocessing, spreadsheet and database programs. His work sure saved us a lot of typing.

David Bond, editor of Aviation Daily, Jim Baumgarner, Grier Graham, Frank Jackman and Avery Vise of the staff of that publication, and Richard Tuttle, editor, and Frank Morring, managing editor, of Aerospace Daily contributed their expertise developed over so many years in the aerospace/aviation publishing business.

This first Almanac is a living document. It will grow year after year as new sections and new material are added. As with everything we do at the Aviation Week Newsletters, we solicit your suggestions about format and what other information successive volumes should contain. If you have suggestions, please write to me at Aviation Week Group Newletters, 1200 G St., N.W., Suite 200, Washington, D.C. 20005.

Edmund Pinto
Publisher

ACKNOWLEDGEMENTS

With special thanks to:

Aerospace Industries Association
Airports Council International
Air Transport Association
Aviation Data Service
Aviation Forecasting & Economics
Avitas
AvStat Associates
Bureau of Census
Bureau Of Economic Analysis
Bureau Of Labor Statistics
Department Of The Air Force
Department Of The Army
Department Of Defense
Department Of Commerce
Department Of The Navy
Department Of Transportation
Export-Import Bank Of The U.S.
Federal Aviation Administration
General Aviation Manufacturing Association
International Air Transport Association
International Civil Aviation Organization
International Trade Administration
McGraw-Hill Aerospace Daily
McGraw-Hill Airports Weekly
McGraw Hill Aviation Daily
McGraw-Hill Aviation Week And Space Technology
National Aeronautics And Space Administration
National Science Foundation
Official Airline Guides
Office Of Management And Budget
Regional Airline Association
Strategic Defense Initiative Organization

Note: Statistics may differ slightly due to varying methods and criteria of calculation.

Cover illustration by O'Connell Meier, Alexandria, Va.
Artist, Khiri Rattanasamay
Design work coordinated by Rob Petrie, of McGraw-Hill's Aviation Week Newsletters.

TABLE OF CONTENTS

SECTION II. REGIONAL AIRLINES

SECTION III. GENERAL AVIATION AND HELICOPTERS

SECTION IV. AIRPORTS

SECTION V. THE AEROSPACE INDUSTRY

SECTION VI. DEPARTMENT OF DEFENSE

SECTION VII. U.S.AIR FORCE

SECTION VIII. U.S. ARMY

SECTION IX. U.S.NAVY

SECTION X.   NATIONAL AERONAUTICS AND SPACE ADMINISTRATION (NASA)

SECTION XI. AEROSPACE EXPORT-IMPORT TRADE

Section I.

Airlines

AMERICAN AIRLINES (AMR CORP.)

American Airlines is the largest and arguably most innovative and influential of the U.S. airlines. It was the first to develop the frequent flyer program and the two-tier wage level and was among the first to recognize the strategic importance of computer reservations systems and yield management. American was the catalyst of the industry's phenomenal growth in the 1980s and, in a dratmatic about face, has been among the most aggressive in slashing capital spending and overall downsizing in the 1990s. American is led by probably the most-recognizable figure in commercial aviation — Robert Crandall.

American parent AMR Corp. suffered a net loss of $935 million (including a huge charge to cover an accounting standards change related to employee retirement benefits — FAS 106) and an operating loss of $25 million in 1992, and in the past three years has lost $1.215 billion. In early 1993, American had 102,400 employees and operated a fleet of 673 jet aircraft through a system of seven hubs. Traditionally a domestic carrier, American is currently developing transatlantic and Latin American route networks, but has been unable to get much going in the Asia/Pacific region. American's domestic network is anchored by major hubs at Dallas/Fort Worth and Chicago O'Hare, and smaller hubs at Raleigh/Durham, Nashville, Miami, San Jose, and San Juan, Puerto Rico. The carrier is fed domestically by the American Eagle carriers, most of which are owned by AMR. American is based in Dallas/Fort Worth.

UNITED AIRLINES (UAL CORP.)

United is generally considered the second largest of the U.S. airlines and is almost always measured against the American barometer. Like American, United grew aggressively in the 1980s and has been forced to cut spending and growth plans drastically in the past couple of years. Unlike American, which seemed intent for several years to grow solely on its own, United has been aggressive in buying the foreign assets of failing carriers (such as Pan Am's Pacific route network) and thus has a more well developed international route structure than its rival. United is also led by a relatively high-profile executive, Stephen Wolf.

In 1992, United parent UAL Corp. had a net loss of $956.8 million (including an FAS 106 charge) and an operating loss of $537.8 million, and in the past three years has lost $1.19 billion. At the end of 1992, United had a fleet of 536 jet aircraft and 83,900 employees. United is the dominant carrier at one of the world's busiest airports — Chicago O'Hare, and also hubs at Denver, Washington Dulles, and San Francisco, and has international hubs at London Heathrow and Tokyo Narita. The carrier is fed domestically its United Express operation, which comprises a number of mostly independently-owned regionals. United is based in Chicago.

DELTA AIR LINES

Delta, the third and final member of the so-called Big Three, is often perceived as having a relatively conservative management and — rightly or wrongly— as under pressure to keep pace with its two larger rivals. Like both American and United, it acquired European assets from a failing competitor, but unlike the other two, its Europeans operations are based at Frankfurt instead of London Heathrow. Delta is not usually considered an industry leader in such areas as pricing, but a move by Delta to match a competitor's offering will usually result in the other members of the Big Three following suit. Delta, which still holds its annual stockholders meeting in its birthplace of Monroe, La., is led by Ronald Allen.

Delta suffered a net loss of $564.8 million and an operating loss of $822.1 million in calendar 1992, has lost $958.3 million in the past three years calendar years and has seen its financial situtation severely eroded by its acquisition of Pan Am's transatlantic operations. At the beginning of 1993, Delta, which is the only major to work on a fiscal year of July 1-June 30, had 75,000 employees and a fleet of 561 aircraft. Delta is by far the dominant carrier at its Atlanta hub. The carrier also hubs at Dallas/Fort Worth, Salt Lake City, and Cincinnati, and considers Orlando, Los Angeles and New York Kennedy mini-hubs. The airline has an extensive transatlantic and beyond network based in Frankfurt and a smaller presence in the Pacific. The carrier has an international alliance with Swissair and Singapore Airlines that involves small equity investments by the three in each other. Delta is based in Atlanta.

NORTHWEST AIRLINES (NWA)

Northwest was taken private in a leverage buyout largely financed by Dutch carrier KLM and led Alfred Checchi and his Wings Holdings in 1989. Since then, the carrier has struggled to pay down huge debt while trying to remain on or near the same level as the Big Three. A financial restructuring being worked on at presstime is crucial to the carrier's chances for long-term survival. Another key factor is the success of the operational merger of Northwest and KLM Checchi and Co-Chairman Gary Wilson lead Northwest. KLM, which controls at least 20% of Northwest, has written down to zero the value of its investment. Northwest's financial results for 1992 were not available.

USAIR (USAIR GROUP)

The current USAir is the product of a merger of USAir, Piedmont and Pacific Southwest, which propelled the carrier from a well-thought of East Coast operator to a would-be nationwide competitor. Despite a year of study and preparation, the merger was such an enormous task and represented the mix of such divergent corporate cultures that the airline is still suffering its after-affects. Possibly more significant for USAir and the industry, however, was the investment in USAir of $300 million by British

Airways in early 1993. The deal, one of the most controversial in airline history, may not only result in BA investing another $450 million in USAir in return for an even greater degree of control, but could be the catalyst that forces the U.S. and the U.K. to radically alter Bermuda 2. The chairman of USAir is Seth Schofield. Three BA executives sit on the USAir Group board.

USAir parent USAir Group suffered an operating loss of $1.23 billion (including an FAS 106 charge) and had an operating loss of $336.5 million. In the past three years, USAir has lost $1.99 billion. The carrier has 47,000 employees and a fleet of 449 jet aircraft, all but 11 of which are narrowbodies. USAir has hubs at Charlotte, Pittsburgh, Philadelphia, Baltimore/Washington, and Indianapolis, and operates almost exclusively within North America. The company is based in Arlington, Va., near Washington National.

CONTINENTAL AIRLINES

Continental Airlines, renowned as much for having been the airline of the infamous Frank Lorenzo as for anything else, fell into bankruptcy for the second time in less than a decade in December 1990, but emerged again in the spring of 1993 on the strength of a $450 million invesmtment by an Air Canada-led investor group. The investment deal gave Air Canada a 28.7% stake in Continental, making it one of a growing number of U.S. carriers supported and partially owned by non-U.S. airline interests. SAS, an early pioneer in cross-border investments, lost its holding in Continental, but the two plan to continue their cooperation at Newark. Continental however, is still saddled with an aging fleet and with a reputation for less-than-stellar service that in the past has kept high-revenue business travelers from booking Continental flights.

Continental Airlines Holdings, which disappeared as an entity when the company emerged from Chapter 11, had a net loss of $125.3 million on revenues of $5.58 billion in 1992 and an operating loss of $107.6 million. When the carrier emerged from bankrutpcy in late April 1993, it had 42,000 employees and a fleet of 325 aircraft. The carrier has hubs at Houston Intercontinental, Denver, Newark, Cleveland, Guam and Honolulu. Continental emerged from Chapter 11 as a parent company with three operating operating subsidiaries: Continental Express, the System One CRS and Continental Micronesia, an independently certificated airline that operates mainly in the Pacific Rim. Continental is based in Houston and its chief executive is Robert Ferguson and its new chairman is David Bonderman, one of the principal of Air Partners, which teamed with Air Canada to come up with the $450 million.

TRANS WORLD AIRLINES

TWA, one of world's most widely recognized airlines, has been struggling for years not to end up like another world renowned U.S. airline pioneer — Pan Am. TWA was taken behind the protective screen of Chapter

11 by former Chairman Carl Icahn in early 1992 and has yet to emerge. A new management team took over the airline in 1992 and brought back many former TWA executives who had left during the Icahn years, but the carrier is a longshot for reorganization.

TWA had a net loss of $317.7 million and an operating loss of $404.6 million in 1992. TWA's primary domestic hub is St. Louis, but the carrier has a weak U.S. network relative to the other major carriers. Over the last several years, the airline has sold numerous foreign routes to its competitors in order to stay afloat, and thus its presence in the transatlantic market has declined dramatically. The carrier's European hub is based in Paris. TWA is based in St. Louis, and its co-chief executives are Robin Wilson and Glenn Zander.

SOUTHWEST AIRLINES

Southwest was the only U.S. major airline to post a profit for 1992 and is virtually guaranteed yet another profitable year in 1993. One of the newest majors, Southwest is a study in exploiting and staying within a niche. It is a high-frequency, low fare, low-cost, no-frills carrier that bursts into markets with multiple daily departures and fares so low that Southwest claims to compete with the automobile in the city-to-city travel market. Southwest expands cautiously, but is aggressive once it enters a market. The carrier is a reflection of Chairman Herb Kelleher.

In 1992, Southwest had net earnings of $103.6 million and an operating profit of $182.6 million. In the past three years, Southwest has made $177.6 million. At the end of 1992, Southwest had a fleet of 141 jets — all of them 737s of various models — and 11,397 employees. The carrier's home base is Dallas Love Field and it also has a large operation at Phoenix, where it competes directly against bankrupt America West. To date, Southwest's operations have been strictly domestic, but there has been speculation the carrier is considering expanding into the Mexican market.

AMERICA WEST

America West, which was elevated to major carrier status along with Southwest in 1990, tried to expand too quickly in the late 1980s and early 1990s and was forced into Chapter protection on June 27, 1991. Since then, the airline has slashed unprofitable routes — including its ill-advised 747 service to Japan — and reduced the size of its jet fleet from 123 planes to just 85. The airline was twice able to attract debtor-in-possession financing to say afloat, but as of presstime was still searching for a long-term investor that would allow it to emerge from Chapter 11. The chairman of America West is Phoenix businessman William Franke, who at the behest of Arizona Gov. Fife Symington, lead the effort to secure the second round of DIP financing for the airline.

America West suffered a net loss of $131.8 million in 1992 on revenues of $1.29 billion. In the past three years, the airline has lost $428.4 million.At the end of 1992, America West had the equivalent of 10,233 full-time employees. Its primary hub is Phoenix, but the carrier has smaller operations at Las Vegas and Columbus.

Major World Airlines
International

BRITISH AIRWAYS

British Airways, the self-proclaimed "World's Favourite Airline" has managed to remain profitable, albeit at a lower level, while most of its major international competitors have been losing millions of dollars. In the year ended March 31, 1993, BA posted net earnings of 178 million pounds (US$269 million), down from a profit of 395 million pounds in the previous year. BA has aggressively sought to extend its reach and influence worldwide. In early 1993, BA invested heavily in both Qantas and USAir, ending up with a 25% stake in the former and a 21% interest in the latter. Over an 18-month period to mid-year 1993, BA acquired the assets of failing U.K. regional Dan-Air; acquired 49% of German regional Delta Air and promptly renamed it Deutsche BA, acquired a 49% stake in French regional TAT, and spun off a new low-cost subsidiary, British Airways Regional. But the carrier has been rocked with seemingly endless bad publicity over its alleged dirty tricks campaign against Virgin Atlantic Airways. BA carried 28.1 million passengers during the year, up 10.5% from the previous year.

JAPAN AIRLINES

Japan Airlines, long a model, money-making airline, lost $372.4 million in the fiscal year ended March 31, 1993, and in 1991 and 1992 found itself in the unusual position of having to slash planned capital spending and dramatically restructure its operations to cut costs. The airline even went so far as to suspend its dividend payments. JAL has been rocked by the worldwide recession and by the bursting of Japan's so-called bubble economy, which drastically impacted outbound Japanese traffic. The maturing of the Asia/pacific market has resulted in greater price competition and lower yields. JAL carried 8.2 million passengers on its international routes, a decline of 2% from the previous year, and 15.1 million, an increase of 1%, on its domestic routes during the year.

LUFTHANSA

Lufthansa has been among the hardest hits of the major European airlines, finding itself buffeted not only be the worldwide recession of the early 1990s, but by the adverse affects of reunification on Germany's economy. The airline, which is about 52% government owned, lost 297 million Deutsche marks in 1992 following a 1991 loss of 331 million marks. In addition to the bad economy, Lufthansa blamed the results on fierce price competition and currency fluctuations. In an effort to cut costs, the airline has slashed thousands of jobs, implemented a wage freeze, dropped unprofitable routes and farmed out its domestic operations to a subsidiary.

ALL NIPPON AIRWAYS

All Nippon Airways, likes it main Japanese rival Japan Airlines, has been severely impacted by Japan's economic problems. But unlike JAL, which is primarily an international operator, ANA, has been somewhat insulated from the perils of the international markets, where it is a relative newcomer, by its extensive domestic network. ANA managed a net profit of $20.9 million (2.42 billion yen) in the fiscal year ended March 31, 1993, and projected fiscal 1994 net earnings of $22.4 million (2.6 billion yen). Despite its profitability, ANA said it will be restructuring its routes with an eye to maintaining sound margins in both domestic and international operations. ANA carried 33.3 million passengers on domestic routes and 1.6 million on international routes in fiscal 1993.

AIR FRANCE

The French government revealed in mid-1993 that Air France is on a list of 21 state-held companies targeted for privatization. The carrier will have to return to profitability first, and that, according to Chairman Bernard Attali, is likely to take at least two years. In 1992, Air France lost US$595 million (3.2 billion francs). The airline, like many of its European and U.S. competitors, has been forced to layoff thousands and deny salary increases in an effort to control costs. In 1992, UTA French Airlines, which had been previously purchased was merged into Air France. The Groupe Air France also holds a 75% stake in Air Inter, the country's dominant domestic carrier. Air France has moved to increase its foreign reach by buying into both Sabena and CSA, the airline of the Czech Republic. Air France has a marketing alliances with Air Canada and Aeromexico, and could eventually sign a similar deal with Continental, which is 24% owned by Air Canada. In 1992, the Groupe Air France airlines carried 32.7 million passengers, up 4% from the previous year.

SINGAPORE AIRLINES

Singapore Airlines, one of the world's most consistently profitable airlines, posted an after-tax profit of US$518.5 million (S$850.6 million) for the year ended March 31, 1993, a decline of about 8.4%. The airline attributed its decline in profitability to the worldwide recession, intense airline industry competition, and the strong Singapore dollar. Singapore Airlines is in a three-way alliance with Delta and Swissair. The three partners have taken small equity stakes in each other to cement the alliance. SIA lost out to British Airways in the bidding for a minority stake in Qantas, but is said to be interested in buying a piece of privately held Ansett Airlines of Australia, which will launch its first international operations within the new few years. SIA carried 8.6 million passengers in fiscal 1992-93, an increase of 6.3% from the previous year.

KLM

KLM lost US$313 million (562 million guilders) in the year ended March 31, 1993, largely because of its investment in Northwest. The struggling U.S. carrier's continuing losses became such a drag on KLM's earnings during 1992 that the Dutch carrier wrote down the value of the $400 million investment to zero. KLM is not only involved in an ongoing operational merger with Northwest, it is talking to the three European Quality Alliance partners — SAS, Swissair and Austrian Airlines — about the formation of a jointly owned management company to run the operations of the four airlines. A four-way merger is likely down the road.

QANTAS

Qantas had an after-tax profit of A$137.7 million in the year ended June 30, 1992. In late calendar 1992, the government-owned international carrier acquired 100% of domestic carrier Australian Airlines. In early 1993, British Airways acquired a 25% stake in the merged entity and the government injected A$1.35 billion into the carrier to pay down debt. The remaining 75% of the government's holding in Qantas is scheduled to sold to the public by late 1993 or early 1994. The privatization is part of an overall liberalization of the Australian air transport market that will see domestic carrier Ansett eventually be given international authority. Qantas carried 4.5 million passengers in fiscal 1992.

CATHAY PACIFIC AIRWAYS

Cathay Pacific Airways posted a 1992 profit of US$388.8 million (HK$3.01 billion), a slight increase over the previous year. Cathay is renowned for its service, but its reputation was tarnished in early 1993 by a flight attendants' strike during the high season. The carrier's future is somewhat murky because of Hong Kong's scheduled mid-1997 reversion to Chinese possession and a bitter dispute between China and the U.K. over the financing of the huge Chek Lap Kok Airport project.

ALITALIA

Alitalia, the Italian-flag carrier, lost 16.8 billion lira in 1992, but that was an improvement over 1991, when the airline lost 43.7 billion lira. Like most other European airlines, Alitalia spent much of 1991 and 1992 cutting costs in order to remain competitive with more efficient U.S. and Asian carriers. In 1992, Alitalia acquired a 30% share in Malev Hungarian Airline.

IBERIA

Iberia is developing a hub in Miami to connect its European operations Latin America, where it has extensive holdings in local airline companies. Iberia lost 34.8 billion pestas in 1992, and was forced to freeze some employee wages in 1993.

AIR CANADA

Air Canada lost US$360.1 million (C$454 million) after a 1991 net loss of C$218 million. Air Canada, as do many observers, believes that Canada cannot support two international airlines. Air Canada has suggested a merger with rival Canadian Airlines International, but as of mid-1993, that had not happened. Air Canada underwent a major restructuring in 1992, and signed commercial agreements with United and Air France. In late April 1993, Air Canada bought a 24% interest in Continental. Air Canada projects that it will post a net profit in 1994.

KOREAN AIR

For a long time South Korea's only international carrier, Korean Air now has to share the spotlight with Asiana Airlines. In 1990, the most recent year for which statistics were available, Korean Air had a net loss of $10.9 million, and carried 11.3 million passengers.

CANADIAN AIRLINES INTERNATIONAL

Canadian Airlines International parent company PWA Corp. lost US$430.9 million (C$543.3 million) in 1992, including C$333 million in restructuring costs related to the company's efforts to restructure is financial obligations and complete a transaction with American Airlines that would see the U.S. carrier take a significant minority interest in Canadian. Indeed, the prospects for Canadian's survival largely depend on whether the American transaction is completed. Without it, Canadian will likely be forced into a merger with Air Canada or will fold. Conversely, if the American deal does goes through, Air Canada may well have to find itself a U.S. backer.

VARIG

Varig is one of the oldest airlines in the world, having been founded in May 1927, and is South America's largest and arguably most well-known. In 1991, the most recent year for which results were available, the Brazilian carrier had a net loss of $137.6 million, and a larger loss was expected for 1992. In late 1992, however, airline officials said they expected Varig to return to profitability again in 1993.

SAS

SAS, which had a pre-tax loss of 743 million Swedish kronor in 1992, has undergone a rigorous cost-cutting campaign since the beginning of the 1990s and is fast becoming one of Europe's most efficient carriers. SAS is 50% owned by the governments of Sweden, Norway and Denmark, and 50% held by private investors. SAS and its chief executive, Jan Carlzon, have long been advocates of multi-carrier alliances and at presstime were in the midst of negotiations to build what would be one of Europe's largest airlines through the aligning of SAS, KLM, Swissair and Austrian Airlines. SAS carried 14.5 million passengers in 1992, up 4.5% from the previous year.

ATA MEMBER AIRLINES - 1993

Alaska Airlines
P.O. Box 68900
Seattle-Tacoma Int'l Airport
Seattle, Washington 98168-0900
(206) 433-3200

Aloha Airlines
P.O. Box 30028
Honolulu, Hawaii 96820-0228
(808) 836-4101

American Airlines
P.O. Box 619616
DFW Airport, Texas 75261-9616
(817) 963-1234

American Trans Air
P.O. Box 51609
Indianapolis International Airport
Indianapolis, Indiana 46251-0609
(317) 247-4000

Continental Airlines
Suite 1501
2929 Allen Parkway
Houston, Texas 77019
(713) 834-5000

Delta Air Lines
Hartsfield Atlanta International Airport
Atlanta, Georgia 30320-9998
(404) 715-2600

DHL Airways
P.O. Box 75122
Cincinnati, Ohio 45275
(606) 283-2232

Evergreen International Airlines
3850 Three Mile Lane
McMinnville, Oregon 97128-9496
(503) 472-0011

Federal Express
P.O. Box 727
Memphis, Tennessee 38116
(901) 369-3600

Hawaiian Airlines
P.O. Box 30008
Honolulu International Airport
Honolulu, Hawaii 96820-0008
(808) 525-5511

Northwest Airlines
Minneapolis-St. Paul Int'l Airport
St. Paul, Minnesota 55111-3075
(612) 726-2111

Reeve Aleutian Airways
4700 West International Airport Road
Anchorage, Alaska 99502-1091
(907) 243-1112

Southwest Airlines
Box 36611, Love Field
Dallas, Texas 75235-1625
(214) 904-4000

Trans World Airlines
100 South Bedford Road
Mt. Kisco, New York 10549
(914) 242-3000

United Airlines
P.O. Box 66100
Chicago, Illinois 60666-0100
(708) 952-4000

United Parcel Service
400 Perimeter Center
Terraces North
Atlanta, Georgia 30346
(404) 913-6000

USAir
2345 Crystal Drive
Crystal Park 4
Arlington, Virginia 22227
(703) 418-7000

Associate Members

Air Canada
P.O. Box 14000
Air Canada Center
St. Laurent, Quebec
Canada H4Y 1H4
(514) 422-5000

Canadian Airlines International
Suite 2800, 700 2nd St., S.W.
Calgary, Alberta
Canada T2P 2W2
(403) 294-2000

AIRCRAFT OPERATING STATISTICS - 1992
Figures are averages for most commonly used models

	Number of Seats	Speed Airborne	Flight Length	Fuel (Gallons Per Hour)	Aircraft Operating Costs Per Hour
B747-400	403	534	4,375	3,361	$7,098
B747-100	399	520	3,094	3,507	5,905
L-1011-100/200	293	495	1,519	2,362	4,072
DC-10-10	282	488	1,387	2,189	4,056
A300-600	266	474	1,206	1,686	3,917
DC-10-30	265	522	2,962	2,607	4,595
B767-300	224	489	2,087	1,485	3,384
B757-200	188	458	1,105	996	2,293
B767-200	187	483	2,064	1,434	2,956
A320-100/200	149	441	944	757	1,868
B727-200	148	430	689	1,242	2,263
B737-400	145	400	564	769	1,743
MD-80	142	419	662	889	1,842
B737-300	131	415	633	739	1,768
DC-9-50	122	387	453	890	1,640
B727-100	117	429	638	1,137	2,220
B737-100/200	112	387	444	793	1,735
DC-9-30	102	381	441	794	1,658
F-100	99	361	388	690	1,445
DC-9-10	77	361	359	727	1,439

LEADING U.S. AIRPORTS -- 1992
PASSENGERS (Arriving & Departing)

Chicago O'Hare	64,441,087	Minneapolis/St. Paul	22,907,490
Dallas/Ft. Worth	51,943,567	Detroit	22,840,643
Los Angeles	46,964,555	Honolulu	22,608,188
Atlanta	42,032,988	Phoenix	22,111,555
San Francisco	31,789,021	Orlando	21,147,888
Denver	30,877,180	St. Louis	20,984,782
New York (JFK)	27,760,912	Las Vegas	20,913,054
Miami	26,483,717	New York (LGA)	19,662,521
Newark	24,286,986	Houston	19,354,013
Boston	22,988,717	Pittsburgh	18,748,884

CARGO TONS (Enplaned & Deplaned)

Memphis	1,413,042	Dayton	450,057
New York (JFK)	1,317,961	Indianapolis	436,834
Los Angeles	1,238,198	Honolulu	380,851
Chicago O'Hare	1,115,116	Boston	371,536
Miami	1,002,242	Philadelphia	366,283
Atlanta	649,284	Seattle	361,607
San Francisco	602,741	Oakland	361,439
Newark	587,753	Denver	327,264
Anchorage	580,369	Minneapolis/St. Paul	301,607
Dallas/Ft. Worth	577,318	Baltimore/Washington	284,818

Source: Airports Council International

OPERATING FLEET - ATA AIRLINES
As of December 31, 1992

	AS	TS	AA	TZ	CO	DL	DH	EZ	FM	HA	NW	RV	WN	TW	UA	UP	US	AC	CA	TOTAL
B-747					7			16	12		49			11	55	11		9	3	173
MD-11			11			9			8											28
DC-10			59		20				30		29				54				8	200
L-1011				13		56				5				19						93
B-767			58			51								9	34		11	21	12	196
A300			34		21															55
A310						21														21
DC-8								3		6						49		5		63
B-757			69	6		80					33				72	30	11			301
A320											40							31	10	81
B-727	17		142	7	89	153	17	11	156		68	2		58	104	48	8			880
MD-80	38		260		64	110					8			33			31			544
DC-9					34	3		8		14	139			46			73	35		352
B-737	17	22			89	71							141		217		243		48	848
F-100			39														40			79
F-28																	33			33
F-27									32											32
L-188												3								3
YS-11												3								3
DHC-7										4										4
SA-227						10														10
Cessna 208						18			216											234
TOTALS	**72**	**22**	**672**	**26**	**324**	**554**	**45**	**38**	**454**	**29**	**366**	**8**	**141**	**176**	**536**	**138**	**450**	**101**	**81**	**4233**

AIRLINE CODES:

AS	Alaska	EZ	Evergreen	UA	United
TS	Aloha	FM	Federal Express	UP	United Parcel Service
AA	American	HA	Hawaiian	US	USAir
TZ	American Trans Air	NW	Northwest	AC	Air Canada
CO	Continental	RV	Reeve Aleutian	CA	Canadian
DL	Delta	WN	Southwest		
DH	DHL	TW	Trans World		

1992 AIR TRAVEL SURVEY

	1982	1991	1992
Percent of adults who have:			
Ever Flown	65%	76%	76%
Flown in past 12 months	24	32	33
Purpose of trips taken:			
Business Trips	52	46	37
Pleasure/Personal Trips	48	54	63

Source: Gallup/ATA Survey

FAA AVIATION FORECASTS
Commercial Air Carriers FY 1993-2004

Year	Passengers (millions)	RPMs (billions)	Jet Aircraft	Departures (millions)
1993	483.2	494.7	4,265	6.4
1994	508.7	525.4	4,311	6.5
1995	533.6	555.5	4,387	6.6
1996	557.2	584.9	4,496	6.8
1997	579.4	613.4	4,553	6.9
1998	601.4	642.4	4,690	7.1
1999	623.9	671.4	4,843	7.3
2000	646.3	700.9	5,016	7.5
2001	669.1	731.0	5,180	7.7
2002	691.8	761.6	5,381	7.8
2003	715.0	793.1	5,566	8.0
2004	738.0	824.6	5,747	8.1

BALANCE SHEET ($000)
U.S. Scheduled Airlines

	1991	1992
Assets		
Current Assets	14,883,193	15,638,459
Investments and special funds	5,152,466	4,888,603
Flight equipment owned	42,896,537	46,995,108
Ground equipment & property	14,285,452	15,066,677
Reserve for depreciation (Owned)	(22,008,987)	(24,357,643)
Leased equipment & property capitalized	6,938,853	6,782,829
Reserve for depreciation (Leased)	(2,669,286)	(2,563,799)
Other property	4,796,321	6,536,571
Deferred charges	6,057,893	6,180,148
Total assets	**75,166,953**	**0**
Liabilities		
Current liabilities	24,022,967	23,920,983
Long-term debt	11,949,986	13,893,377
Other non-current	13,619,278	19,689,746
Deferred credit	7,521,409	7,729,689
Stockholders' equity--net of Treasury Stock	13,218,802	9,933,158
Preferred stock	269,072	270,394
Common stock	259,632	318,637
Other paid-in capital	8,326,857	9,616,906
Retained earnings	4,713,130	55,419
Less: Treasury Stock	349,889	328,198
Total liabilities and stockholders' equity	**70,332,442**	**75,166,953**

U.S. SCHEDULED AIRLINES

Majors	**Nationals**	**Regionals**	
(Annual revenues of over $1 billion)	**(Annual revenues of $100 million to $1 billion)**	Aerial Air Transport Int'l Airline of the Americas	MGM Grand Miami Air Millon
America West	Air Wisconsin	Airmark	North American
American	Alaska	American Int'l	Northern Air
Continental	Aloha	Amerijet	Patriot
Delta	American Trans Air	Arrow	Private Jet
Federal Express	Emery	AV Atlantic	Reeve
Northwest	Evergreen	Braniff Int'l	Reno
Southwest	Hawaiian	Buffalo	Rich
Trans World	Horizon Air	Carnival	Ryan International
United	Markair	Casino Express	Sierra Pacific
USAir	Midwest Express	Challenge Air Cargo	Simmons
	Southern Air	Executive Airlines	Spirit Air
	Tower	Express One	Sun Country
	United Parcel Service	Fine Airlines	Trans Air Link
	USAir Shuttle	Florida West	Trans Continental
	Westair	Great American	Trans States
	World	Int'l Cargo Xpress	Wilbur's
		Jet Fleet	Worldwide
		Key	Wrangler
		Kiwi	Zantop

EMPLOYMENT
U.S. Scheduled Airlines

	1982	1991	1992
Pilots and Copilots	28,144	49,232	50,852
Other Flight Personnel	6,900	8,033	7,235
Flight Attendants	50,860	81,794	84,071
Mechanics	43,393	58,819	58,670
Aircraft and Traffic Service Personnel	87,813	237,292	240,518
Office Employees	66,997	44,304	43,245
All Other	46,388	54,091	55,821
Total Employment	330,495	533,565	540,412

Average Compensation per Employee

	1982	1991	1992
Salaries and Wages	$31,561	$39,395	$40,576
Benefits and Pensions	5,245	7,202	8,583
Payroll Taxes	2,015	2,921	3,045
Total Compensation	$38,821	$49,518	$52,204

PASSENGER YIELD
Revenue per Passenger Mile (¢)

	1982	1991	1992
Domestic	12.8	13.2	12.9
International	10.2	11.3	11.5
Total	12.3	12.7	12.5

FREIGHT AND EXPRESS YIELD
Revenue per Freight & Express Ton Mile (¢)

	1982 *	1991	1992
Domestic	46.8	64.8	64.7
International	36.8	43.6	43.7
Total	42.7	53.9	53.6

* Does not include Federal Express

ATA AIRLINE STATISTICS -- 1992

	Number of Aircraft	Employees	Aircraft Departures	Passengers (000)	Revenue Passenger Miles (000)	Passenger Revenues ($000)	Cargo Revenues ($000)	Total Operating Revenues ($000)	Operating Profit/ (Loss) ($000)	Net Profit/ (Loss) ($000)
Alaska	72	6,729	116,117	6,167	5,477,411	777,432	69,076	922,393	(108,941)	(81,101)
Aloha	22	2,015	79,022	4,662	627,578	170,675	29,697	206,335	9	(6,687)
American	672	91,185	931,098	85,955	97,109,980	11,891,244	576,802	13,581,102	(77,204)	(734,865)
American Trans Air	26	2,359	3,608	480	751,376	385,126	--	421,790	3,246	(2,140)
Continental	324	34,975	481,065	38,359	43,071,177	4,636,720	244,456	5,209,639	(194,523)	(299,429)
Delta	554	79,157	1,010,058	82,911	80,498,837	10,726,879	661,841	11,639,080	(825,508)	(564,826)
DHL	45	NA	NA	NA	NA	NA	NA	NA	NA	NA
Evergreen *	38	729	10,621	--	--	--	213,965	332,666	27,056	(7,692)
Federal Express	454	84,161	262,198	--	--	--	2,963,242	7,712,065	269,751	(83,915)
Hawaiian	29	2,840	67,122	4,645	3,319,254	343,387	15,287	399,725	(84,679)	32,866
Northwest	366	44,695	536,666	43,443	58,225,038	6,903,441	736,172	7,963,785	(308,804)	(386,206)
Reeve Aleutian	8	279	3,510	57	44,437	13,841	8,465	24,846	(5,626)	(2,678)
Southwest	141	10,975	438,190	31,023	13,788,124	1,623,828	33,088	1,685,250	181,783	103,559
Trans World	176	27,830	272,504	22,292	28,882,501	2,953,216	159,176	3,569,923	(369,496)	(317,653)
United	536	78,566	720,592	66,596	92,486,440	11,204,332	793,343	12,724,577	(496,470)	(933,285)
United Parcel Service *	138	2,855	80,254	--	--	--	1,173,398	1,174,204	14,669	7,110
USAir	450	45,692	916,371	54,654	35,096,736	5,785,830	169,534	6,235,622	(375,545)	(1,223,384)

Associate Members

	Number of Aircraft	Employees	Aircraft Departures	Passengers (000)	Revenue Passenger Miles (000)	Passenger Revenues ($000)	Cargo Revenues ($000)	Total Operating Revenues ($000)	Operating Profit/ (Loss) ($000)	Net Profit/ (Loss) ($000)
Air Canada	101	19,448	148,218	9,900	14,391,000	2,813,000	342,000	3,501,000	(145,000)	(454,000)
Canadian	81	15,000	NA	8,137	13,324,018	2,061,000	209,700	2,441,500	(212,000)	(396,200)

* Includes non-scheduled service   NA = Not Available

AIR TRANSPORT 1993
The Annual Report of the U.S. Scheduled Airline Industry
1992 HIGHLIGHTS

TRAFFIC	1991	1992	Percent Change
Passengers Enplaned (000)	452,301	473,305	4.6
Domestic Service	412,360	429,900	4.3
International Service	39,941	43,405	8.7
Revenue Passenger Miles (000)	447,954,829	478,081,118	6.7
Available Seat Miles (000)	715,199,140	751,815,137	5.1
Passenger Load Factor (%)	62.6	63.6	
Aircraft Departures	6,782,782	6,866,325	1.2
Cargo Revenue Ton Miles (000)	12,129,963	13,053,681	7.6
Freight and Express Revenue Ton Miles (000)	10,225,199	10,989,344	7.5
Mail Revenue Ton Miles (000)	1,904,764	2,064,337	8.4
Total Revenue Ton Miles (000)	56,925,444	60,861,792	6.9

FINANCIAL	1991	1992	Percent Change
Passenger Revenue ($000)	57,091,675	59,747,693	4.7
Domestic Service ($000)	44,032,884	44,684,618	1.5
International Service ($000)	13,058,791	15,063,075	15.3
Freight and Express Revenues ($000)	5,508,572	5,891,390	6.9
Mail Revenues ($000)	957,077	1,174,063	22.7
Total Operating Revenues ($000)	75,158,493	77,949,558	3.7
Total Operating Expenses ($000)	76,943,234	80,333,091	4.4
Operating Profit ($000)	(1,784,741)	(2,383,533)	
Net Profit ($000)	(1,940,157)	(4,028,452)	
Rate of Return on Investment (%)	-0.5	-6.4	
Operating Profit Margin (%)	-2.4	-3.1	
Net Profit Margin (%)	-2.6	-5.2	

Customized airline data services are available from the ATA Office of Industry Data.

INCOME STATEMENT ($000)
U.S. Scheduled Airlines

	1991			1992		
	Domestic	International	Total	Domestic	International	Total
Operating Revenues:						
Passenger	44,032,884	13,058,791	57,091,675	44,684,618	15,063,075	59,747,693
Freight and Express	3,206,194	2,302,378	5,508,572	3,358,957	2,532,433	5,891,390
Mail	734,376	222,701	957,077	930,417	243,646	1,174,063
Charter	1,841,604	1,875,754	3,717,358	1,760,406	976,011	2,736,417
Public Service Revenue	3,083	132	3,215	3,220	296	3,516
Other	6,412,292	1,468,304	7,880,596	6,814,993	1,581,486	8,396,479
Total Operating Revenues	56,230,433	18,928,060	75,158,493	57,552,611	20,396,947	77,949,558
Operating Expenses:						
Flying Operations	16,830,585	5,636,309	22,466,894	17,116,921	5,813,071	22,929,992
Maintenance	6,682,050	2,152,099	8,834,149	6,858,443	2,132,783	8,991,226
Passenger Service	5,068,080	1,861,276	6,929,356	5,339,652	2,198,051	7,537,703
Aircraft & Traffic Servicing	9,140,387	2,831,394	11,971,781	9,780,522	3,236,958	13,017,480
Promotion and Sales	8,855,936	4,602,325	13,458,261	8,930,482	5,220,857	14,151,339
Administrative	2,874,630	977,086	3,851,716	2,966,533	876,804	3,843,337
Transport Related	4,089,505	1,232,561	5,322,066	4,330,086	1,170,695	5,500,781
Depreciation and Amortization	3,216,986	892,025	4,109,011	3,336,398	1,024,835	4,361,233
Total Operating Expenses	56,758,159	20,185,075	76,943,234	58,659,037	21,674,054	80,333,091
Operating Income or (Loss)	(527,726)	(1,257,015)	(1,784,741)	(1,106,426)	(1,277,107)	(2,383,533)
Other Income or (Expense)						
Interest Expense	(1,349,103)	(427,891)	(1,776,994)	(1,346,591)	(379,422)	(1,726,013)
Income Taxes	66,209	317,643	383,852	193,970	577,350	771,320
Other	529,452	708,274	1,237,726	(463,212)	(227,014)	(690,226)
Net Profit or (Loss)	(1,281,168)	(658,989)	(1,940,157)	(2,722,259)	(1,306,193)	(4,028,452)
Operating Profit Margin (%)	-0.9	-6.6	-2.4	-1.9	-6.3	-3.1
Net Profit Margin (%)	**-2.3**	-3.5	-2.6	-4.7	-6.4	-5.2

TOP 30 DOMESTIC AIRLINE MARKETS*
Passengers--Outbound plus Inbound
(Twelve Months Ended December, 1992)

1	New York	Los Angeles	2,904,060	16	Honolulu	Lihue, Kauai	1,273,460
2	New York	Boston	2,350,240	17	Los Angeles	Las Vegas	1,217,420
3	New York	Chicago	2,330,750	18	Chicago	Los Angeles	1,188,980
4	New York	Washington	2,282,480	19	New York	West Palm Beach	1,187,110
5	Los Angeles	San Francisco	2,153,360	20	Los Angeles	Honolulu	1,165,470
6	New York	Miami	2,142,690	21	San Francisco	San Diego	1,058,870
7	Dallas/Ft.Worth	Houston	2,090,390	22	Honolulu	Kona, Hawaii	1,055,760
8	Honolulu	Kahului, Maui	2,035,100	23	Los Angeles	Oakland	1,042,880
9	New York	San Francisco	2,010,350	24	New York	Dallas/Ft. Worth	1,029,860
10	New York	Orlando	1,727,260	25	Honolulu	Hilo, Hawaii	1,019,250
11	New York	Ft. Lauderdale	1,547,970	26	Chicago	St. Louis	973,580
12	New York	San Juan	1,524,400	27	Boston	Washington	952,450
13	Los Angeles	Phoenix	1,402,160	28	Los Angeles	Washington	925,220
14	Chicago	Detroit	1,339,600	29	San Francisco	Honolulu	880,390
15	New York	Atlanta	1,330,730	30	Chicago	Washington	875,740

* Includes all commercial airports in a metropolitan area. Does not include connecting passengers.

Source: DOT Origin/Destination Survey.

AIRCRAFT ON ORDER - ATA AIRLINES
As of December 31, 1992

| | NUMBER | | FIRM ORDER DELIVERY DATES | | | |
AIRCRAFT TYPE	FIRM	OPTIONS	1993	1994	1995	1996 & Beyond
Airbus						
A300	26	50	1	5	6	14
A310	9		1	8		
A320	73	70	23	21	15	14
A330	36	20				36
Boeing						
B-737	272	396	47	53	32	140
B-747	18	51	6			12
B-757	169	212	44	29	33	63
B-767	69	96	22	7	10	30
B-777	34	34			11	23
Fokker						
100	36	75	21	15		
McDonnell Douglas						
MD-80	27	18	14	6	7	
MD-90	60	130		4	6	50
MD-11	21	55	15	2	4	
TOTALS	850	1207	194	150	124	382

The value of firm aircraft orders was $42.8 billion.

SAFETY
U.S. Air Carriers
Scheduled Service

	Departures (Millions)	Fatal Accidents	Fatalities	Fatal Accidents Per 100,000 Departures
1982	5.0	5	235	0.081*
1983	5.0	4	15	0.079
1984	5.4	1	4	0.018
1985	5.8	7	197	0.120
1986	6.4	3	8	0.031*
1987	6.6	5	232	0.061*
1988	6.7	3	285	0.030*
1989	6.6	11	278	0.166
1990	6.9	6	39	0.087
1991	6.8	4	62	0.059
1992	6.9	4	33	0.058

* Sabotage caused accidents are included in Accidents and Fatalities but not in the Accident Rates.

Source: National Transportation Safety Board

TRAFFIC AND OPERATIONS DATA
U.S. Scheduled Airlines

	1991			1992		
	Domestic	International	Total	Domestic	International	Total
Passenger Traffic -- Scheduled Service						
Revenue Passengers Enplaned (000)	412,360	39,941	452,301	429,900	43,405	473,305
Revenue Passenger Miles (000)	332,565,881	115,388,948	447,954,829	347,502,877	130,578,241	478,081,118
Available Seat Miles (000)	543,637,976	171,561,164	715,199,140	557,103,106	194,712,031	751,815,137
Revenue Passenger Load Factor (%)	61.2	67.3	62.6	62.4	67.1	63.6
Average Length of Haul (Miles)	806	2,889	990	808	3,008	1,010
Cargo Traffic (Revenue Ton Miles) -- Scheduled Service						
Total (000)	6,357,695	5,772,268	12,129,963	6,759,130	6,294,551	13,053,681
Freight and Express (000)	4,946,052	5,279,147	10,225,199	5,191,094	5,798,250	10,989,344
U.S. Mail (000)	1,411,643	493,121	1,904,764	1,568,036	496,301	2,064,337
Overall Traffic and Operations Data						
Total revenue ton miles - Charter service (000)	3,053,970	2,499,935	5,553,905	3,625,080	1,975,303	5,600,383
Total revenue ton miles - All services (000)	42,668,248	19,811,099	62,479,347	45,134,497	21,327,680	66,462,177
Total Available ton miles - All services (000)	80,879,199	35,495,307	116,374,506	83,695,875	38,146,924	121,842,799
Ton mile load factor - All services (%)	52.8	55.8	53.7	53.9	55.9	54.5
Revenue Aircraft Departures - Scheduled services	6,419,407	363,375	6,782,782	6,477,140	389,185	6,866,325
Revenue Aircraft Miles - Scheduled services (000)	3,708,477	707,137	4,415,614	3,793,522	825,595	4,619,117
Revenue Aircraft Hours - Scheduled services	9,140,777	1,433,052	10,573,829	9,283,984	1,659,760	10,943,744

TOP U.S. 25 AIRLINES IN 1992

	Passengers			Revenue Passenger Miles			Freight Ton Miles			Total Operating Revenues
	(000)			(000)			(000)			($000)
1	American	85,955	1	American	97,109,980	1	Federal Express	3,953,227	1	American	13,581,102
2	Delta	82,911	2	United	92,486,440	2	Northwest	1,845,580	2	United	12,724,577
3	United	66,596	3	Delta	80,498,837	3	United	1,321,732	3	Delta	11,639,080
4	USAir	54,654	4	Northwest	58,225,038	4	American	1,169,714	4	Northwest	7,963,785
5	Northwest	43,443	5	Continental	43,071,177	5	Delta	819,836	5	Federal Express	7,712,065
6	Continental	38,359	6	USAir	35,096,736	6	Continental	546,702	6	USAir	6,235,622
7	Southwest	31,023	7	Trans World	28,882,501	7	Trans World	366,046	7	Continental	5,209,639
8	Trans World	22,292	8	Southwest	13,788,124	8	United Parcel Service	356,760	8	Trans World	3,569,923
9	America West	15,118	9	America West	11,654,954	9	Challenge Air Cargo	157,317	9	Southwest	1,685,250
10	Alaska	6,167	10	Alaska	5,477,411	10	USAir	142,396	10	America West	1,302,632
11	Aloha	4,662	11	Hawaiian	3,319,254	11	Arrow	51,977	11	United Parcel Service	1,174,204
12	Hawaiian	4,645	12	Tower	1,605,320	12	America West	50,456	12	Alaska	922,393
13	Air Wisconsin	2,864	13	American Trans Air	751,376	13	Alaska	49,436	13	American Trans Air	421,790
14	Horizon Air	2,380	14	Midwest Express	661,566	14	Zantop	37,557	14	Hawaiian	399,725
15	Simmons	2,341	15	Aloha	627,578	15	Tower	31,417	15	Evergreen	332,666
16	Westair	1,947	16	Air Wisconsin	581,777	16	Southwest	18,491	16	Tower	221,420
17	USAir Shuttle	1,472	17	Carnival	550,733	17	Hawaiian	15,470	17	Air Wisconsin	217,100
18	Trans States	1,322	18	Braniff Int'l	509,403	18	Amerijet	15,246	18	Horizon Air	208,149
19	Executive Airlines	912	19	Horizon Air	485,336	19	Aloha	8,256	19	Aloha	206,335
20	Midwest Express	808	20	Markair	476,032	20	Markair	8,019	20	World	200,286
21	Carnival	575	21	Simmons	387,122	21	Northern Air	7,793	21	Emery	187,180
22	Markair	549	22	Westair	384,574	22	Midwest Express	4,767	22	Simmons	184,675
23	Braniff Int'l	518	23	USAir Shuttle	318,889	23	MGM Grand	4,336	23	Westair	176,090
24	American Trans Air	480	24	Trans States	278,057	24	Ryan International	2,041	24	Sun Country	151,115
25	Tower	463	25	Key	159,372	25	Reeve Aleutian	2,007	25	Southern Air	138,317

TOP 25 U.S. AIRLINES IN 1991

	Passengers (000)			Revenue Passenger Miles (000)			Freight Ton Miles (000)			Total Operating Revenues ($000)
1	American	75,892	1	United	81,853,018	1	Federal Express	3,622,660	1	American	12,096,075
2	Delta	74,125	2	American	81,585,861	2	Northwest	1,634,455	2	United	11,660,131
3	United	61,891	3	Delta	67,272,534	3	United	1,214,343	3	Delta	10,062,781
4	USAir	55,600	4	Northwest	53,197,868	4	American	883,862	4	Federal Express	7,555,246
5	Northwest	41,096	5	Continental	41,431,838	5	Delta	668,392	5	Northwest	7,533,713
6	Continental	36,969	6	USAir	34,119,846	6	Continental	563,751	6	USAir	6,049,170
7	Southwest	25,211	7	Trans World	27,961,852	7	Pan American	376,694	7	Continental	5,283,494
8	Trans World	20,523	8	Pan American	18,898,150	8	Trans World	368,940	8	Trans World	3,657,492
9	America West	15,844	9	America West	12,912,368	9	United Parcel Service	209,634	9	Pan American	2,093,896
10	Pan American	10,559	10	Southwest	11,274,334	10	USAir	136,665	10	America West	1,420,069
11	Alaska	5,810	11	Alaska	4,907,299	11	Challenge Air Cargo	129,072	11	Southwest	1,313,677
12	Aloha	4,915	12	Midway	3,400,910	12	America West	72,117	12	United Parcel Service	1,123,460
13	Midway	4,314	13	Hawaiian	2,019,681	13	Arrow	64,192	13	Alaska	933,629
14	Hawaiian	3,764	14	Tower	1,433,928	14	Air Transport Int'l.	49,069	14	American Trans Air	421,855
15	West Air	3,062	15	Aloha	664,939	15	Alaska	47,436	15	Hawaiian	364,339
16	Air Wisconsin	2,365	16	West Air	649,265	16	Zantop	32,617	16	Evergreen	358,858
17	Horizon Air	2,022	17	Eastern	620,056	17	Southwest	12,964	17	West Air	255,580
18	Trump Shuttle	1,430	18	Midwest Express	610,733	18	Midway	11,087	18	Midway	250,816
19	Executive Airlines	944	19	American Trans Air	519,626	19	Northern Air	10,054	19	Tower Air	245,810
20	Trans States	915	20	Air Wisconsin	493,868	20	Markair	9,201	20	World	226,595
21	Eastern	818	21	Horizon Air	404,671	21	Amerijet	8,773	21	Aloha	214,022
22	Midwest Express	768	22	Carnival	353,711	22	Aloha	8,056	22	Air Wisconsin	211,306
23	Simmons	546	23	Trump Shuttle	307,120	23	Hawaiian	5,242	23	Simmons	184,825
24	Carnival	446	24	Trans States	189,707	24	Midwest Express	3,994	24	Horizon	183,142
25	Markair	374	25	Markair	174,222	25	Eastern	3,881	25	Southern Air	174,851

U.S. INDUSTRY TRAFFIC MARKET SHARE
12 Months 1992

#		RPMs	Share (%)		#		ASMs	Share (%)		#		LF (%)
1.	American	97,396,091	20.743		1.	American	152,945,589	20.708		1.	United	67.4
2.	United	92,488,299	19.698		2.	United	137,210,480	18.578		2.	Northwest	65.3
3.	Delta	80,494,988	17.144		3.	Delta	131,389,184	17.789		3.	TWA	64.7
4.	Northwest	58,226,333	12.401		4.	Northwest	89,141,911	12.069		4.	Southwest	64.5
5.	Continental	43,071,907	9.173		5.	Continental	67,880,679	9.191		5.	American	63.7
6.	USAir	35,097,275	7.475		6.	USAir	59,666,766	8.079		6.	Continental	63.5
7.	TWA	28,882,400	6.151		7.	TWA	44,651,200	6.046		7.	Delta	61.3
8.	Southwest	13,787,005	2.936		8.	Southwest	21,366,642	2.893		8.	America West	61.1
9.	America West	11,780,568	2.509		9.	America West	19,271,353	2.609		9.	USAir	58.8
10.	Alaska	5,537,000	1.179		10.	Alaska	9,617,000	1.302		10.	Alaska	57.6
11.	Aloha	628,932	0.134		11.	Air Wis	1,204,123	0.163		11.	Aloha	57.5
12.	Air Wis	586,666	0.125		12.	Comair	1,157,988	0.157		12.	Horizon	53.7
13.	Atlantic SE	547,068	0.117		13.	Aloha	1,093,364	0.148		13.	Atlantic SE	50.8
14.	Comair	521,106	0.111		14.	Atlantic SE	1,077,154	0.146		14.	Air Wis	48.7
15.	Horizon	485,600	0.103		15.	Horizon	904,900	0.123		15.	Comair	45.0
16.	Midwest Express	---			16.	Midwest Express	---			16.	Metro	---
17.	Metro	---			17.	Metro	---			17.	Midwest Express	---
	Total	**469,531,238**				**Total**	**738,578,333**					**63.6**

1982-1992 SUMMARY
U.S. Scheduled Airlines

	1982	1983	1984	1985	1986	1987	1988	1989	1990	1991	1992
Traffic--Scheduled Service											
Revenue passengers enplaned (000)	294,102	318,638	344,683	382,022	418,946	447,678	454,614	453,692	465,560	452,301	473,305
Revenue passenger miles (000)	259,643,870	281,829,148	305,115,855	336,403,021	366,545,855	404,471,484	423,301,559	432,714,309	457,926,286	447,954,829	478,081,118
Available seat miles (000)	440,119,206	464,537,979	515,323,339	547,788,432	607,435,847	648,720,938	676,802,328	684,375,876	733,374,893	715,199,140	751,815,137
Revenue passenger load factor (%)	59.0	60.7	59.2	61.4	60.3	62.3	62.5	63.2	62.4	62.6	63.6
Average passenger trip length (miles)	883	884	885	881	875	903	931	954	984	990	1,010
Freight and express ton miles (000)	5,482,198	6,092,932	6,566,571	6,030,543	7,344,054	8,260,278	9,632,219	10,275,002	10,546,329	10,225,199	10,989,344
Aircraft departures	4,963,794	5,033,906	5,448,150	5,835,474	6,426,970	6,581,309	6,699,564	6,622,080	6,923,593	6,782,782	6,866,325
Financial											
Passenger revenue ($000)	30,549,719	32,744,618	36,939,345	39,235,809	40,056,093	44,940,391	50,295,686	53,802,067	58,453,215	57,091,675	59,747,693
Freight and express revenue ($000)	2,437,703	2,592,567	2,859,419	2,680,715	5,627,996	6,398,156	7,477,731	6,892,754	5,431,627	5,508,572	5,891,390
Mail revenue ($000)	688,675	653,129	712,070	889,575	838,278	923,022	971,807	955,455	970,475	957,077	1,174,063
Charter revenue ($000)	1,085,537	1,075,428	1,112,050	1,279,812	1,268,899	1,611,673	1,697,793	2,051,883	2,876,581	3,717,358	2,736,417
Total operating revenue ($000)	36,407,635	38,953,672	43,825,047	46,664,414	50,524,933	56,985,709	63,748,886	69,315,854	76,141,739	75,158,493	77,949,558
Total operating expense ($000)	37,141,070	38,643,262	41,673,536	45,238,150	49,201,832	54,516,820	60,312,383	67,504,587	78,054,094	76,943,234	80,333,091
Operating profit ($000)	(733,435)	310,410	2,151,511	1,426,264	1,323,101	2,468,889	3,436,503	1,811,267	(1,912,355)	(1,784,741)	(2,383,533)
Interest expense ($000)	1,384,084	1,482,352	1,540,377	1,588,306	1,692,548	1,695,388	1,845,762	1,944,388	1,978,163	1,776,994	1,726,013
Net profit ($000)	(915,814)	(188,051)	824,668	862,715	(234,909)	593,398	1,685,599	127,902	(3,921,002)	(1,940,157)	(4,028,452)
Revenue per passenger mile (¢)	11.8	11.6	12.1	11.7	10.9	11.1	11.9	12.4	12.8	12.7	12.5
Rate of return on investment (%)	2.1	6.0	9.9	9.6	4.9	7.2	10.8	6.3	(6.0)	(0.5)	(6.4)
Operating profit margin (%)	(2.0)	0.8	4.9	3.1	2.6	4.3	5.4	2.6	(2.5)	(2.4)	(3.1)
Net profit margin (%)	(2.5)	(0.5)	1.9	1.8	(0.5)	1.0	2.6	0.2	(5.1)	(2.6)	(5.2)
Employees	330,495	328,648	345,079	355,113	421,686	457,349	480,553	506,728	545,809	533,565	540,412

Note: Federal Express began reporting as a Section 401 carrier in 1986 and is included in 1986 and later years.

MAJOR AND NATIONAL AIRLINE PASSENGER TRAFFIC 1987

	Air California	Air Wisconsin	Alaska Airlines	Aloha Airlines	West Airlines	American Airlines	American Trans Air	Aspen Airways
Enplaned Passengers - Sch+NSch Serv.	1,332,844	2,160,010	3,963,395	3,056,877	11,231,607	55,598,871	1,531,970	842,171
RPM's - Sch. + NSch. Serv. (000's)	663,115	271,676	2,987,462	403,018	5,785,814	56,793,989	3,692,280	153,771
ASM's - Sch. + NSch. Serv. (000's)	1,190,204	608,119	5,398,290	740,780	10,318,096	88,742,505	4,738,230	299,376
Departures Performed - Sch+NSch Serv.	25,878	87,296	81,621	56,031	177,214	663,834	12,368	30,333
Block Hours	37,637	53,832	125,434	32,162	230,417	1,398,545	43,278	26,676
Employment	NA	1,139	4,351	0	6,132	57,275	1,578	456
Total Operating Revenue	$ 101,923,000	150,580,092	541,188,139	118,152,648	576,637,590	7,124,481,000	0	49,007,473
Total Operating Expenses	108,261,000	136,464,368	504,170,737	105,876,547	611,992,801	6,651,297,000	244,606,198	48,795,846
Operating Profit or Loss	-6,338,000	14,115,724	37,017,402	12,276,101	-35,355,211	473,184,000	-425,357	211,627
Net Income	$ -6,252,000	6,068,913	32,472,569	6,251,374	-45,674,993	213,828,000	0	228,735

	Braniff	Continental Airlines	Delta Air Lines	Eastern Air Lines	Evergreen Intl. Airlines	Federal Express	Flying Tiger Line	Hawaiian Airlines
Enplaned Passengers - Sch+NSch Serv.	3,388,574	40,324,913	54,129,209	44,671,712	0	0	81,834	5,000,554
RPM's - Sch. + NSch. Serv. (000's)	3,696,947	39,853,433	43,819,490	36,125,576	0	0	1,133,717	3,539,385
ASM's - Sch. + NSch. Serv. (000's)	5,586,866	64,433,905	78,060,637	56,076,791	0	0	1,273,693	5,117,798
Departures Performed - Sch+NSch Serv.	39,971	555,767	733,349	528,463	23,497	149,588	25,931	81,531
Block Hours	91,525	1,131,356	1,283,333	946,712	56,575	207,552	98,514	78,947
Employment	1,837	30,763	50,039	43,776	588	45,151	6,072	2,344
Total Operating Revenue	$ 293,058,000	4,020,598,000	6,093,331,000	4,529,208,249	126,213,214	3,572,699,000	1,175,210,000	298,777,098
Total Operating Expenses	310,836,000	3,993,299,000	5,659,088,000	4,470,310,098	108,505,148	3,191,421,000	1,023,798,000	310,076,523
Operating Profit or Loss	-17,778,000	27,299,000	434,243,000	58,898,151	17,708,066	381,278,000	151,412,000	-11,299,425
Net Income	$ -10,364,000	-257,971,000	233,297,000	-181,675,947	8,015,510	166,226,000	81,710,000	-8,384,593

Abbreviations

RPM = Revenue Passenger Miles
NSch = Non Scheduled
ASM = Available Seat Miles
Rev = Revenue
Sch = Scheduled
Serv = Service

MAJOR AND NATIONAL AIRLINE PASSENGER TRAFFIC 1987 (CONTINUED)

	Horizon Air	Markair	Midway Airlines	Midwest Express Airlines	Northwest Airlines	Pan Am. World Airways	Piedmont Aviation	Presidential Airways
Enplaned Passengers - Sch+NSch Serv.	1,403,842	307,861	3,632,441	291,232	37,297,303	15,028,996	25,315,162	850,317
RPM's - Sch. + NSch. Serv. (000's)	289,199	134,401	2,576,177	191,096	39,788,540	26,363,373	11,465,889	408,309
ASM's - Sch. + NSch. Serv. (000's)	575,991	279,495	4,492,299	287,382	61,695,839	41,785,589	18,939,921	843,445
Departures Performed - Sch+NSch Serv.	141,081	17,309	66,004	8,424	517,572	144,481	451,410	23,640
Block Hours	107,778	19,809	121,630	13,871	1,003,134	368,492	550,253	41,642
Employment	134,120	626	2,304	295	34,172	21,883	17,818	658
Total Operating Revenue	$ 107,669,235	83,640,338	340,688,716	44,189,447	5,073,726,000	3,121,574,000	1,963,706,247	42,683,557
Total Operating Expenses	104,733,499	70,822,820	315,674,401	38,914,187	4,868,611,000	3,292,013,000	1,803,466,642	75,387,388
Operating Profit or Loss	2,935,736	12,817,518	25,014,315	5,275,260	205,115,000	-170,439,000	160,239,605	-32,703,831
Net Income	$ 887,057	2,448,198	19,759,410	3,222,823	140,716,000	-274,595,000	98,549,523	-29,969,613

	Southwest Airlines	Southwest Airlines	Tower Air	Trans World Airlines	United Air Lines	USAir	World Airways
Enplaned Passengers - Sch+NSch Serv.	0	14,236,201	251,003	24,720,341	55,225,491	24,982,503	184,605
RPM's - Sch. + NSch. Serv. (000's)	0	6,773,820	1,909,927	33,013,985	66,347,593	13,295,502	1,333,676
ASM's - Sch. + NSch. Serv. (000's)	0	11,487,684	2,183,195	52,078,868	101,453,732	20,307,834	1,479,258
Departures Performed - Sch+NSch Serv.	20,308	246,320	649	311,430	670,695	394,970	1,092
Block Hours	12,711,790	276,719	11,782	682,347	1,449,767	535,420	13,872
Employment	588	5,281	421	30,089	59,669	15,768	472
Total Operating Revenue	$ 119,513,565	698,675,051	137,992,937	4,056,434,547	7,862,794,507	2,070,312,000	101,218,000
Total Operating Expenses	104,338,820	657,397,127	129,008,335	3,815,999,379	7,711,577,440	1,806,855,000	86,269,000
Operating Profit or Loss	15,174,745	41,277,924	8,984,602	240,435,168	151,217,067	263,457,000	14,950,000
Net Income	$ 7,625,666	19,691,931	2,608,507	106,200,426	33,337,922	164,113,000	7,846,000

Abbreviations

RPM	=	Revenue Passenger Miles	ASM	=	Available Seat Miles
NSch	=	Non Scheduled	Rev	=	Revenue

Sch = Scheduled
Serv = Service

MAJOR AND NATIONAL AIRLINE PASSENGER TRAFFIC 1988

	Air Wisconsin	Alaska Airlines	Aloha Airlines	America West Airlines	American Airlines	American Trans Air	Aspen Airways	Braniff
Enplaned Passengers - Sch+NSch Serv.	2,419,851	4,867,555	3,435,009	12,790,877	64,309,483	1,270,626	893,935	4,669,927
RPM's - Sch. + NSch. Serv. (000's)	421,334	3,656,347	451,109	7,120,084	64,770,369	3,712,126	170,535	4,462,248
ASM's - Sch. + NSch. Serv. (000's)	852,171	6,682,225	722,708	12,200,104	102,045,221	4,852,470	341,500	6,724,528
Departures Performed - Sch+NSch Serv.	88,453	98,743	56,911	190,562	763,789	9,654	30,683	54,100
Block Hours	73,116	156,930	51,630	253,466	1,600,249	50,515	22,338	118,221
Employment	1,266	4,475	1,306	7,554	65,340	1,684	503	2,537
Total Operating Revenue	$ 184,565,659	708,855,000	134,213,092	781,460,405	8,550,586,000	0	58,301,244	412,438,000
Total Operating Expenses	161,889,145	646,908,000	124,041,205	763,319,309	7,749,591,000	233,026,534	60,587,085	424,462,000
Operating Profit or Loss	22,676,514	61,947,000	10,171,887	18,141,096	800,995,000	-16,013,359	-2,285,841	-12,024,000
Net Income	$ 17,824,553	37,600,000	8,937,868	9,361,386	449,445,000	-23,025,950	-2,421,732	-25,603,000

	Continental Airlines	Delta Air Lines	Eastern Air Lines	Evergreen Intl. Airlines	Federal Express	Flying Tiger Line	Hawaiian Airlines	Horizon Air
Enplaned Passengers - Sch+NSch Serv.	37,862,496	60,009,084	35,655,757	0	1	84,243	5,685,902	1,413,624
RPM's - Sch. + NSch. Serv. (000's)	40,766,521	51,686,213	28,871,226	0	0	1,148,294	3,617,097	290,460
ASM's - Sch. + NSch. Serv. (000's)	66,829,428	88,784,293	46,894,934	0	0	1,284,978	5,295,858	576,673
Departures Performed - Sch+NSch Serv.	523,583	802,902	450,605	9,682	176,800	33,115	85,576	149,459
Block Hours	1,147,622	1,440,697	804,317	72,317	249,568	121,957	20,756,927	112,145
Employment	30,217	56,901	32,542	689	53,055	6,287	2,655	1,440
Total Operating Revenue	$ 4,553,349,000	7,393,275,000	3,888,120,535	208,909,076	4,301,551,000	1,294,894,000	353,445,449	113,385,000
Total Operating Expenses	4,474,887,000	6,868,638,000	4,097,557,217	164,630,976	3,880,193,000	1,147,928,000	358,199,125	110,523,000
Operating Profit or Loss	78,462,000	524,637,000	-209,436,682	44,278,100	421,358,000	146,966,000	-4,753,676	2,862,000
Net Income	$ -315,520,000	344,523,000	-335,351,403	19,631,981	210,107,000	74,406,000	-8,484,288	2,565,000

Abbreviations

RPM = Revenue Passenger Miles   ASM = Available Seat Miles   Sch = Scheduled

NSch = Non Scheduled   Rev = Revenue   Serv = Service

MAJOR AND NATIONAL AIRLINE PASSENGER TRAFFIC 1988 (CONTINUED)

	Markair	Midway Airlines	Midwest Express Airlines	Northwest Airlines	Pan Am. World Airways	Piedmont Aviation	Presidential Airways	Southern Air Transport
Enplaned Passengers - Sch+NSch Serv.	366,635	4,266,045	411,187	35,948,992	16,958,893	27,125,160	874,755	0
RPM's - Sch. + NSch. Serv. (000's)	159,661	2,996,027	268,181	40,850,210	29,715,277	13,057,649	328,884	0
ASM's - Sch. + NSch. Serv. (000's)	317,017	5,355,027	408,358	62,025,503	46,544,820	22,106,273	718,834	0
Departures Performed - Sch+NSch Serv.	21,616	78,359	12,211	488,669	163,653	485,246	38,253	24,535
Block Hours	25,133	132,844	19,986	983,990	430,370	618,101	45,843	65,531
Employment	693	3,559	480	34,973	22,826	19,913	747	652
Total Operating Revenue	$ 99,117,752	388,001,883	61,367,830	5,587,783,000	3,592,617,000	2,363,177,797	69,156,688	140,815,730
Total Operating Expenses	83,667,701	374,509,269	55,622,378	5,392,145,000	3,697,910,000	2,085,816,721	83,075,941	119,472,118
Operating Profit or Loss	15,450,051	13,492,614	5,745,452	195,638,000	-105,293,000	277,361,076	-13,919,253	21,343,612
Net Income	$ 3,160,432	6,544,924	3,762,152	162,789,000	-118,254,000	159,440,419	-15,399,271	10,329,377

	Southwest Airlines	Tower Air	Trans World Airlines	United Air Lines	United Parcel Service	USAir	World Airways
Enplaned Passengers - Sch+NSch Serv.	16,878,979	329,190	25,193,436	56,370,488	0	32,644,603	157,540
RPM's - Sch. + NSch. Serv. (000's)	7,751,718	1,832,016	34,803,192	69,101,362	0	17,506,131	1,162,756
ASM's - Sch. + NSch. Serv. (000's)	13,426,120	2,133,710	56,298,424	101,721,040	0	28,483,032	2,099,976
Departures Performed - Sch+NSch Serv.	275,125	970	322,092	626,726	13,324	523,954	1,489
Block Hours	316,745	10,324	727,134	1,413,989	28,857	732,005	15,352
Employment	6,448	488	30,518	60,139	1,854	21,581	317
Total Operating Revenue	$ 860,446,611	132,831,381	4,361,144,616	8,796,352,228	118,706,853	2,802,994,000	126,589,000
Total Operating Expenses	774,367,173	121,998,772	4,101,725,920	8,127,784,098	126,332,497	2,659,266,000	107,463,000
Operating Profit or Loss	86,079,438	10,832,609	259,418,696	668,568,130	-7,625,644	143,728,000	19,127,000
Net Income	$ 57,400,227	3,440,657	249,742,519	589,236,000	-5,807,913	76,169,000	25,245,000

Abbreviations

RPM	=	Revenue Passenger Miles	ASM	=	Available Seat Miles
NSch	=	Non Scheduled	Rev	=	Revenue

Sch	=	Scheduled
Serv	=	Service

MajOR AND NATIONAL AIRLINE PASSENGER TRAFFIC 1989

	Air Wisconsin	Alaska Airlines	Aloha Airlines	America West Airlines	American Airlines	American Trans Air	Aspen Airways	Braniff
Enplaned Passengers - Sch+NSch Serv.	2,194,975	5,017,087	4,012,489	13,420,437	72,380,711	1,346,473	802,495	4,664,769
RPM's - Sch. + NSch. Serv. (000's)	434,346	4,060,900	534,451	7,933,884	73,502,831	3,967,162	189,540	4,028,842
ASM's - Sch. + NSch. Serv. (000's)	918,434	7,324,135	846,512	13,724,514	115,222,063	5,293,886	356,403	10,368,644
Departures Performed - Sch+NSch Serv.	83,833	101,126	64,088	205,115	811,845	9,716	29,192	64,728
Block Hours	75,805	171,401	38,849	300,342	1,774,716	48,084	26,624	131,916
Employment	1,360	5,041	1,554	9,435	75,086	1,982	583	0
Total Operating Revenue	$ 178,090,889	794,075,000	155,085,066	998,312,758	9,960,947,000	0	59,818,027	425,661,000
Total Operating Expenses	176,450,296	736,238,000	146,612,038	950,196,461	9,230,151,000	261,004,534	65,251,973	489,639,000
Operating Profit or Loss	1,640,593	57,837,000	8,473,028	48,116,297	730,796,000	-11,110,002	-5,433,946	-63,978,000
Net Income	$ 355,396	42,323,000	2,245,704	20,017,616	423,100,000	-13,112,729	-5,471,846	-155,444,000

	Continental Airlines	Delta Air Lines	Eastern Air Lines	Evergreen Int'l. Airlines	Federal Express	Flying Tiger Line	Hawaiian Airlines	Horizon Air
Enplaned Passengers - Sch+NSch Serv.	35,652,330	68,254,294	14,544,818	0	202,376	45,001	5,537,193	1,598,432
RPM's - Sch. + NSch. Serv. (000's)	39,527,610	59,387,458	11,627,748	0	331,268	581,239	3,566,317	319,097
ASM's - Sch. + NSch. Serv. (000's)	63,753,431	93,654,291	19,162,253	0	493,797	652,420	5,243,156	607,490
Departures Performed - Sch+NSch Serv.	489,379	837,327	182,386	15,170	216,360	20,425	80,741	156,235
Block Hours	1,088,972	1,541,585	332,668	49,519	352,125	71,041	82,379	113,435
Employment	31,725	61,025	15,468	638	70,460	0	2,808	1,553
Total Operating Revenue	$ 4,944,226,000	8,648,315,000	1,551,678,305	211,334,984	5,833,824,000	730,148,000	348,268,164	132,521,000
Total Operating Expenses	4,788,009,000	7,971,765,000	2,416,399,523	186,835,519	5,399,977,000	717,374,000	392,689,331	126,527,000
Operating Profit or Loss	156,217,000	676,550,000	-864,721,218	24,499,465	433,849,000	12,774,000	-44,421,167	5,994,000
Net Income	$ 3,059,000	473,174,000	-852,315,516	4,233,335	124,633,000	10,294,000	-42,688,117	1,865,000

Abbreviations

RPM = Revenue Passenger Miles	ASM = Available Seat Miles	Sch = Scheduled
NSch = Non Scheduled	Rev = Revenue	Serv = Service

MAJOR AND NATIONAL AIRLINE PASSENGER TRAFFIC 1989 (CONTINUED)

	Markair	Midway Airlines	Midwest Express Airlines	Northwest Airlines	Pan Am. World Airways	Piedmont Aviation	Presidential Airways	Southern Air Transport
Enplaned Passengers - Sch+NSch Serv.	396,347	4,702,069	632,660	38,733,751	17,325,637	16,669,803	531,760	0
RPM's - Sch. + NSch. Serv. (000's)	169,162	3,492,916	408,462	46,450,144	29,658,669	8,585,883	244,293	0
ASM's - Sch. + NSch. Serv. (000's)	338,916	6,155,382	675,877	71,066,959	46,613,520	14,586,030	490,869	0
Departures Performed - Sch+NSch Serv.	21,932	83,999	17,695	509,790	172,384	291,036	34,335	25,885
Block Hours	26,734	163,981	30,244	1,087,539	473,707	391,559	38,167	71,777
Employment	709	4,685	667	37,481	27,769	0	0	718
Total Operating Revenue	$ 108,559,387	463,042,834	91,872,348	6,553,827,000	3,611,853,000	1,653,710,000	55,184,135	145,599,221
Total Operating Expenses	92,391,372	476,574,820	84,398,272	6,263,766,000	3,930,672,000	1,422,938,000	65,354,784	131,969,302
Operating Profit or Loss	16,168,015	-13,531,986	7,474,076	290,061,000	-318,819,000	230,772,000	-10,170,649	13,629,919
Net Income	$ 2,679,693	-21,684,761	4,819,224	355,247,000	-414,730,000	139,741,000	-15,644,447	5,906,350

	Southwest Airlines	Tower Air	Trans World Airlines	United Air Lines	United Parcel Service	USAir	World Airways
Enplaned Passengers - Sch+NSch Serv.	20,283,864	326,602	25,241,349	54,947,148	0	44,659,743	105,732
RPM's - Sch. + NSch. Serv. (000's)	9,348,456	1,878,167	35,159,543	69,638,708	0	25,315,332	1,427,580
ASM's - Sch. + NSch. Serv. (000's)	14,884,021	2,109,902	57,430,561	104,547,873	0	41,288,851	1,823,929
Departures Performed - Sch+NSch Serv.	304,765	990	322,607	621,002	60,537	747,198	1,457
Block Hours	350,804	10,442	747,407	1,454,640	127,099	1,082,997	26,070
Employment	7,330	492	32,577	65,099	2,101	34,807	545
Total Operating Revenue	$ 1,015,124,354	141,438,001	4,507,348,412	9,641,887,601	650,377,814	4,204,661,000	173,639,000
Total Operating Expenses	917,549,416	131,202,630	4,482,970,858	9,185,032,148	617,450,539	4,432,755,000	149,434,000
Operating Profit or Loss	97,574,938	10,235,371	24,377,554	456,855,453	32,927,275	-228,094,000	24,205,000
Net Income	$ 71,391,399	3,405,852	-298,547,053	358,088,000	41,458,278	-137,652,000	28,029,000

Abbreviations

RPM	= Revenue Passenger Miles	ASM = Available Seat Miles	Sch = Scheduled
NSch	= Non Scheduled	Rev = Revenue	Serv = Service

MAJOR AND NATIONAL AIRLINE PASSENGER TRAFFIC 1990

	Air Wisconsin	Alaska Airlines	Aloha Airlines	America West Airlines	American Airlines	American Trans Air	Aspen Airways	Continental Airlines
Enplaned Passengers - Sch+NSch Serv.	2,247,710	5,457,142	4,637,499	15,624,060	73,254,556	2,520,491	403,621	36,204,620
RPM's - Sch. + NSch. Serv. (000's)	466,954	4,494,012	658,815	11,114,287	76,961,369	4,600,462	80,056	40,301,812
ASM's - Sch. + NSch. Serv. (000's)	1,025,379	8,379,611	1,064,833	18,286,161	123,574,738	6,754,809	148,649	66,141,975
Departures Performed - Sch+NSch Serv.	87,497	110,899	77,043	224,560	828,149	17,220	12,259	498,665
Block Hours	79,571	193,793	47,042	366,586	1,932,746	56,594	11,426	1,119,478
Employment	1,358	5,822	1,808	12,764	85,680	2,127	374	33,553
Total Operating Revenue	$ 196,648,826	895,745,000	188,852,257	1,321,642,030	11,008,677,000	0	33,251,329	5,202,234,000
Total Operating Expenses	195,978,317	878,593,000	180,939,053	1,353,272,972	10,940,704,000	359,420,196	35,612,194	5,444,049,000
Operating Profit or Loss	670,509	17,152,000	7,913,240	-31,630,942	67,973,000	-8,500,768	-2,360,865	-241,815,000
Net Income	$ 1,288,981	15,046,000	1,374,193	-74,670,994	-76,777,000	-12,643,897	1,897,251	-1,236,387,000

	Delta Air Lines	Eastern Air Lines	Evergreen Intl. Airlines	Federal Express	Hawaiian Airlines	Horizon Air	Markair	Midway Airlines
Enplaned Passengers - Sch+NSch Serv.	65,788,168	21,435,146	17,963	387,647	5,050,972	1,829,738	438,733	6,508,433
RPM's - Sch. + NSch. Serv. (000's)	59,039,397	17,362,075	38,016	885,891	3,204,660	364,497	194,900	4,862,140
ASM's - Sch. + NSch. Serv. (000's)	99,903,062	28,026,506	66,475	1,176,294	4,793,594	727,679	421,853	8,310,753
Departures Performed - Sch+NSch Serv.	878,468	280,549	7,236	245,652	74,898	176,169	25,903	122,687
Block Hours	1,649,090	516,597	21,336	420,254	73,597	142,011	29,881	229,758
Empoyment	64,791	19,075	555	80,006	2,917	1,828	801	5,171
Total Operating Revenue	$ 8,746,083,000	2,181,796,438	217,608,222	7,612,986,000	340,024,283	163,820,000	128,302,682	614,809,293
Total Operating Expenses	8,981,211,000	2,715,149,964	194,857,250	7,189,297,000	432,751,719	157,979,000	112,448,871	699,339,127
Operating Profit or Loss	-235,128,000	-533,353,526	22,750,972	423,689,000	-92,727,436	5,841,000	15,853,811	-84,529,834
Net Income	$ -154,033,000	-1,115,907,931	2,942,579	127,367,000	-114,253,402	1,052,000	7,866,185	-139,215,049

Abbreviations

RPM	=	Revenue Passenger Miles	ASM	=	Available Seat Miles
NSch	=	Non Scheduled	Rev	=	Revenue

Sch	=	Scheduled
Serv	=	Service

MAJOR AND NATIONAL AIRLINE PASSENGER TRAFFIC 1990 (CONTINUED)

	Midwest Express Airlines	Northwest Airlines	Pan Am. World Airways	Southern Air Transport	Southwest Airlines	Tower Air	Trans World Airlines	United Air Lines
Enplaned Passengers - Sch+NSch Serv.	771,472	41,049,622	17,927,920	0	22,108,794	454,242	24,414,112	57,613,631
RPM's - Sch. + NSch. Serv. (000's)	585,036	52,198,424	31,597,436	0	10,007,151	2,088,363	34,651,673	76,136,447
ASM's - Sch. + NSch. Serv. (000's)	1,036,314	78,123,005	48,223,671	0	16,489,152	2,397,733	55,614,793	114,992,321
Departures Performed - Sch+NSch Serv.	21,524	518,758	177,483	25,350	338,561	1,708	303,797	654,510
Block Hours	39,989	1,152,065	490,827	62,161	392,796	13,037	728,798	1,565,542
Employment	835	42,283	28,823	742	8,267	466	33,189	70,179
Total Operating Revenue	$ 125,824,731	7,257,113,000	3,930,722,000	152,502,559	1,186,831,234	172,263,787	4,606,082,103	10,956,058,538
Total Operating Expenses	120,749,639	7,398,781,000	4,421,181,000	133,475,772	1,105,215,047	160,526,112	4,768,287,673	11,010,310,090
Operating Profit or Loss	5,075,092	-141,668,000	-490,459,000	19,026,787	81,616,187	11,737,675	-162,205,570	-54,251,552
Net Income	$ 2,959,958	-10,414,000	-638,074,000	13,631,910	47,082,866	5,614,201	-237,563,948	95,755,167

	United Parcel Service	USAir	World Airways
Enplaned Passengers - Sch+NSch Serv.	0	60,250,238	515,420
RPM's - Sch. + NSch. Serv. (000's)	0	36,329,927	1,043,469
ASM's - Sch. + NSch. Serv. (000's)	0	60,663,392	1,525,139
Departures Performed - Sch+NSch Serv.	71,014	1,094,833	3,720
Block Hours	152,339	1,548,593	18,627
Empoyment	2,768	50,464	559
Total Operating Revenue	$ 891,089,744	6,084,704,000	243,681,000
Total Operating Expenses	878,945,645	6,627,939,000	225,400,000
Operating Profit or Loss	12,144,099	-543,235,000	18,281,000
Net Income	$ 38,448,230	-410,748,000	11,972,000

Abbreviations

RPM = Revenue Passenger Miles	ASM = Available Seat Miles	Sch = Scheduled
NSch = Non Scheduled	Rev = Revenue	Serv = Service

MAJOR AND NATIONAL AIRLINE PASSENGER TRAFFIC 1991

	Air Wisconsin	Alaska Airlines	Aloha Airlines	America West Airlines	American Airlines	American Trans Air	Aspen Airways	Braniff
Enplaned Passengers - Sch+NSch Serv.	2,365,449	5,865,658	4,924,999	16,906,839	75,925,925	2,575,666	129,274	169,818
RPM's - Sch. + NSch. Serv. (000's)	495,011	4,948,473	666,559	13,030,686	81,702,863	4,602,473	32,070	127,814
ASM's - Sch. + NSch. Serv. (000's)	1,095,821	8,789,602	1,060,781	20,628,041	132,578,577	7,088,914	51,104	211,537
Departures Performed - Sch+NSch Serv.	91,134	111,715	76,952	227,707	853,892	19,083	2,293	2,099
Block Hours	81,632	200,156	49,134	400,713	2,036,274	47,405	2,267	1,566
Employment	1,544	6,070	1,915	11,862	88,130	2,218	0	160
Total Operating Revenue	$ 199,106,634	933,629,000	214,022,075	1,420,069,253	12,098,076,000	0	12,199,157	11,425,273
Total Operating Expenses	226,994,537	914,000,000	196,214,065	1,524,726,356	12,080,570,000	410,552,589	11,077,743	17,061,649
Operating Profit or Loss	-27,887,903	19,629,000	17,808,010	-104,657,103	17,506,000	-24,185,219	1,121,414	-5,636,376
Net Income	$ -37,738,830	10,932,000	5,177,826	-222,016,114	-165,422,000	-29,751,428	477,220	-7,363,356

	Continental Airlines	Delta Air Lines	Evergreen Intl. Airlines	Federal Express	Hawaiian Airlines	Horizon Air	Markair	Midway Airlines
Enplaned Passengers - Sch+NSch Serv.	37,813,760	74,186,854	35,735	353,020	4,059,067	2,036,910	384,139	3,863,894
RPM's - Sch. + NSch. Serv. (000's)	42,379,528	67,345,074	89,120	843,861	2,623,480	411,285	177,557	3,059,147
ASM's - Sch. + NSch. Serv. (000's)	67,429,774	111,596,285	164,371	1,281,014	4,146,294	793,683	487,587	5,084,205
Departures Performed - Sch+NSch Serv.	492,488	939,158	14,670	252,222	62,391	181,550	22,569	67,452
Block Hours	1,115,724	1,788,255	22,754	430,067	63,382	127,366	24,969	129,161
Employment	34,672	71,755	759	83,067	2,767	1,954	974	0
Total Operating Revenue	$ 5,283,494,000	10,062,781,000	358,657,945	7,555,246,000	402,698,435	183,142,000	120,673,984	352,437,000
Total Operating Expenses	5,552,639,000	10,329,134,000	324,762,790	7,234,643,000	512,840,397	175,178,000	116,478,413	412,007,000
Operating Profit or Loss	-269,145,000	-266,353,000	33,895,155	320,603,000	-110,141,962	7,964,000	4,195,572	-59,570,000
Net Income	$ -340,915,000	-239,475,000	10,005,218	-34,016,000	-109,012,547	3,600,000	3,085,426	-82,635,000

Abbreviations

RPM = Revenue Passenger Miles	ASM = Available Seat Miles	Sch = Scheduled
NSch = Non Scheduled	Rev = Revenue	Serv = Service

MAJOR AND NATIONAL AIRLINE PASSENGER TRAFFIC 1991 (CONTINUED)

	Midwest Express Airlines	Northwest Airlines	Pan Am. World Airways	Southern Air Transport	Southwest Airlines	Tower Air	Trans World Airlines	United Air Lines
Enplaned Passengers - Sch+NSch Serv.	790,190	41,239,211	9,873,980	0	25,244,552	556,937	20,777,849	62,039,973
RPM's - Sch. + NSch. Serv. (000's)	656,235	53,931,257	17,585,547	0	11,335,055	2,518,479	28,660,738	82,290,499
ASM's - Sch. + NSch. Serv. (000's)	1,196,945	81,690,316	27,741,708	0	18,545,946	3,195,802	44,506,523	124,100,862
Departures Performed - Sch+NSch Serv.	22,640	517,927	104,304	21,308	382,910	2,361	259,426	691,402
Block Hours	42,042	1,167,072	303,002	44,829	442,625	18,271	613,034	1,665,984
Empoyment	895	43,082	0	718	9,277	528	28,620	75,050
Total Operating Revenue	$ 125,262,110	7,533,711,000	2,093,898,000	174,851,157	1,313,676,850	245,809,775	3,657,491,517	11,660,131,000
Total Operating Expenses	124,694,263	7,593,827,000	2,562,650,000	156,573,993	1,251,633,551	226,824,938	4,005,854,196	12,150,737,000
Operating Profit or Loss	567,847	-60,116,000	-468,752,000	18,277,164	62,043,299	18,984,837	-348,362,679	-490,606,000
Net Income	$ 102,832	-3,099,000	-283,056,000	15,146,668	26,919,066	6,963,745	34,550,578	-335,344,000

	United Parcel Service	USAir	World Airways
Enplaned Passengers - Sch+NSch Serv.	0	55,838,814	480,681
RPM's - Sch. + NSch. Serv. (000's)	0	34,398,677	1,023,133
ASM's - Sch. + NSch. Serv. (000's)	0	58,574,720	1,742,220
Departures Performed - Sch+NSch Serv.	76,779	948,009	3,610
Block Hours	84,998	1,462,773	19,750
Employment	2,694	46,600	495
Total Operating Revenue	$ 1,123,460,000	6,049,170,000	226,595,000
Total Operating Expenses	1,083,683,000	6,251,271,000	211,348,000
Operating Profit or Loss	39,777,000	-202,101,000	15,247,000
Net Income	$ 27,610,000	-259,998,000	20,482,000

Abbreviations

RPM = Revenue Passenger Miles	ASM = Available Seat Miles	Sch = Scheduled
NSch = Non Scheduled	Rev = Revenue	Serv = Service

MAJOR AND NATIONAL AIRLINE OPERATING EXPENSE 1987

	Air California	Air Wisconsin	Alaska Airlines	Aloha Airlines	America West Airlines	American Airlines	American Trans Air	Aspen Airways
Mtls- Aircraft Fuel & Oil	$ 15,110,000	11,747,244	77,327,940	16,989,679	111,298,591	1,032,191,000	57,558,803	3,207,102
Mtls- Passenger Food	1,966,000	398,761	26,525,858	369,760	11,582,015	289,407,000	10,463,735	71,887
Svcs- Advertising & Other Promotion	3,525,000	662,999	18,806,671	3,218,719	14,764,665	168,868,000	1,417,916	45,611
Landing Fees	2,066,000	2,993,959	9,385,212	853,868	15,517,242	123,540,000	4,823,316	561,606
Rentals	16,414,000	10,100,218	43,244,299	7,772,170	116,437,073	262,498,000	7,563,971	3,396,262
Total Salaries & Benefits	32,136,539	33,086,451	160,263,558	35,401,279	143,333,333	2,522,797,440	59,407,624	8,533,019
Maintenance Labor	1,485,539	5,139,503	5,627,926	2,729,982	0	214,287,440	9,956,081	1,786,217
Bfts- Personnel Expense	3,021,000	1,597,578	9,489,923	798,847	9,077,435	129,434,000	13,975,742	412,053
LABOR (Less Maint Labor)	27,630,000	26,349,370	145,145,709	31,872,450	134,255,898	2,179,076,000	35,475,801	6,334,749
Exp- Maintenance	10,380,000	24,234,995	41,904,018	15,181,102	29,160,438	746,136,000	47,899,285	11,072,384
Svcs- Traffic Commissions - Passenger	7,320,000	8,588,598	42,831,555	3,464,193	40,085,053	567,836,000	3,241,581	1,688,418
Nonop- Interest Income	0	0	0	0	0	0	0	0
Nonop- Int. on L-T-D & Cap.Leases	403,000	6,890,105	18,457,616	4,718,990	28,894,960	182,737,000	8,044,561	169,269
Nonop- Int. Exp. Other	-88,000	-1,531,188	353,076	72,312	-5,714,115	-18,725,000	93,916	260,818
INTEREST EXPENSE	315,000	5,358,917	18,810,692	4,791,302	23,180,845	164,012,000	8,138,477	430,087
Total Operating Expenses	$ 108,261,000	103,648,454	504,171,149	105,876,333	611,992,801	6,651,297,000	244,606,197	23,742,837
Rev- Passengers	89,790,000	140,889,018	459,272,859	90,831,563	559,180,837	6,147,715,000	0	46,719,607

Abbreviations

Mtls	=	Materials	Bfts	=	Benefits
Rev	=	Revenue	L-T-D & Cap Leases	=	Long Term Debt and Capital Leases

Nonop = Non -operating

MAJOR AND NATIONAL AIRLINE OPERATING EXPENSE 1987 (CONTINUED)

	Braniff	Continental Airlines	Delta Air Lines	Eastern Air Lines	Evergreen Int'l. Airlines	Federal Express	Flying Tiger Line	Hawaiian Airlines
Mtls- Aircraft Fuel & Oil	$ 69,166,000	729,089,000	910,137,000	695,521,139	7,491,103	117,023,000	188,176,000	65,572,719
Mtls- Passenger Food	9,506,000	136,892,000	222,188,000	158,028,375	5,341	0	2,239,000	6,775,367
Svcs- Advertising & Other Promotion	11,886,000	87,460,000	103,423,000	102,018,099	143,139	60,776,000	4,933,000	7,762,128
Landing Fees	6,129,000	87,017,000	100,462,000	91,068,976	241,121	10,644,000	32,400,000	3,448,221
Rentals	42,641,000	448,878,000	366,437,000	119,309,993	8,137,382	275,574,266	65,931,000	30,552,744
Total Salaries & Benefits	61,624,000	945,245,000	2,436,990,000	1,694,589,705	19,473,978	1,707,423,004	317,580,031	88,496,844
Maintenance Labor	0	55,328,000	147,712,000	191,946,883	4,954,792	67,094,484	34,849,031	6,442,021
Bfts- Personnel Expense	6,215,000	74,868,000	112,244,000	83,175,309	3,510,142	45,143,521	31,353,000	7,190,623
LABOR (Less Maint Labor)	55,409,000	815,049,000	2,177,034,000	1,419,467,513	11,009,044	1,595,184,999	251,378,000	74,864,200
Exp- Maintenance	36,469,000	428,991,000	512,861,000	640,396,970	26,666,222	242,626,000	122,746,000	54,588,633
Svcs- Traffic Commissions - Passenger	19,627,000	330,203,000	504,802,000	343,020,636	0	0	0	9,166,881
Nonop- Interest Income	0	0	0	0	0	0	0	0
Nonop- Int. on L-T-D & Cap.Leases	0	230,057,000	91,577,000	281,141,062	9,559,533	69,109,000	48,547,000	7,171,939
Nonop- Int. Exp. Other	12,000	45,271,000	-28,319,000	26,911,613	0	0	5,174,000	42,187
INTEREST EXPENSE	12,000	275,328,000	63,258,000	308,052,675	9,559,533	69,109,000	53,721,000	7,214,126
Total Operating Expenses	$ 315,836,000	3,993,299,000	5,659,088,000	4,470,310,098	69,069,872	3,191,421,000	1,030,928,000	310,076,523
Rev- Passengers	248,421,000	3,627,227,000	5,607,370,000	4,051,640,987	0	0	0	232,308,657

Abbreviations

Mtls	=	Materials	Bfts	=	Benefits
Rev	=	Revenue	L-T-D & Cap Leases	=	Long Term Debt and Capital Leases

Nonop = Non -operating

MAJOR AND NATIONAL AIRLINE OPERATING EXPENSE 1987 (CONTINUED)

	Horizon Air	Markair	Midway Airlines	Midwest Express Airlines	Northwest Airlines	Pan Am. World Airways	Piedmont Aviation	Presidential Airways
Mtls- Aircraft Fuel & Oil	$ 5,801,297	10,731,707	57,683,315	3,327,434	834,613,000	506,351,000	262,333,084	5,981,561
Mtls- Passenger Food	260,368	1,531,857	7,286,441	1,537,960	145,537,000	96,511,000	58,299,159	387,145
Svcs- Advertising & Other Promotion	85⁻,669	2,050,910	10,525,704	816,791	78,693,000	77,084,000	40,962,947	1,207,096
Landing Fees	1,124,599	546,462	5,431,728	337,041	107,244,000	78,763,000	33,947,283	965,291
Rentals	10,075,318	4,171,566	31,337,136	357,863	188,709,000	235,137,000	125,786,956	10,940,268
Total Salaries & Benefits	18,142,513	24,818,805	72,848,917	5,231,473	1,623,246,000	1,156,594,000	710,242,961	8,722,669
Maintenance Labor	2,062,759	2,975,128	7,741,327	455,128	137,941,000	80,001,000	32,926,230	1,109,594
Bfts- Personnel Expense	1,340,951	677,789	6,116,883	1,384,354	105,715,000	72,874,000	42,364,140	732,636
LABOR (Less Maint Labor)	14,738,803	21,165,888	58,990,707	3,391,991	1,379,590,000	1,003,719,000	634,952,591	6,880,439
Exp- Maintenance	15,828,328	9,982,097	38,644,195	5,611,010	510,893,000	303,855,000	181,010,607	7,895,112
Svcs- Traffic Commissions - Passenger	4,428 061	2,387,699	29,138,870	2,277,069	653,149,000	244,400,000	123,052,029	1,510,481
Nonop- Interest Income	0	0	0	0	0	0	0	0
Nonop- Int. on L-T-D & Cap.Leases	702,224	11,600,503	6,514,012	0	79,559,000	96,275,000	28,825,085	1,096,037
Nonop- Int. Exp. Other	0	0	133,888	23,115	-17,200,000	11,677,000	-16,835,591	394,103
INTEREST EXPENSE	702,224	11,600,503	6,647,900	23,115	62,359,000	107,952,000	11,989,494	1,490,140
Total Operating Expenses	$ 53,817,482	70,822,820	315,674,401	21,212,812	4,868,611,000	3,292,013,000	1,803,464,822	49,875,494
Rev- Passengers	101,811,489	43,599,225	329,674,559	42,390,599	4,371,624,000	2,613,015,000	1,867,304,357	39,972,917

Abbreviations

Mtls	=	Materials	Bfts	=	Benefits
Rev	=	Revenue	L-T-D & Cap Leases	=	Long Term Debt and Capital Leases

Nonop = Non -operating

MAJOR AND NATIONAL AIRLINE OPERATING EXPENSE 1987 (CONTINUED)

	Southern Air Transport	Southwest Airlines	Tower Air	Trans World Airlines	United Air Lines	USAir	World Airways
Mtls- Aircraft Fuel & Oil	$ 5,547,385	121,900,144	15,922,702	618,518,448	1,228,180,039	260,740,000	15,746,000
Mtls- Passenger Food	0	3,128,350	4,483,536	125,734,912	271,494,973	49,114,000	2,067,704
Svcs- Advertising & Other Promotion	151,900	21,143,636	91,551	92,865,943	174,913,540	32,027,000	1,000
Landing Fees	374,651	16,567,350	1,681,273	71,192,358	131,594,437	33,309,000	1,464,000
Rentals	6,366,295	31,859,978	5,381,407	190,459,933	211,278,148	58,582,000	12,391,069
Total Salaries & Benefits	16,576,753	220,827,884	11,092,989	1,305,148,802	2,870,584,289	793,185,000	22,966,818
Maintenance Labor	2,936,007	8,339,586	0	140,723,620	327,259,101	71,192,000	785,000
Bfts- Personnel Expense	3,641,716	15,727,105	3,347,622	81,909,719	159,123,400	39,831,000	4,887,223
LABOR (Less Maint Labor)	9,999,030	196,761,193	7,745,367	1,082,515,463	2,384,201,788	682,162,000	17,294,595
Exp- Maintenance	36,048,655	70,678,080	21,986,094	460,728,472	915,141,451	208,470,000	14,325,000
Svcs- Traffic Commissions - Passenger	0	50,630,929	6,874,513	322,221,764	835,311,497	144,712,000	557,000
Nonop- Interest Income	0	0	0	0	0	0	0
Nonop- Int. on L-T-D & Cap.Leases	3,427,210	25,648,889	5,649,126	239,996,373	118,143,012	37,458,000	6,241,000
Nonop- Int. Exp. Other	0	-7,740,785	0	15,462,586	46,431,861	-18,648,000	518,000
INTEREST EXPENSE	3,427,210	17,908,104	5,649,126	255,458,959	164,574,873	18,810,000	6,759,000
Total Operating Expenses	$ 56,510,185	657,397,127	80,038,991	3,801,239,217	7,711,577,440	1,806,855,000	86,266,000
Rev- Passenger	0	676,268,440	56,898,390	3,414,807,616	6,774,077,606	1,948,858,000	0

Abbreviations

Mtls = Materials	Bfts = Benefits	Nonop = Non -operating
Rev = Revenue	L-T-D & Cap Leases = Long Term Debt and Capital Leases	

MAJOR AND NATIONAL AIRLINE OPERATING EXPENSE 1988

	Air Wisconsin	Alaska Airlines	Aloha Airlines	America West Airlines	American Airlines	American Trans Air	Aspen Airways	Braniff
Mtls- Aircraft Fuel & Oil	$ 15,646,450	88,924,000	16,594,113	114,985,382	1,100,956,000	49,685,306	6,995,243	77,938,000
Mtls- Passenger Food	1,064,395	34,065,000	423,703	15,164,355	369,533,000	7,819,272	217,093	12,409,000
Svcs- Advertising & Other Promotion	1,110 662	25,594,000	4,049,808	19,516,000	206,327,000	490,525	217,511	14,826,000
Landing Fees	4,480,929	11,932,000	1,055,893	17,436,881	144,056,000	3,690,221	1,513,575	7,355,000
Rentals	15,361,078	63,956,000	11,399,071	151,913,457	421,742,000	6,910,508	9,165,504	59,928,000
Total Salaries & Benefits	51,302,301	210,555,000	44,705,416	194,927,860	2,817,394,000	62,879,963	19,955,431	89,997,000
Maintenance Labor	7,678,314	6,801,000	4,344,974	9,329,011	249,835,000	10,490,564	3,997,119	82,000
Bfts- Personnel Expense	2,804,347	12,135,000	977,044	10,535,487	145,144,000	12,462,368	1,204,286	9,997,000
LABOR (Less Maint Labor)	40,819,140	191,619,000	39,383,398	175,063,362	2,422,415,000	39,927,031	14,754,026	79,918,000
Exp- Maintenance	32,415,231	57,298,000	19,327,592	46,547,206	842,512,000	55,424,087	14,155,067	51,208,000
Svcs- Traffic Commissions - Passenger	14,242,831	53,741,000	3,996,313	54,215,701	726,575,000	2,529,126	4,197,788	28,795,000
Nonop- Interest Income	0	0	0	0	0	0	0	0
Nonop- Int. on L-T-D & Cap.Leases	6,661,513	16,348,000	6,211,264	41,278,850	176,695,000	8,771,425	213,991	875,000
Nonop- Int. Exp. Other	-230,653	104,000	138,316	-3,263,000	-17,504,000	42,077	257,989	688,000
INTEREST EXPENSE	6,430,860	16,452,000	6,349,580	38,015,850	159,191,000	8,813,502	471,980	1,563,000
Total Operating Expenses	$ 161,889,145	646,908,000	126,490,626	763,319,309	7,749,591,000	226,789,255	60,587,085	424,462,000
Rev- Passenger s	172,916,937	611,445,000	101,551,479	731,297,176	7,553,296,000	0	54,283,559	367,349,000

Abbreviations

Mtls	= Materials	Bfts	=	Benefits
Rev	= Revenue	L-T-D & Cap Leases	=	Long Term Debt and Capital Leases

Nonop = Non -operating

MAJOR AND NATIONAL AIRLINE OPERATING EXPENSE 1988 (CONTINUED)

	Continental Airlines	Delta Air Lines	Eastern Air Lines	Evergreen Int'l. Airlines	Federal Express	Flying Tiger Line	Hawaiian Airlines	Horizon Air
Mtls- Aircraft Fuel & Oil	$ 700,644,000	974,068,000	565,568,229	18,056,866	123,143,000	208,399,000	68,443,194	9,597,000
Mtls- Passenger Food	138,185,000	276,752,000	116,195,184	9,303	0	2,919,000	7,279,545	589,000
Svcs- Advertising & Other Promotion	105,075,000	106,363,000	104,547,711	59,965	81,548,000	8,195,000	5,608,535	1,611,000
Landing Fees	95,840,000	113,659,000	74,987,472	1,492,962	13,241,000	41,429,000	5,670,278	2,262,000
Rentals	529,983,000	494,823,000	127,212,806	16,532,805	341,517,000	84,841,000	35,662,557	19,430,000
Total Salaries & Benefits	1,080,236,000	2,876,243,000	1,541,950,843	41,337,830	2,063,013,000	347,106,000	108,922,743	40,019,000
Maintenance Labor	112,700,000	166,684,000	169,384,652	9,426,582	73,682,000	39,785,000	7,791,112	4,363,000
Bfts- Personnel Expense	83,227,000	140,042,000	78,053,121	8,111,052	55,141,000	36,269,000	8,677,588	3,226,000
LABOR (Less Maint Labor)	884,309,000	2,569,517,000	1,294,513,070	23,800,196	1,934,190,000	271,052,000	92,454,043	32,430,000
Exp- Maintenance	608,871,000	630,364,000	583,910,555	36,227,346	289,938,000	152,043,000	69,635,581	17,851,000
Svcs- Traffic Commissions - Passenger	387,256,000	706,871,000	302,445,514	0	0	0	10,064,402	8,074,000
Nonop- Interest Income	0	0	0	0	0	0	0	0
Nonop- Int. on L-T-D & Cap.Leases	265,826,000	80,448,000	277,858,779	14,590,428	77,204,000	43,789,000	6,641,327	1,255,000
Nonop- Int. Exp. Other	13,989,000	-24,949,000	22,852,334	0	0	5,036,000	745,265	0
INTEREST EXPENSE	279,815,000	55,499,000	300,711,113	14,590,428	77,204,000	48,825,000	7,386,592	1,255,000
Total Operating Expenses	$ 4,474,887,000	6,868,638,000	4,097,557,217	164,630,977	3,880,193,000	1,161,475,000	358,199,129	110,523,000
Rev- Passenger s	4,074,316,000	6,839,290,000	3,481,832,049	0	0	0	266,682,289	106,706,000

Abbreviations

Mtls	= Materials	B⁼ts	= Benefits	Nonop	= Non -operating
Rev	= Revenue	L-T-D & Cap Leases	= Long Term Debt and Capital Leases		

MAJOR AND NATIONAL AIRLINE OPERATING EXPENSE 1988 (CONTINUED)

	Markair	Midway Airlines	Midwest Express Airlines	Northwest Airlines	Pan Am. World Airways	Piedmont Aviation	Presidential Airways	Southern Air Transport
Mtls- Aircraft Fuel & Oil	$ 12,365,628	63,906,279	8,427,701	810,013,000	545,852,000	280,914,801	10,831,677	11,164,667
Mtls- Passenger Food	1,687,632	8,338,080	4,459,338	155,789,000	115,445,000	71,149,890	1,194,029	0
Svcs- Advertising & Other Promotion	1,907,139	13,366,474	3,549,759	85,144,000	84,626,000	39,634,612	869,418	332,866
Landing Fees	854,114	5,120,434	1,035,620	121,187,000	93,985,000	38,646,713	1,245,166	882,301
Rentals	5,249,939	36,873,349	1,363,328	234,151,000	298,695,000	200,773,605	23,391,649	6,790,966
Total Salaries & Benefits	30,608,942	88,239,071	13,046,717	1,781,118,000	1,212,383,000	806,937,910	20,741,825	39,113,550
Maintenance Labor	3,602,107	7,955,907	1,438,269	160,185,000	96,283,000	37,655,992	2,860,497	6,121,591
Bfts- Personnel Expense	618,208	7,605,276	1,120,609	116,887,000	82,276,000	50,795,865	1,175,953	9,345,362
LABOR (Less Maint Labor)	26,388,627	72,677,888	10,487,839	1,504,046,000	1,033,824,000	718,486,053	16,705,375	23,646,597
Exp- Maintenance	13,581,952	51,802,573	8,290,242	607,097,000	376,286,000	221,519,396	6,456,206	43,363,374
Svcs- Traffic Commissions - Passenger	3,107,243	29,689,685	4,638,898	853,817,000	278,357,000	164,716,843	4,990,963	0
Nonop- Interest Income	0	0	0	0	0	0	0	0
Nonop- Int. on L-T-D & Cap.Leases	11,594,811	8,324,424	0	58,330,000	109,586,000	29,951,170	1,276,707	9,695,953
Nonop- Int. Exp. Other	0	155,190	0	-34,696,000	13,834,000	-17,235,748	1,394,813	0
INTEREST EXPENSE	11,594,811	8,479,614	0	23,634,000	123,420,000	12,715,422	2,671,520	9,695,953
Total Operating Expenses	$ 83,667,701	374,509,269	55,622,378	5,392,145,000	3,697,910,000	2,085,816,720	83,055,462	119,472,116
Rev- Passenger s	52,716,492	367,727,005	58,245,951	4,815,771,000	3,069,766,000	2,259,185,462	60,542,927	0

Abbreviations

Mtls	=	Materials	Bfts	=	Benefits	Nonop = Non -operating
Rev	=	Revenue	L-T-D & Cap Leases	=	Long Term Debt and Capital Leases	

MAJOR AND NATIONAL AIRLINE OPERATING EXPENSE 1988 (CONTINUED)

	Southwest Airlines	Tower Air	Trans World Airlines	United Air Lines	United Parcel Service	USAir	World Airways
Mtls- Aircraft Fuel & Oil	$ 130,320,436	21,595,195	623,170,600	1,152,723,832	24,108,498	335,186,000	23,075,000
Mtls- Passenger Food	1,846,942	6,374,263	141,479,148	288,419,203	0	76,159,000	2,658,000
Svcs- Advertising & Other Promotion	28,007,958	631,482	98,660,360	209,175,578	0	56,285,000	0
Landing Fees	19,634,928	2,505,024	81,987,548	137,871,435	2,710,461	45,947,000	3,142,000
Rentals	43,431,633	5,557,731	212,503,427	276,476,805	470,644	175,775,000	18,409,000
Total Salaries & Benefits	271,694,911	19,438,010	1,459,239,941	2,925,852,211	30,460,403	1,107,190,000	24,301,000
Maintenance Labor	11,190,998	0	156,855,733	331,717,166	9,114,991	101,589,000	1,006,000
Bfts- Personnel Expense	16,732,471	4,819,219	85,055,273	158,350,707	3,067,001	63,962,000	5,455,000
LABOR (Less Maint Labor)	243,771,442	14,618,791	1,217,328,935	2,435,784,338	18,278,411	941,639,000	17,840,000
Exp- Maintenance	77,749,445	21,042,730	513,613,038	937,525,322	27,501,530	342,910,000	14,703,000
Svcs- Traffic Commissions - Passenger	52,976,067	11,466,218	372,396,705	1,026,820,797	0	195,350,000	602,000
Nonop- Interest Income	0	0	0	0	0	0	0
Nonop- Int. on L-T-D & Cap.Leases	29,221,179	5,972,300	331,708,895	194,951,126	55,665	55,594,000	2,168,000
Nonop- Int. Exp. Other	-9,278,823	0	12,484,308	5,572,981	-34,425	-24,974,000	-105,000
INTEREST EXPENSE	19,942,356	5,972,300	344,193,203	200,524,107	21,240	30,620,000	2,063,000
Total Operating Expenses	$ 774,367,173	121,998,771	4,101,725,920	8,127,784,098	126,332,497	2,659,266,000	107,465,000
Rev- Passengers	828,343,642	72,383,693	3,700,454,425	7,650,565,215	0	2,654,484,000	0

Abbreviations

Mtls	=	Materials	Bfts	=	Benefits	Nonop	=	Non -operating
Rev	=	Revenue	L-T-D & Cap Leases	=	Long Term Debt and Capital Leases		

MAJOR AND NATIONAL AIRLINE OPERATING EXPENSE 1989

	Air Wisconsin	Alaska Airlines	Aloha Airlines	America West Airlines	American Airlines	American Trans Air	Aspen Airways	Braniff
Mtls- Aircraft Fuel & Oil	$ 18,514,555	111,769,000	20,807,697	147,297,599	1,379,219,000	50,762,676	7,964,657	53,391,953
Mtls- Passenger Food	1,169,033	37,891,000	508,885	21,484,689	436,018,000	6,629,895	288,684	7,876,552
Svcs- Advertising & Other Promotion	1,040,849	25,277,000	4,720,845	30,224,553	194,055,000	967,649	266,443	7,940,581
Landing Fees	5,651,258	12,969,000	1,286,906	21,291,143	167,045,000	4,135,487	1,623,086	4,736,581
Rentals	23,527,146	71,251,000	15,225,706	176,276,665	586,250,000	8,747,483	8,990,273	44,385,928
Total Salaries & Benefits	55,582,364	239,666,000	52,303,911	262,647,991	3,343,283,000	74,365,277	20,931,793	77,077,821
Maintenance Labor	8,502,192	7,782,000	5,032,213	16,143,762	299,885,000	12,042,855	3,925,223	5,739,809
Bfts- Personnel Expense	3,424,362	14,239,000	1,204,968	14,811,565	174,705,000	17,414,808	1,136,606	7,525,369
LABOR (Less Maint Labor)	43,656,310	217,645,000	46,066,730	231,692,664	2,868,693,000	44,907,614	15,869,964	63,812,643
Exp- Maintenance	36,084,035	69,262,000	26,604,817	53,837,499	1,013,598,000	66,306,626	16,018,269	80,190,000
Svcs- Traffic Commissions - Passenger	13,505,553	64,346,000	4,864,695	71,346,418	843,941,000	2,564,888	4,463,022	21,533,999
Nonop- Interest Income	0	0	0	0	0	0	0	0
Nonop- Int. on L-T-D & Cap.Leases	5,035,192	13,865,000	7,458,482	51,018,566	212,986,000	9,775,970	369,838	10,566,000
Nonop- Int. Exp. Other	-923,091	-3,274,000	554,888	-7,249,685	-59,569,000	61,588	357,091	4,251,000
INTEREST EXPENSE	4,112,101	10,591,000	8,013,370	43,768,881	153,417,000	9,837,558	726,929	14,817,000
Total Operating Expenses	$ 176,450,296	736,238,000	146,612,038	950,196,461	9,230,151,000	261,004,534	65,251,973	316,217,093
Rev- Passengers	166,341,995	695,026,000	123,037,864	923,448,931	8,836,152,000	0	57,341,563	385,818,000

Abbreviations

Mtls	=	Materials	Bfts	=	Benefits
Rev	=	Revenue	L-T-D & Cap Leases	=	Long Term Debt and Capital Leases

Nonop = Non -operating

MAJOR AND NATIONAL AIRLINE OPERATING EXPENSE 1989 (CONTINUED)

	Continental Airlines	Delta Air Lines	Eastern Air Lines	Evergreen Int'l. Airlines	Federal Express	Flying Tiger Line	Hawaiian Airlines	Horizon Air
Mtls- Aircraft Fuel & Oil	$ 769,919,000	1,166,166,000	264,596,134	27,472,640	279,022,000	132,710,000	72,359,508	11,504,000
Mtls- Passenger Food	167,741,000	322,414,000	63,332,250	2,080	1,324,000	1,569,000	7,920,266	739,000
Svcs- Advertising & Other Promotion	126,227,000	137,014,000	61,979,975	323,676	101,669,000	4,631,000	6,476,967	2,081,000
Landing Fees	100,804,000	127,313,000	37,141,269	2,689,932	38,759,000	24,300,000	5,436,809	2,318,000
Rentals	603,537,000	546,169,000	161,613,328	21,311,141	512,059,000	59,454,000	43,390,742	20,728,000
Total Salaries & Benefits	1,176,821,000	3,308,703,000	779,678,802	39,972,125	2,744,269,000	201,562,000	121,963,810	44,959,000
Maintenance Labor	229,680,000	185,871,000	47,597,308	8,263,841	89,721,000	24,607,000	9,467,902	4,608,000
Bfts- Personnel Expense	83,992,000	148,570,000	53,593,134	9,458,107	88,251,000	18,836,000	8,864,667	3,483,000
LABOR (Less Maint Labor)	863,149,000	2,974,262,000	678,488,360	22,250,177	2,566,297,000	158,119,000	103,631,241	36,868,000
Exp- Maintenance	659,903,000	679,940,000	299,665,760	52,747,773	413,850,000	91,519,000	77,591,367	21,806,000
Svcs- Traffic Commissions - Passenger	447,841,000	863,856,000	138,209,466	0	0	0	9,980,432	8,938,000
Nonop- Interest Income	0	0	0	0	0	0	0	0
Nonop- Int. on L-T-D & Cap.Leases	247,542,000	65,571,000	275,743,025	24,776,141	81,258,000	22,336,000	3,943,389	2,464,000
Nonop- Int. Exp. Other	7,307,000	-43,317,000	47,190,068	0	11,932,000	2,053,000	1,195,059	0
INTEREST EXPENSE	254,849,000	22,254,000	322,933,093	24,776,141	93,190,000	24,389,000	5,138,448	2,464,000
Total Operating Expenses	$ 4,788,009,000	7,971,765,000	2,416,399,523	186,835,519	5,456,755,000	723,397,000	392,689,330	126,525,000
Rev- Passengers	4,334,429,000	8,048,222,000	1,358,557,546	0	0	0	260,259,369	125,096,000

Abbreviations

Mtls	=	Materials	Bfts	=	Benefits	Nonop	=	Non-operating
Rev	=	Revenue	L-T-D & Cap Leases	=	Long Term Debt and Capital Leases			

MAJOR AND NATIONAL AIRLINE OPERATING EXPENSE 1989 (CONTINUED)

	Markair	Midway Airlines	Midwest Express Airlines	Northwest Airlines	Pan Am. World Airways	Piedmont Aviation	Presidential Airways	Southern Air Transport
Mtls- Aircraft Fuel & Oil	$ 14,517,731	86,863,260	14,773,651	1,021,561,000	599,347,000	195,541,000	8,202,371	10,553,397
Mtls- Passenger Food	1,721,575	10,491,559	6,519,063	187,377,000	120,041,000	43,987,000	672,483	0
Svcs- Advertising & Other Promotion	1,716,255	24,958,035	4,414,829	123,222,000	87,845,000	18,931,000	623,244	180,030
Landing Fees	885,296	8,053,590	1,597,227	132,852,000	91,760,000	25,441,000	986,093	902,552
Rentals	5,199,711	47,764,581	2,294,456	293,374,000	368,733,000	140,426,000	17,106,079	6,889,081
Total Salaries & Benefits	31,549,282	118,708,053	20,108,860	2,011,960,000	1,302,668,000	527,860,000	17,275,004	47,150,768
Maintenance Labor	3,700,662	11,010,357	2,316,927	156,870,000	114,458,000	26,345,000	2,824,156	7,446,457
Bfts- Personnel Expense	993,323	10,266,323	1,422,916	131,572,000	88,077,000	32,133,000	1,168,029	11,980,385
LABOR (Less Maint Labor)	26,855,292	97,431,373	16,369,017	1,723,518,000	1,100,133,000	469,382,000	13,282,819	27,723,926
Exp- Maintenance	15,853,780	59,081,044	11,509,400	679,671,000	442,195,000	167,341,000	8,280,115	52,212,523
Svcs- Traffic Commissions - Passenger	3,435,900	36,112,244	6,948,820	974,252,000	260,159,000	113,181,000	3,723,173	0
Nonop- Interest Income	0	0	0	0	0	0	0	0
Nonop- Int. on L-T-D & Cap.Leases	12,432,662	7,862,585	0	56,401,000	103,586,000	24,812,000	2,209,854	9,909,215
Nonop- Int. Exp. Other	0	175,872	0	-41,042,000	12,923,000	-5,726,000	3,218,783	0
INTEREST EXPENSE	12,432,662	8,038,457	0	15,359,000	116,509,000	19,086,000	5,428,637	9,909,215
Total Operating Expenses	$ 92,391,372	476,574,820	84,398,272	6,263,766,000	3,930,672,000	1,422,938,000	67,495,849	131,969,302
Rev- Passengers	58,837,635	441,574,054	85,056,758	5,635,239,000	3,016,711,000	1,580,094,000	50,980,253	0

Abbreviations

Mtls	=	Materials	Bfts	=	Benefits	Nonop =	Non -operating
Rev	=	Revenue	L-T-D & Cap Leases	=	Long Term Debt and Capital Leases		

MAJOR AND NATIONAL AIRLINE OPERATING EXPENSE 1989 (CONTINUED)

	Southwest Airlines	Tower Air	Trans World Airlines	United Air Lines	United Parcel Service	USAir	World Airways
Mtls- Aircraft Fuel & Oil	$ 169,026,484	24,826,406	715,103,075	1,356,227,352	134,058,979	569,372,000	21,530,000
Mtls- Passenger Food	2,334,249	6,827,054	155,080,118	320,381,054	0	146,796,000	1,843,000
Svcs- Advertising & Other Promotion	26,250,726	260,768	126,994,895	206,412,025	0	132,587,000	1,000
Landing Fees	25,425,908	2,666,535	89,917,463	146,970,813	13,207,074	66,850,000	2,643,000
Rentals	47,810,273	3,334,144	219,725,373	429,985,456	2,370,792	345,779,000	33,641,000
Total Salaries & Benefits	322,276,901	22,467,718	1,636,322,651	3,264,265,618	139,532,316	1,793,417,000	32,923,000
Maintenance Labor	13,886,235	0	176,035,852	426,981,528	38,608,327	162,000,000	1,693,000
Bfts- Personnel Expense	21,226,806	6,236,430	94,964,158	184,999,577	12,704,786	114,965,000	8,037,000
LABOR (Less Maint Labor)	287,163,860	16,231,288	1,365,322,641	2,652,284,513	88,219,203	1,516,452,000	23,193,000
Exp- Maintenance	102,666,576	21,045,231	605,325,570	1,163,132,059	158,279,946	576,010,000	36,284,000
Svcs- Traffic Commissions - Passenger	61,005,859	11,662,343	386,646,813	1,245,507,364	0	301,328,000	70,000
Nonop- Interest Income	0	0	0	0	0	0	0
Nonop- Int. on L-T-D & Cap.Leases	33,550,038	6,755,647	391,195,232	165,206,418	-216,389	87,221,000	1,291,000
Nonop- Int. Exp. Other	-10,226,529	0	22,808,417	-6,202,374	80,359	-59,252,000	476,000
INTEREST EXPENSE	23,323,509	6,755,647	414,003,649	159,004,044	-136,030	27,969,000	1,767,000
Total Operating Expenses	$ 917,549,416	131,200,628	4,482,970,858	9,185,032,148	617,450,539	4,432,755,000	149,434,000
Rev- Passengers	973,568,083	75,792,964	3,835,929,178	8,457,744,566	0	3,979,539,000	0

Abbreviations

Mtls = Materials	Bfts = Benefits	Nonop = Non -operating
Rev = Revenue	L-T-D & Cap Leases = Long Term Debt and Capital Leases	

MAJOR AND NATIONAL AIRLINE OPERATING EXPENSE 1990

	Air Wisconsin	Alaska Airlines	Aloha Airlines	America West Airlines	American Airlines	American Trans Air	Aspen Airways	Continental Airlines
Mtls- Aircraft Fuel & Oil	$ 25,219,544	106,416,000	31,763,272	250,116,820	1,913,504,000	79,489,670	3,943,780	1,033,112,000
Mtls- Passenger Food	1,406,198	30,761,000	664,433	42,752,635	504,318,000	14,688,729	192,553	194,355,000
Svcs- Advertising & Other Promotion	969,010	22,859,000	6,126,342	35,178,157	204,782,000	2,746,916	159,208	115,542,000
Landing Fees	5,669,639	11,913,000	1,519,953	25,584,543	188,452,000	5,996,039	785,246	109,918,000
Rentals	25,842,310	67,025,000	18,859,595	232,610,512	686,842,000	22,372,676	6,840,773	631,183,000
Total Salaries & Benefits	60,451,858	196,488,000	65,064,698	365,540,857	3,798,232,000	88,350,529	11,777,165	1,337,505,000
Maintenance Labor	9,276,135	10,161,000	6,944,499	24,598,600	340,185,000	15,499,153	1,670,646	268,489,000
Bfts- Personnel Expense	3,280,297	11,836,000	1,459,699	21,326,657	194,444,000	19,544,898	595,654	84,708,000
LABOR (Less Maint Labor)	47,895,426	174,491,000	56,660,500	319,615,600	3,263,603,000	53,306,478	9,510,865	984,308,000
Exp- Maintenance	41,510 236	93,310,000	29,128,154	95,280,017	1,214,528,000	78,828,030	5,813,898	712,076,000
Svcs- Traffic Commissions - Passenger	15,803,309	52,779,000	6,041,981	97,533,204	962,089,000	3,794,772	2,300,060	479,286,000
Nonop- Interest Income	0	0	0	0	0	0	0	0
Nonop- Int. on L-T-D & Cap.Leases	6,864,015	13,002,000	7,650,805	67,109,973	257,683,000	11,003,977	64,482	218,900,000
Nonop- Int. Exp. Other	-1,127,147	-8,630,000	890,117	-6,374,550	-91,935,000	117,019	726,558	6,885,000
INTEREST EXPENSE	5,736,368	4,372,000	8,540,922	60,735,423	165,748,000	11,120,996	791,040	225,785,000
Total Operating Expenses	$ 198,978,317	642,318,000	180,939,017	1,353,272,972	10,940,704,000	359,420,195	35,612,194	5,444,049,000
Rev- Passenger s	188,567,329	781,995,000	153,068,544	1,232,058,708	9,729,319,000	0	29,273,140	4,540,771,000

Abbreviations

Mtls = Materials	Bfts = Benefits	Nonop = Non -operating
Rev = Revenue	L-T-D & Cap Leases = Long Term Debt and Capital Leases	

MAJOR AND NATIONAL AIRLINE OPERATING EXPENSE 1990 (CONTINUED)

	Delta Air Lines	Eastern Air Lines	Evergreen Int'l. Airlines	Federal Express	Hawaiian Airlines	Horizon Air	Markair	Midway Airlines
Mtls- Aircraft Fuel & Oil	$ 1,612,935,000	469,593,018	33,551,718	496,272,000	85,850,634	17,806,000	20,290,802	154,631,647
Mtls- Passenger Food	340,900,000	95,638,923	55,287	2,778,000	7,792,166	1,003,000	1,674,717	23,226,111
Svcs- Advertising & Other Promotion	179,200,000	100,229,561	87,561	44,031,000	7,204,119	2,419,000	1,693,154	23,553,056
Landing Fees	154,391,000	45,571,150	2,041,045	61,764,000	5,000,990	3,274,000	856,267	7,680,413
Rentals	569,300,000	149,739,515	22,417,906	422,587,000	55,873,371	22,584,000	6,819,055	87,180,854
Total Salaries & Benefits	3,606,795,000	789,037,589	46,634,761	1,094,551,000	122,519,991	52,071,000	36,246,320	180,432,352
Maintenance Labor	208,134,000	64,874,085	14,372,842	124,048,000	11,379,478	5,767,000	4,861,130	23,932,323
Bfts- Personnel Expense	167,973,000	56,626,916	10,508,203	71,987,000	9,254,729	4,091,000	1,282,782	14,712,025
LABOR (Less Maint Labor)	3,230,688,000	667,536,588	21,753,716	898,516,000	101,885,784	42,213,000	30,102,408	141,788,004
Exp- Maintenance	785,979,000	401,518,722	60,016,318	462,912,000	91,650,147	28,900,000	21,838,242	93,517,408
Svcs- Traffic Commissions - Passenger	847,799,000	180,439,905	0	0	12,821,044	11,498,000	4,360,946	51,732,827
Nonop- Interest Income	0	0	0	0	0	0	0	0
Nonop- Int. on L-T-D & Cap.Leases	94,347,000	209,035,559	40,422,523	190,506,000	7,870,517	3,857,000	9,984,178	13,444,492
Nonop- Int. Exp. Other	-36,961,000	16,419,210	0	5,810,000	882,560	0	0	77,704
INTEREST EXPENSE	57,386,000	225,454,769	40,422,523	196,316,000	8,753,077	3,857,000	9,984,178	13,522,196
Total Operating Expenses	$ 8,981,211,000	2,715,149,964	194,857,248	7,189,297,000	432,751,719	157,990,000	112,533,758	699,339,127
Rev- Passenger s	8,150,766,000	1,939,279,902	0	0	249,019,806	155,005,000	71,994,976	585,979,772

Abbreviations

Mtls	=	Materials	Bfts	=	Benefits
Rev	=	Revenue	L-T-D & Cap Leases	=	Long Term Debt and Capital Leases

Nonop = Non -operating

MAJOR AND NATIONAL AIRLINE OPERATING EXPENSE 1990 (CONTINUED)

	Midwest Express Airlines	Northwest Airlines	Pan Am. World Airways	Southern Air Transport	Southwest Airlines	Tower Air	Trans World Airlines	United Air Lines
Mtls- Aircraft Fuel & Oil	$ 25,195,570	1,408,958,000	788,690,000	16,619,330	242,421,298	39,540,313	903,003,895	1,815,702,955
Mtls- Passenger Food	8,648,989	234,257,000	136,245,000	0	2,759,500	8,744,592	163,887,040	355,128,346
Svcs- Advertising & Other Promotion	5,661,157	131,656,000	72,968,000	60,960	28,979,827	276,688	119,793,076	213,661,666
Landing Fees	2,005,717	145,471,000	100,598,000	1,416,891	31,526,158	3,529,732	94,627,506	180,843,789
Rentals	8,020,263	359,422,000	415,538,000	5,914,744	55,725,803	3,996,921	225,735,438	648,466,580
Total Salaries & Benefits	26,605,452	2,302,795,000	1,361,012,000	48,483,864	381,541,262	32,788,251	1,638,389,667	3,670,024,290
Maintenance Labor	2,058,324	181,687,000	119,373,000	8,598,649	16,654,879	0	177,270,154	495,316,667
Bfts- Personnel Expense	2,212,796	165,052,000	93,494,000	10,214,406	23,347,984	7,304,183	92,705,035	202,176,019
LABOR (Less Maint Labor)	22,334,332	1,956,056,000	1,148,145,000	29,670,809	341,538,399	25,484,068	1,368,414,478	2,972,531,604
Exp- Maintenance	11,602,639	764,437,000	489,455,000	48,495,656	114,266,949	21,693,325	588,914,254	1,367,725,352
Svcs- Traffic Commissions - Passenger	9,133,107	1,133,452,000	297,252,000	0	71,877,722	12,648,271	399,230,934	1,601,481,438
Nonop- Interest Income	0	0	0	0	0	0	0	0
Nonop- Int. on L-T-D & Cap.Leases	560	110,386,000	137,988,000	6,773,489	32,003,234	6,521,265	349,591,455	145,458,502
Nonop- Int. Exp. Other	0	-22,451,000	16,029,000	0	-13,737,724	0	19,874,903	-24,727,573
INTEREST EXPENSE	560	87,935,000	154,017,000	6,773,489	18,265,510	6,521,265	369,466,358	120,730,929
Total Operating Expenses	$ 120,749,639	7,398,773,000	4,421,181,000	133,475,772	1,105,215,047	160,526,112	4,768,287,673	11,010,310,090
Rev- Passengers	115,670,908	6,337,961,000	3,346,701,000	0	1,144,421,193	76,478,954	3,848,870,957	9,534,189,086

Abbreviations

Mtls	=	Materials	Bfts	=	Benefits	
Rev	=	Revenue	L-T-D & Cap Leases	=	Long Term Debt and Capital Leases	

Nonop = Non -operating

MAJOR AND NATIONAL AIRLINE OPERATING EXPENSE 1990 (CONTINUED)

	United Parcel Service	USAir	World Airways
Mtls- Aircraft Fuel & Oil	$ 219,109,348	976,494,000	45,977,000
Mtls- Passenger Food	0	213,878,000	3,635,000
Svcs- Advertising & Other Promotion	0	134,238,000	42,000
Landing Fees	16,405,286	103,808,000	3,158,000
Rentals	30,091,882	572,037,000	37,720,000
Total Salaries & Benefits	194,867,246	2,701,064,000	49,737,000
Maintenance Labor	40,338,193	192,699,000	2,325,000
Bfts- Personnel Expense	19,296,045	171,217,000	13,942,000
LABOR (Less Maint Labor)	135,233,008	2,337,148,000	33,470,000
Exp- Maintenance	250,362,278	797,874,000	57,949,000
Svcs- Traffic Commissions - Passenger	25,058,719	431,651,000	23,000
Nonop- Interest Income	0	0	0
Nonop- Int. on L-T-D & Cap.Leases	-81,798	139,641,000	5,115,000
Nonop- Int. Exp. Other	274,496	-42,165,000	1,965,000
INTEREST EXPENSE	192,698	97,476,000	7,080,000
Total Operating Expenses	$ 878,945,645	6,627,939,000	225,400,000
Rev- Passengers	0	5,753,353,000	0

Abbreviations

Mtls = Materials	Bfts = Benefits	Nonop = Non -operating
Rev = Revenue	L-T-D & Cap Leases = Long Term Debt and Capital Leases	

MAJOR AND NATIONAL AIRLINE OPERATING EXPENSE 1991

	Air Wisconsin	Alaska Airlines	Aloha Airlines	America West Airlines	American Airlines	American Trans Air	Aspen Airways	Braniff
Mtls- Aircraft Fuel & Oil	$ 23,540,354	135,538,000	28,726,999	234,040,484	1,794,003,000	92,857,818	1,255,242	4,116,986
Mtls- Passenger Food	1,348,851	46,203,000	768,288	45,508,930	575,463,000	16,374,295	168,615	519,271
Svcs- Advertising & Other Promotion	3,089,644	30,443,000	7,220,151	31,198,192	249,709,000	3,159,983	51,674	666,790
Landing Fees	6,739,528	19,457,000	1,649,883	30,321,461	220,710,000	7,092,657	187,251	1,911,187
Rentals	34,447,497	98,125,000	20,683,821	319,241,535	848,474,000	27,778,828	1,309,305	1,804,576
Total Salaries & Benefits	70,307,136	292,348,000	74,297,220	402,443,028	4,249,838,000	100,093,714	2,632,811	3,560,536
Maintenance Labor	10,092,240	14,611,000	6,993,697	28,776,190	382,368,000	18,847,969	230,945	602,933
Bfts- Personnel Expense	4,094,819	17,671,000	1,426,324	24,800,878	220,344,000	21,078,798	273,303	0
LABOR (Less Maint Labor)	56,120,077	260,066,000	65,877,199	348,865,960	3,647,126,000	60,166,947	2,128,563	2,957,603
Exp- Maintenance	48,507,369	100,945,000	33,633,454	99,619,621	1,315,113,000	89,123,967	1,411,067	2,414,005
Svcs- Traffic Commissions - Passenger	16,127,604	71,061,000	6,599,653	127,462,589	1,109,180,000	4,114,814	767,564	188,515
Nonop- Interest Income	0	0	0	0	0	0	0	0
Nonop- Int. on L-T-D & Cap.Leases	10,957,917	31,220,000	6,540,865	68,576,470	339,156,000	9,625,386	-2,874	0
Nonop- Int. Exp. Other	-1,463,041	-6,900,000	550,347	-6,664,437	-146,841,000	138,729	273,912	208,393
INTEREST EXPENSE	9,494,876	24,320,000	7,091,212	61,912,033	192,315,000	9,764,115	271,038	208,393
Total Operating Expenses	$ 226,994,537	914,000,000	196,214,066	1,524,726,356	12,080,570,000	410,552,589	11,077,743	17,061,648
Rev- Passengers	192,272,166	807,867,000	176,760,245	1,310,037,031	10,694,110,000	0	11,403,464	8,858,766

Abbreviations

Mtls	=	Materials	Bfts	=	Benefits	Nonop = Non -operating
Rev	=	Revenue	L-T-D & Cap Leases	=	Long Term Debt and Capital Leases	

MAJOR AND NATIONAL AIRLINE OPERATING EXPENSE 1991 (CONTINUED)

	Continental Airlines	Delta Air Lines	Evergreen Int'l. Airlines	Federal Express	Hawaiian Airlines	Horizon Air	Markair	Midway Airlines
Mtls- Aircraft Fuel & Oil	$ 916,314,000	1,490,686,000	54,860,041	427,411,000	64,817,717	15,909,000	18,752,474	55,925,000
Mtls- Passenger Food	242,175,000	403,496,000	322,826	3,905,000	6,954,382	1,215,000	1,822,719	7,294,000
Svcs- Advertising & Other Promotion	108,571,000	229,423,000	1,017,363	56,358,000	3,108,593	2,132,000	2,453,880	7,271,000
Landing Fees	121,103,000	189,349,000	4,205,280	58,604,000	3,869,561	4,718,000	1,178,724	5,951,000
Rentals	664,349,000	819,299,000	36,524,276	444,656,000	75,636,221	25,443,000	8,748,597	43,404,000
Total Salaries & Benefits	1,408,237,000	4,017,137,000	70,454,254	1,206,440,000	117,246,523	60,269,000	35,105,884	82,764,000
Maintenance Labor	260,327,000	230,188,000	23,538,480	128,539,000	10,559,432	6,619,000	4,964,339	10,890,000
Bfts- Personnel Expense	83,935,000	206,176,000	16,930,677	70,564,000	12,880,708	4,568,000	1,248,490	4,733,000
LABOR (Less Maint Labor)	1,063,975,000	3,580,773,000	29,985,097	1,007,337,000	93,806,383	49,082,000	28,893,055	67,141,000
Exp- Maintenance	755,439,000	960,996,000	112,407,092	478,700,000	87,839,023	33,442,000	22,120,489	54,902,000
Svcs- Traffic Commissions - Passenger	541,366,000	1,044,227,000	0	0	42,628,607	13,284,000	4,734,082	19,833,000
Nonop- Interest Income	0	0	0	0	0	0	0	0
Nonop- Int. on L-T-D & Cap.Leases	127,261,000	170,200,000	41,183,110	188,740,000	8,260,322	1,183,000	9,285,306	4,491,000
Nonop- Int. Exp. Other	3,868,000	-52,614,000	0	6,658,000	1,674,329	668,000	0	0
INTEREST EXPENSE	131,129,000	117,586,000	41,183,110	195,398,000	9,934,651	1,851,000	9,285,306	4,491,000
Total Operating Expenses	$ 5,552,639,000	10,329,134,000	324,762,793	7,234,643,000	459,145,280	175,177,000	116,437,540	282,153,000
Rev- Passengers	4,656,484,000	9,358,412,000	0	0	285,234,459	173,651,000	64,450,933	335,074,000

Abbreviations

Mtls = Materials Bfts = Benefits Nonop = Non -operating

Rev = Revenue L-T-D & Cap Leases = Long Term Debt and Capital Leases

MAJOR AND NATIONAL AIRLINE OPERATING EXPENSE 1991 (CONTINUED)

	Midwest Express Airlines	Northwest Airlines	Pan Am. World Airways	Southern Air Transport	Southwest Airlines	Tower Air	Trans World Airlines	United Air Lines
Mtls- Aircraft Fuel & Oil	$ 24,572,317	1,239,058,000	415,486,000	20,254,714	226,510,372	59,460,068	647,159,886	1,678,040,000
Mtls- Passenger Food	8,700,300	246,999,000	75,162,000	0	3,332,454	13,384,693	125,783,840	418,541,000
Svcs- Advertising & Other Promotion	5,760,164	100,429,000	47,527,000	283,908	37,347,195	1,501,357	62,965,893	206,738,000
Landing Fees	2,213,320	160,955,000	57,464,000	1,844,262	39,801,647	5,212,591	85,671,781	217,089,000
Rentals	10,331,960	401,700,000	292,776,000	10,927,545	92,546,127	4,627,368	199,678,678	867,525,000
Total Salaries & Benefits	32,012,129	2,480,186,000	844,886,000	52,875,670	435,162,837	39,851,686	1,468,764,213	4,204,694,000
Maintenance Labor	3,233,890	189,294,000	81,325,000	8,268,694	19,638,735	0	167,759,256	504,525,000
Bfts- Personnel Expense	1,931,592	176,039,000	56,453,000	13,627,907	27,201,950	9,010,915	69,483,999	238,860,000
LABOR (Less Maint Labor)	26,846,647	2,114,853,000	707,108,000	30,979,069	388,322,152	30,840,771	1,231,520,958	3,461,309,000
Exp- Maintenance	11,641,415	771,447,000	277,069,000	53,923,617	135,035,512	33,211,029	516,058,575	1,422,996,000
Svcs- Traffic Commissions - Passenger	10,677,356	1,288,442,000	155,424,000	0	81,244,596	15,109,941	319,427,442	1,864,641,000
Nonop- Interest Income	0	0	0	0	0	0	0	0
Nonop- Int. on L-T-D & Cap.Leases	5,121	191,207,000	50,464,000	4,047,456	43,942,086	8,717,857	313,339,975	173,414,000
Nonop- Int. Exp. Other	41,485	-17,776,000	8,862,000	0	-15,300,969	0	20,385,180	-54,829,000
INTEREST EXPENSE	46,606	173,431,000	59,326,000	4,047,456	28,641,117	8,717,857	333,725,155	118,585,000
Total Operating Expenses	$ 124,694,263	7,593,826,000	2,562,650,000	156,573,993	1,251,633,549	226,052,025	4,005,854,196	12,150,737,000
Rev- Passengers	116,212,451	6,499,364,000	1,720,053,000	0	1,267,896,738	109,572,472	2,914,858,872	10,160,804,000

Abbreviations

Mtls	=	Materials	Bfts	=	Benefits	
Rev	=	Revenue	L-T-D & Cap Leases	=	Long Term Debt and Capital Leases	

Nonop = Non -operating

MAJOR AND NATIONAL AIRLINE OPERATING EXPENSE 1991 (CONTINUED)

	United Parcel Service	USAir	World Airways
Mtls- Aircraft Fuel & Oil	$ 217,172,000	775,359,000	50,433,000
Mtls- Passenger Food	0	258,379,000	4,351,000
Svcs- Advertising & Other Promotion	12,000	106,801,000	5,000
Landing Fees	21,094,000	106,390,000	4,013,000
Rentals	92,229,000	590,228,000	35,081,000
Total Salaries & Benefits	257,639,000	2,523,703,000	41,103,000
Maintenance Labor	61,542,000	193,859,000	2,314,000
Bfts- Personnel Expense	17,992,000	131,212,000	10,533,000
LABOR (Less Maint Labor)	178,105,000	2,198,632,000	28,256,000
Exp- Maintenance	335,680,000	768,929,000	45,469,000
Svcs- Traffic Commissions - Passenger	0	488,368,000	91,000
Nonop- Interest Income	0	0	0
Nonop- Int. on L-T-D & Cap.Leases	-9,000	178,113,000	853,000
Nonop- Int. Exp. Other	55,000	-9,883,000	1,299,000
INTEREST EXPENSE	46,000	168,230,000	2,152,000
Total Operating Expenses	$ 1,083,631,000	6,251,271,000	211,348,000
Rev- Passenger s	0	5,687,892,000	0

Abbreviations

Mtls	=	Materials	Bfts	=	Benefits
Rev	=	Revenue	L-T-D & Cap Leases	=	Long Term Debt and Capital Leases

Nonop = Non -operating

OPERATING EXPENSES OF U.S. AIR CARRIERS
DOMESTIC AND INTERNATIONAL OPERATIONS
Calendar Years 1977 - 1991 (Millions of Dollars)

Year	TOTAL Operating Expenses	Flying Operations	Main-tenance	Passenger Service	Aircraft & Traffic Servicing	Promotion and Sales	Depreciation & Amorti-zation	Other
DOMESTIC OPERATIONS								
1977	$15,166	$ 5,288	$2,001	$1,461	$2,728	$1,713	$ 967	$1,008
1978	17,172	5,669	2,155	1,711	3,120	2,040	1,231	1,246
1979	21,523	7,998	2,457	2,091	3,702	2,564	1,373	1,337
1980	26,409	11,029	2,758	2,329	4,051	3,096	1,560	1,586
1981	29,051	12,037	2,822	2,522	4,497	3,708	1,723	1,742
1982	29,478	11,529	2,709	2,668	4,665	4,160	1,876	1,869
1983	31,186	11,370	2,878	2,983	5,104	4,764	2,107	1,980
1984	33,812	12,161	3,176	3,192	5,369	5,310	2,223	2,380
1985	36,611	12,684	3,604	3,464	5,781	6,089	2,318	2,670
1986	39,934	11,368	4,475	3,793	7,680	6,820	2,652	3,171
1987	43,925	12,509	4,951	4,169	8,575	7,399	2,855	3,468
1988	47,739	13,176	5,643	4,444	9,527	8,235	2,977	3,737
1989	52,460	14,749	6,184	4,775	9,449	8,718	3,078	5,507
1990	58,983	18,166	6,921	5,220	9,094	9,102	3,273	7,207
1991 p	56,596	16,766	6,654	5,063	9,131	8,847	3,175	6,960
INTERNATIONAL OPERATIONS								
1977	$3,852	$ 1,303	$ 450	$ 351	$ 668	$ 526	$ 253	$ 301
1978	4,355	1,351	498	427	768	623	323	363
1979	5,505	1,960	571	538	922	774	352	388
1980	6,766	2,775	616	600	1,049	917	385	423
1981	6,574	2,757	540	583	932	945	382	435
1982	6,452	2,596	512	577	893	954	396	525
1983	6,693	2,490	548	664	936	1,162	389	505
1984	7,485	2,629	677	749	975	1,308	446	701
1985	7,984	2,738	768	852	1,069	1,414	482	662

(Continued on next page)

OPERATING EXPENSES OF U.S. AIR CARRIERS (Continued)
DOMESTIC AND INTERNATIONAL OPERATIONS
Calendar Years 1977 - 1991 (Millions of Dollars)

Year	TOTAL Operating Expenses	Flying Operations	Main-tenance	Passenger Service	Aircraft & Traffic Servicing	Promotion and Sales	Depreciation & Amorti-zation	Other
1986	8,458	2,402	901	877	1,386	1,665	518	711
1987	10,226	2,836	1,096	1,059	1,749	2,094	533	860
1988	12,403	3,230	1,332	1,280	2,193	2,742	618	1,009
1989	14,954	3,919	1,724	1,454	2,483	3,108	746	1,520
1990	18,878	5,454	2,051	1,738	2,657	3,833	887	2,295
1991 p	19,872	5,609	2,121	1,657	2,822	4,573	886	2,204

Source: Department of Transportation, Office of Aviation Statistics, "Air Carrier Financial Statistics Quarterly" (Quarterly).
NOTE: Detail may not add to totals because of rounding.
Scheduled and non-scheduled service for all certificated route air carriers.
Excludes supplemental air carriers, commuters, and air taxis.
General and administrative and other transport-related expenses.
p Preliminary.

OPERATING REVENUES AND EXPENSES OF U.S. AIR CARRIERS
DOMESTIC AND INTERNATIONAL OPERATIONS
Calendar Years 1964 - 1991 (Millions of Dollars)

Year	TOTAL OPERATIONS			Domestic Operations			International Operations		
	Oper-ating Reve-nues	Oper-ating Ex-penses	Oper-ating Profit (or Loss)	Oper-ating Reve-nues	Oper-ating Ex-penses	Oper-ating Profit (or Loss)	Oper-ating Reve-nues	Oper-ating Ex-penses	Oper-ating Profit (or Loss)
1964	$ 4,251	$ 3,781	$ 470	$ 3,169	$ 2,849	$ 320	$ 1,082	$ 932	$ 150
1965	4,958	4,286	672	3,691	3,239	452	1,267	1,047	220
1966	5,745	4,970	775	4,171	3,670	502	1,574	1,300	274
1967	6,865	6,157	708	4,981	4,560	421	1,884	1,597	287
1968	7,753	7,248	505	5,691	5,397	295	2,062	1,852	210
1969	8,791	8,403	387	6,936	6,613	322	1,855	1,790	65
1970	9,290	9,247	43	7,180	7,181	(1)	2,109	2,066	44
1971	10,046	9,717	328	7,753	7,496	257	2,292	2,221	71
1972	11,163	10,578	584	8,652	8,158	493	2,512	2,420	91
1973	12,419	11,834	585	9,694	9,200	494	2,725	2,633	91
1974	14,703	13,978	725	11,546	10,761	785	3,157	3,218	(60)
1975	15,356	15,229	128	12,020	11,903	117	3,336	3,326	11
1976	17,503	16,781	721	13,899	13,324	575	3,605	3,457	147
1977	19,926	19,018	908	15,822	15,166	657	4,104	3,852	252
1978	22,892	21,527	1,366	18,189	17,172	1,018	4,703	4,355	348
1979	27,227	27,028	199	21,652	21,523	129	5,575	5,505	69
1980	33,728	33,949	(222)	26,404	26,409	(6)	6,543	6,766	(223)
1981	36,211	36,612	(401)	28,788	29,051	(264)	6,390	6,574	(184)
1982	36,066	36,804	(739)	28,728	29,478	(750)	6,435	6,452	(17)
1983	38,593	38,231	362	31,014	31,186	(171)	7,163	6,693	470
1984	44,060	41,946	2,114	35,394	33,812	1,582	7,975	7,485	490
1985	48,580	47,207	1,372	37,629	36,611	1,018	8,302	7,984	319
1986	50,086	48,855	1,231	41,001	39,984	1,060	8,621	8,458	163
1987	56,787	54,339	2,448	45,658	43,925	1,733	10,925	10,226	698
1988	63,679	60,236	3,443	50,187	47,739	2,448	13,402	12,403	998

(Continued on next page)

OPERATING REVENUES AND EXPENSES OF U.S. AIR CARRIERS (Continued)
DOMESTIC AND INTERNATIONAL OPERATIONS
Calendar Years 1964 - 1991 (Millions of Dollars)

Year	TOTAL OPERATIONS			Domestic Operations			International Operations		
	Oper-ating Reve-nues	Oper-ating Ex-penses	Oper-ating Profit (or Loss)	Oper-ating Reve-nues	Oper-ating Ex-penses	Oper-ating Profit (or Loss)	Oper-ating Reve-nues	Oper-ating Ex-penses	Oper-ating Profit (or Loss)
1989	69,225	67,413	1,812	54,314	52,460	1,855	14,911	14,954	(43)
1990	75,984	77,898	(1,913)	57,994	58,983	(989)	17,990	18,914	(924)
1991 p	74,942	76,669	(1,727)	56,119	56,596	(477)	18,823	20,073	(1,250)

Source: Department of Transportation, Office of Aviation Statistics, "Air Carrier Financial Statistics Quarterly" (Quarterly).
NOTE: Detail may not add to totals because of rounding.
Scheduled and non-scheduled service for all certificated route air carriers.
Excludes supplemental air carriers, commuters, and air taxis.
For 1980 and subsequent years, includes 'Other' operations not reported as
'Domestic' or 'International.'
p Preliminary.

SOURCES OF OPERATING REVENUES OF U.S. AIR CARRIERS
DOMESTIC AND INTERNATIONAL OPERATIONS
Calendar Years 1977 - 1991 (Millions of Dollars)

Year	TOTAL Operating Revenues	Passenger Service	Mail	Freight & Air Express	Excess Baggage	Other
DOMESTIC OPERATIONS						
1977	$15,822	$13,773	$355	$1,109	$21	$ 564
1978	18,189	15,753	336	1,347	23	730
1979	21,652	18,931	417	1,485	28	791
1980	26,404	23,317	446	1,582	32	1,027
1981	28,788	25,504	497	1,659	36	1,091
1982	28,728	25,440	524	1,505	42	1,218
1983	31,014	27,519	516	1,602	52	1,326
1984	35,393	31,437	552	1,716	70	1,618
1985	37,629	33,343	733	1,581	78	1,895
1986	41,001	33,814	679	4,278	85	2,159
1987	45,658	37,492	704	4,952	67	2,443
1988	50,187	41,002	789	5,807	72	2,518
1989	54,314	43,670	767	5,408	70	4,399
1990	57,994	46,282	747	4,276	76	6,613
1991	56,119	44,515	721	4,478	78	6,326
INTERNATIONAL OPERATIONS						
1977	$ 4,104	$ 3,047	$112	$ 710	$21	$ 215
1978	4,703	3,534	117	750	20	282
1979	5,575	4,271	131	837	23	313
1980	6,543	4,984	175	1,011	25	348
1981	6,390	4,916	165	984	25	299
1982	6,435	4,959	177	990	25	283
1983	7,163	5,605	152	999	23	384
1984	7,975	6,074	158	1,169	27	546
1985	8,302	6,451	161	1,130	28	532
1986	8,621	6,551	154	1,451	28	437
1987	10,925	8,374	180	1,783	33	555
1988	13,402	10,357	183	2,150	39	672
1989	14,911	11,181	188	2,417	47	1,078
1990	17,990	13,468	223	2,602	43	1,654
1991 p	18,823	14,063	222	3,069	50	1,419

Source: Department of Transportation, Office of Aviation Statistics, "Air Carrier Financial Statistics Quarterly" (Quarterly).

NOTE: Detail may not add to totals because of rounding.
Scheduled and non-scheduled service for all certificated route air carriers.
Excludes supplemental air carriers, commuters, and air taxis.
Scheduled and charter.
Subsidy included with Mail through 1979, and thereafter included in Other, which also includes revenues not related to transport, plus, beginning in 1981, transport revenues not specifically broken out by category by some small carriers.
p Preliminary.

OPERATING REVENUES AND EXPENSES
OF WORLD SCHEDULED AIRLINES
Calendar Years 1987 - 1991 (Millions of U.S. Dollars)

	1987	1988	1989	1990	1991 p
OPERATING REVENUES:					
Scheduled Services:					
Passenger	$111,600	$127,250	$137,200	$153,290	
Freight	17,350	19,380	19,110	18,510	
Mail	1,980	2,050	2,040	2,250	
Total Scheduled Services	$130,930	$148,680	$158,350	$174,050	NA
Non-Scheduled Services	5,640	6,360	6,650	7,020	
Incidental	10,430	11,160	14,000	17,630	
Total Operating Revenues	**$147,000**	**$166,200**	**$179,000**	**$198,700**	**$203,400**
OPERATING EXPENSES:					
Flight Operations	$ 37,390	$ 39,270	$ 44,520	$ 56,060	
Maintenance & Overhaul	15,960	18,320	19,590	22,790	
Depreciation & Amortization	10,820	12,150	12,520	14,030	
User Charges & Station					NA
Expenses	24,410	28,440	29,080	32,200	
Passenger Services	14,260	15,900	17,880	20,880	
Ticketing, Sales & Promotion	24,570	27,080	30,070	32,960	
General, Administrative & Other	12,390	14,840	17,540	21,280	
Total Operating Expenses	**$139,800**	**$156,000**	**$171,200**	**$200,200**	**$204,400**
OPERATING RESULT	**$ 7,200**	**$ 10,200**	**$ 7,800**	**$ (1,500)**	**$ (1,000)**
Percent of Revenue	4.9%	6.1%	4.4%	-0.8%	-0.5%
NET RESULT	**$ 2,500**	**$ 5,000**	**$ 3,700**	**$ (4,300)**	**NA**
Percent of Revenue	1.7%	3.0%	2.1%	-2.2%	NA

Source: International Civil Aviation Organization, "Civil Aviation Statistics of the World" (Annually).
Excludes domestic operations in the USSR.
Net Result equals Operating Result minus non-operating items, including interest, income taxes, retirement of property and equipment, affiliated companies, and subsidies.
NA Not available.
p Preliminary.
( ) Denotes loss.

JET FUEL COSTS AND CONSUMPTION BY U.S. AIR CARRIERS
Calendar Years 1977 - 1991

Year	Gallons Consumed (Millions)	Total Cost (Millions)	Cost Per Gallon (Cents)	Cost Index (1982 = 100)	Cost of Fuel as Percent of Cash Operating Expenses
1977	10,282.0	$ 3,729.8	36.3	37.0	20.1%
1978	10,627.1	4,178.2	39.3	40.1	19.7
1979	11,278.1	6,503.0	57.7	58.8	24.4
1980	10,874.0	9,769.5	89.8	91.6	29.7
1981	10,087.8	10,498.0	104.1	106.1	29.3
1982	9,942.1	9,755.2	98.1	100.0	27.2
1983	10,214.4	9,073.1	88.8	90.5	24.5
1984	11,050.4	9,361.7	84.7	86.3	23.8
1985	11,675.1	9,326.7	79.9	81.4	22.2
1986	12,643.0	6,995.8	55.3	56.4	16.3
1987	13,629.5	7,593.8	55.7	56.8	16.0
1988	14,204.8	7,557.2	53.2	54.2	14.4
1989	14,103.9	8,472.7	60.1	61.2	14.9
1990	14,841.1	11,465.2	77.3	78.7	17.6
1991	13,798.4	9,329.5	67.6	68.9	14.8

Source: Air Transport Association of America, "Airline Cost Index" (Quarterly) Majors and Nationals excluding Air Florida, Capitol, Transamerica, and World.

US MAJOR CARRIERS JET FLEET ANALYSIS
AVERAGE NUMBER OPERATED DURING JUNE 1992

Aircraft Type	#	Operated Stage	Average Age (Yrs)
American			
A300-B4-600	31	(3)	3.0
B727-100	30	(2)	26.7
B727-200	125	(2)	17.7
B747-SP	2	(3)	12.0
B757-200	57	(3)	1.6
B767-200	30	(3)	6.9
B767-300	20	(3)	2.6
DC-10-10	49	(3)	19.0
DC-10-30	10	(3)	17.7
F-100	19	(3)	0.8
MD-82	234	(3)	5.0
MD-83	20	(3)	2.9
MD-11	10	(3)	1.0
Total Fleet	**637**		**9.2**
America West			
A320-200	18	(3)	2.5
B737-100	1	(2)	23.0
B737-200	25	(2)	12.5
B737-300	40	(3)	5.0
B747-200B	2	(2)	21.0
B757-200	10	(3)	5.8
Total Fleet	**96**		**7.1**
Continental			
A300-B4-100	20	(3)	11.0
B727-100	3	(2)	28.2
B727-200	80	(2)	18.3
B737-100	14	(2)	24.4
B737-200	25	(2)	23.2
B737-300	55	(3)	6.5
B747-100	2	(2)	22.5
B747-200B	6	(2)	20.7
DC-9-30	34	(2)	21.3
DC-10-10	7	(3)	20.1
DC-10-30	10	(3)	16.6
MD-81	5	(3)	11.5
MD-82	55	(3)	7.9
MD-83	1	(3)	5.5
Total Fleet	**317**		**15.0**

Aircraft Type	#	Operated Stage	Average Age (Yrs)
Delta			
A310-200	7	(3)	7.9
A310-300	14	(3)	4.6
B727-200	152	(2)	15.7
B737-200	59	(2)	8.4
B737-300	13	(3)	7.0
B757-200	78	(3)	4.0
B767-200	15	(3)	9.6
B767-300	35	(3)	2.8
DC-9-30	29	(2)	23.6
L1011-1	32	(3)	16.0
L1011-200	1	(3)	14.5
L1011-250	6	(3)	10.2
L1011-500	17	(3)	11.5
MD-88	105	(3)	2.4
MD-11	6	(3)	1.3
Total Fleet	**569**		**9.4**
Federal Express			
B727-100F	89	(2)	25.7
B727-200F	79	(3)	12.8
B747-100	2	(2)	21.5
B747-100F	2	(2)	22.0
B747-200F	8	(3)	13.8
DC-10-10F	11	(3)	17.7
DC-10-30F	18	(2)	9.6
MD-11F	4	(3)	1.4
Total Fleet	**213**		**18.2**
Northwest			
A320-200	32	(3)	1.3
B727-100	3	(2)	28.2
B727-200	70	(2)	16.4
B747-100	12	(2)	22.5
B747-200B	20	(3)	12.9
B747-200F	8	(2)	13.1
B747-400	10	(3)	3.1
B757-200	33	(3)	6.3
DC-9-10/15	26	(2)	26.1
DC-9-10F	4	(2)	25.5
DC-9-30	77	(2)	23.5
DC-9-40	12	(2)	24.3
DC-9-50	28	(2)	13.8
DC-10-30	8	(3)	17.3
DC-10-40	21	(3)	19.3
MD-82	8	(3)	10.6
Total Fleet	**372**		**16.2**
Southwest			
B737-200	49	(2)	12.2
B737-300	63	(3)	4.6
B737-500	25	(3)	1.2
Total Fleet	**137**		**6.7**

Aircraft Type	#	Operated Stage	Average Age (Yrs)
TWA			
B727-100	8	(2)	27.8
B727-200	55	(2)	20.1
B747-100	12	(2)	21.8
B747-200B	3	(2)	20.2
B767-200	9	(3)	9.7
DC-9-10/15	7	(2)	26.1
DC-9-30	35	(2)	23.0
DC-9-30F	1	(2)	23.5
DC-9-40	3	(2)	17.8
L1011-1	15	(3)	19.6
L1011-50	7	(3)	18.4
L1011-100	11	(3)	14.6
MD-82	25	(3)	8.0
MD-83	4	(3)	5.5
Total Fleet	**195**		**18.5**
United			
B727-100	16	(2)	28.6
B727-200	104	(2)	16.6
B737-200	73	(2)	20.5
B737-300	101	(3)	3.9
B737-500	35	(3)	0.9
B747-100	18	(2)	21.4
B747-200B	9	(2)	13.1
B747-400	15	(3)	1.7
B747-SP	11	(3)	15.1
B757-200	60	(3)	1.4
B767-200	19	(3)	9.9
B767-300	9	(3)	0.7
DC-10-10	46	(3)	18.1
DC-10-30	4	(3)	12.8
DC-10-30CF/F	4	(2)	14.0
Total Fleet	**524**		**11.5**
USAir			
B727-200	35	(2)	18.3
B737-200	81	(2)	13.4
B737-300	101	(3)	5.4
B737-400	54	(3)	2.9
B757-200	10	(3)	9.3
B767-200	11	(3)	3.8
DC-9-30	73	(2)	19.8
F28-1000	17	(2)	19.1
F28-4000/6000	24	(2)	9.2
F-100	33	(3)	1.8
MD-81	20	(3)	11.4
MD-82	11	(3)	9.3
Total Fleet	**470**		**10.5**

Grand Total 3,530   11.6

Stage numbers (2) and (3) refer to U.S. Government noise standards.

US NATIONAL CARRIERS FLEET ANALYSIS
AVERAGE NUMBER OPERATED DURING JUNE 1992

Aircraft Type	#	Operated Stage	Average Age (Yrs)
Alaska			
B727-100	1	(2)	24.0
B727-200	23	(2)	13.0
B737-200C	7	(2)	12.9
MD-82	10	(3)	8.5
MD-83	24	(3)	3.7
Total Fleet	**65**		**9.0**
Aloha			
B737-200	17	(2)	14.2
B737-300	3	(3)	4.7
Total Fleet	**20**		**12.8**
Atlantic Southeast			
DHC-7-102	2	(3)	10.5
EMB 110	11	(3)	10.5
EMB 120	51	(3)	4.0
Total Fleet	**64**		**5.3**
Comair			
EMB 110	6	(3)	11.0
EMB 120	36	(3)	1.9
SF 340A	19	(3)	7.0
SA 227	11	(3)	6.6
Total Fleet	**72**		**4.7**
Evergreen			
B727-100	2	(2)	25.0
B727-100F	4	(2)	25.8
B727-100C	5	(2)	25.4
B747-100F	5	(2)	20.4
B747-100C	6	(2)	21.7
B747-200B	2	(2)	19.0
B747-200C	2	(2)	18.5
DC-8-62F	1	(2)	25.0
DC-8-73F	2	(3)	24.0
DC-9-15F	2	(2)	25.5
DC-9-32F	2	(2)	24.0
DC-9-33F	4	(2)	24.3
Total Fleet	**37**		**24.5**

Aircraft Type	#	Operated Stage	Average Age (Yrs)
Hawaiian			
DHC-7-102	4	(3)	11.0
DC-8-62	2	(2)	23.5
DC-8-62F	3	(2)	24.8
DC-8-63	2	(2)	23.5
DC-9-51	13	(2)	15.1
L1011-1	1	(3)	17.5
L1011-50	4	(3)	17.3
Total Fleet	**29**		**17.1**
Horizon			
DHC-8-102	22	(3)	4.0
F28-1000	3	(2)	20.5
SA 227	32	(3)	7.5
Total Fleet	**57**		**6.8**
Metro			
BAe 3101	15	(3)	6.5
SF 340A	16	(3)	4.8
SF 340B	10	(3)	2.9
Total Fleet	**41**		**5.0**
Midwest Express			
DC-9-10	8	(2)	26.6
DC-9-30	4	(2)	24.5
MD-88	2	(3)	3.5
Total Fleet	**14**		**22.7**
Southern Air Transport			
B707-300C	4	(2)	23.0
DC-8-71F	5	(3)	23.4
L-100	16	(3)	22.6
Total Fleet	**25**		**22.8**

Aircraft Type	#	Operated Stage	Average Age (Yrs)
Tower			
B747-100	7	(2)	22.2
Total Fleet	7		22.2
UPS			
B727-100C	39	(2)	25.6
B727-200F	8	(3)	14.3
B747-100	11	(3)	21.2
B757-200	28	(3)	2.9
DC-8-71F	23	(3)	24.6
DC-8-73F	26	(3)	23.7
Total Fleet	**135**		**19.3**
WestAir			
BAe-146-200	5	(3)	4.6
BAe 3101	35	(3)	5.2
BAe 3201	6	(3)	3.5
EMB 120	29	(3)	3.2
SHORTS 360	2	(3)	7.0
Total Fleet	**77**		**4.3**
World			
DC-10-10	1	(3)	19.0
DC-10-30F	5	(3)	14.5
DC-10-30	1	(3)	13.0
Total Fleet	**7**		**14.9**

Grand Total	**648**	**11.7**

Stage numbers (2) and (3) refer to U.S. Government noise standards.

ACTIVE U.S. AIR CARRIER FLEET
By Type of Aircraft, Number of Engines and Model
December 1987 - 1991

	1987	1988	1989	1990	1991
TOTAL	**5,250**	**5,660**	**5,778**	**6,083**	**6,054**
Turbojets -- TOTAL	**3,575**	**3,915**	**3,942**	**4,148**	**4,167**
Four-Engine -- TOTAL	**382**	**427**	**428**	**432**	**410**
Boeing 707	31	31	27	25	27
Boeing 747	156	171	180	190	184
B.Ae. 146	57	57	53	44	17
McDonnell Douglas DC-8	138	168	168	173	182
Three-Engine -- TOTAL	**1,469**	**1,542**	**1,459**	**1,438**	**1,376**
Boeing 727	1,168	1,246	1,167	1,152	1,073
Lockheed L-1011	116	112	107	101	100
McDonnell Douglas DC-10/MD-11	185	184	185	185	203
Twin-Engine -- TOTAL	**1,724**	**1,946**	**2,055**	**2,278**	**2,381**
Airbus A-300	52	57	63	67	63
Airbus A-310	13	19	19	21	42
Airbus A-320	-	-	11	10	35
Boeing 737	633	706	756	812	835
Boeing 757	95	122	146	199	234
Boeing 767	83	126	111	120	136
B.Ae. BAC-111	39	30	-	3	1
Cessna C550	-	-	5	7	-
Dassault Falcon	-	-	-	-	2
Fokker F-28	47	47	53	68	75
Grumman G-1159	-	-	-	1	3
Learjet LR-25	-	1	2	1	2
Learjet LR-35	2	1	1	2	-
McDonnell Douglas DC-9/MD-80	760	837	888	967	953
Turboprops -- TOTAL	**1,241**	**1,375**	**1,476**	**1,595**	**1,598**
Four-Engine -- TOTAL	**102**	**95**	**96**	**88**	**75**
Canadair CL44D	6	6	5	5	-
De Havilland DHC-7	41	39	41	40	33
Lockheed 188 Electra	34	30	30	24	24
Lockheed 382/L-100 Hercules	21	20	20	19	18
Twin-Engine -- TOTAL	**1,139**	**1,280**	**1,380**	**1,507**	**1,523**
Beech BE65	4	1	-	-	-
Beech BE90	4	1	-	-	-
Beech BE99	52	84	53	54	32
Beech BE100	-	1	1	2	1
Beech BE200	5	7	10	16	8
Beech BE1900	48	80	109	147	167
B.Ae. ATP	-	-	-	4	10
B.Ae. Jetstream	113	135	165	222	214
CASA C212 Aviocar	16	18	16	16	13
Cessna C441	2	3	4	2	2
Convair 580/600/640	77	72	58	33	37
DeHavilland DHC-6	71	63	69	67	69
DeHavilland DHC-8	34	44	64	74	81
Dornier DO228	18	33	34	32	31
Embraer EMB110/EMB120	133	139	164	204	190
Fairchild/Fokker F-27/FH-227	47	51	53	58	50
Fairchild Swearingen SA-226	101	90	57	22	31
Fairchild Swearingen SA-227	163	191	212	218	200
Grumman G-73	-	7	5	7	4
Grumman G-159	14	5	6	7	2
Grumman G-500	-	1	-	-	-
Mitsubishi MU-2	1	-	-	1	1

(Continued on next page)

ACTIVE U.S. AIR CARRIER FLEET (Continued)
By Type of Aircraft, Number of Engines, and Model
December 1987 - 1991

	1987	1988	1989	1990	1991
Nihon YS-11	36	22	21	21	22
Nord ND-262/STC-262	12	9	2	1	-
Piper PA31T	6	9	12	8	8
Piper 42	-	-	-	-	1
Rockwell Aero Commander 690	1	1	-	-	-
Saab-Fairchild SF340A	51	68	85	109	153
Shorts SD-3/SD-330	110	110	118	103	93
Shorts SC-7	-	-	-	2	2
Societe Nationale Industrielle Aerospatiale SNAIS ATR-42	20	35	62	77	101
Piston-Engine -- TOTAL	**421**	**362**	**353**	**329**	**283**
Four-Engine -- TOTAL	**38**	**36**	**35**	**31**	**26**
Douglas DC-6	37	35	34	30	25
Douglas DC-7	1	1	1	1	1
Three-Engine -- TOTAL	**3**	**3**	**5**	**6**	**5**
Pilatus Britten-Norman BN2A-MK-3 Turbo Islander	3	3	5	6	5
Twin-Engine -- TOTAL	**380**	**323**	**313**	**292**	**252**
Helicopters -- TOTAL	**13**	**8**	**7**	**11**	**6**

Source: Federal Aviation Administration, "FAA Statistical Handbook of Aviation" (Annually)
NOTE: Effective 1978, includes certificated route air carriers, supplemental air carriers (charters), multi-engine aircraft in passenger service of commuters, and all aircraft over 12,500 pounds operated by air taxis, commercial operators, and travel clubs. "Active aircraft" must have a current U.S. registration and have flown during the calendar year.

CIVIL TRANSPORT AIRCRAFT BACKLOG
December 31, 1987 - 1991

Company and Model	1987	1988	1989	1990	1991
TOTAL AIRCRAFT ON ORDER					
(Domestic and Foreign Orders)	**824**	**1,373**	**1,989**	**2,138**	**1,829**
Value (Millions of Dollars)	**$32,401**	**$58,474**	**$89,069**	**$112,339**	**$108,833**
Boeing--TOTAL	**573**	**937**	**1,440**	**1,563**	**1,456**
B-737	342	488	739	754	615
B-747	120	153	165	250	234
B-757	67	205	344	333	333
B-767	44	91	192	192	188
B-777	-	-	-	34	86
Lockheed--TOTAL	**2**	**1**	**-**	**-**	**-**
L-100	2	1	-	-	-
McDonnell Douglas--TOTAL	**249**	**435**	**549**	**575**	**373**
DC-10	7	1	-	-	-
MD-11	29	88	126	175	138
MD-80	213	346	423	400	235
TOTAL FOREIGN ORDERS	**420**	**840**	**1,092**	**1,205**	**1,073**
Value (Millions of Dollars)	$20,196	$39,504	$54,956	$ 71,213	$ 72,733
Boeing--TOTAL	**293**	**547**	**750**	**872**	**844**
B-737	137	263	359	412	329
B-747	95	124	141	211	205
B-757	28	91	119	125	144
B-767	33	69	131	124	114
B-777	-	-	-	-	52
Lockheed--TOTAL	**2**	**-**	**-**	**-**	**-**
L-100	2	-	-	-	-
McDonnell Douglas--TOTAL	**125**	**293**	**342**	**333**	**229**
DC-10	3	1	-	-	-
MD-11	27	75	96	131	101
MD-80	95	217	246	202	128

Source: Aerospace Industries Association, based on company reports.
Unfilled firm orders for U.S.-manufactured transport aircraft over 30,000 pounds
(including the turboprop-powered Lockheed L-100) excluding options, but including
new transports contracted for lease from the manufacturer.

SPECIFICATIONS OF U.S. CIVIL JET TRANSPORT AIRCRAFT
On Order or in Production as of 1991

Number of Engines and Crew, and Model Designation	Initial Service	Standard Mixed Class	Operating Empty Weight (000's lbs)	Maximum Takeoff Gross Weight (000's lbs)	Range (Nautical Miles)	Engine Manufacturer and Model
FOUR ENGINES/ CREW OF 3						
747-400	1988	412-509	390	870	8,380	GE CF6-80C2
THREE ENGINES/ CREW of 2						
MD-11	1989	293-410	288	618	7,980	GE CF6-80C2-DF1, RR Trent-685, or P&W PW4360
MD-11ER	1989	277	265	603	8,525	GE CF6-80C2-DF1 or P&W PW4360
TWO ENGINES/ CREW OF 2						
737-300	1984	141	70-71	125-139	1,840 -2,950	CFMI CFM56-3-B1 or B2
737-400	1988	159	73-74	139-151	2,250 -2,800	CFMI CFM56-3-B2 or CFM56-3C
737-500	1990	108-132	68	116	2,500	CFMI CFM56-3-B1 or CFM56-3C-1
757-200	1982	186-200	126	240	4,550	RR RB211-535E or P&W PW2037
767-200ER	1984	174-290	180	351	5,942	P&W JT9D-7R4 or GE CF6-80A
767-300	1986	204-290	190	351	4,650	P&W JT9D-7R4 or GE CF6-80A
767-300ER	1987	204-290	196	400	6,650	P&W PW4000 or GE CF6-80C2
777	1995	360-390	295	506	4,200	RR Trent-871, GE GE90-B1, or P&W PW4073
MD-80 series:						
MD-81	1980	155	78	140	1,630	P&W JT8D-209 or P&W JT8D-217A
MD-82	1981	155	79	150	2,176	P&W JT8D-217C
MD-83	1985	155	80	160	2,618	P&W JT8D-219
MD-87	1987	130	74	140	2,405	P&W JT8D-217C
MD-88	1987	155	79	150	2,176	P&W JT8D-217C
MD-90	1994	172	87	156	2,260	IAE V2500-D5

Source: Aerospace Industries Association, based on company reports and Aviation Week & Space Technology, "Aerospace Forecast & Inventory" (Annually).
All jet-powered passenger transport aircraft 33,000 pounds or more empty weight.
The Boeing Company manufacturers models: 737, 747, 757, 767, & 777 and McDonnell Douglas Corporation manufactures models: MD-11, MD-80, and MD-90.
Full passenger load and baggage.
P&W = Pratt & Whitney; GE = General Electric; RR = Rolls-Royce;
CFMI = General Electric/Snecma; IAE = International Aero Engines.
Wide-body aircraft.

TURBINE-ENGINED AIRCRAFT IN THE WORLD AIRLINE FLEET
(By Model, 1987 - 1991)

	1987	1988	1989	1990	1991
TOTAL AIRCRAFT IN SERVICE	**11,711**	**12,575**	**13,514**	**14,651**	**15,181**
Turbojets--TOTAL	**7,600**	**8,085**	**8,587**	**9,426**	**9,819**
Aerospatiale SE-210 Caravelle	60	59	56	49	38
Aerospatiale SN-601 Corvette	11	12	12	7	2
Airbus A300	267	272	294	327	331
Airbus A310	94	116	147	180	193
Airbus A320	-	2	23	130	247
Antonov 124	-	-	-	-	7
B.Ae./Aerospatiale Concorde	14	14	14	14	14
B.Ae. 146	59	82	102	144	166
B.Ae. One-Eleven	166	167	164	132	146
B.Ae. Trident	34	27	27	25	32
B.Ae. (HS) 125	18	16	17	16	17
Beech 400 Beechjet	-	-	-	-	1
Boeing 707/720	273	245	224	210	198
Boeing 727	1,676	1,686	1,684	1,648	1,515
Boeing 737	1,284	1,426	1,585	1,836	2,019
Boeing 747	629	653	676	775	806
Boeing 757	117	167	215	324	380
Boeing 767	163	207	254	345	399
Canadair CL-601 Challenger	-	1	-	-	2
Cessna 500/550/650 Citation I/II/III	28	37	48	43	44
Convair 880/990	12	2	2	-	-
Dassault Falcon 10/20/50	30	39	44	39	43
Dassault Mercure	11	11	11	11	11
Fokker F-28 Fellowship	197	203	203	199	197
Fokker 100	-	1	14	58	93
Gates Learjet	43	56	56	37	34
Gulfstream II/III G-1159	15	14	14	15	16
Ilyushin IL-62	60	66	67	56	39
Ilyushin IL-76	44	55	58	60	61
Israel Aircraft 1121/1124	9	7	3	2	2
Lockheed L-1011 Tristar	230	229	229	228	227
Lockheed L-1329 Jetstar	12	13	13	6	5
MBB Hansa HFB-320	1	1	5	-	-
McDonnell Douglas DC-8	258	282	276	253	257
McDonnell Douglas DC-9	856	853	842	847	741
McDonnell Douglas DC-10	355	361	370	365	361
McDonnell Douglas MD-11	-	-	-	3	36
McDonnell Douglas MD-80	362	462	588	799	908
Mitsubishi MU-300 Diamond	-	1	2	-	-
Rockwell/Sabreliner 60	-	-	3	3	3
Tupolev Tu-134	98	101	97	74	54
Tupolev Tu-154	74	87	95	111	156
Yakolev Yak-40/42	42	52	53	55	48
Turboprops--TOTAL	**3,808**	**4,219**	**4,687**	**5,049**	**5,174**
Aero Spacelines SuperGuppy	-	4	4	-	-
Aerospatiale N.262/Mohawk 298	28	25	23	16	14
Aerospatiale/Aeritalia ATR 42	36	76	122	178	210
Aerospatiale/Aeritalia ATR 72	-	-	-	17	48
Airtech CN-235	-	2	8	18	24
Antonov An-12	14	14	15	19	20
Antonov An-24/26/28/30/32	200	215	251	246	216
B.Ae. ATP	-	-	12	31	41
B.Ae. Vanguard	8	9	7	5	4
B.Ae. Viscount	47	45	40	33	27
B.Ae. (HP-137) Jetstream 31	114	166	201	277	205
B.Ae. (HS) Argosy	5	5	5	-	-

(Continued on next page)

TURBINE-ENGINED AIRCRAFT IN THE WORLD AIRLINE FLEET (Continued)
(By Model, 1987 – 1991)

	1987	1988	1989	1990	1991
Turboprops (Continued)					
B.Ae. HS-748	157	154	152	139	130
Beech 18 Turbo	15	21	24	24	20
Beech 90 King Air	36	44	40	26	28
Beech 99	169	171	173	140	122
Beech 100 King Air	21	24	22	23	24
Beech 200/300 Super King Air	62	70	83	78	76
Beech 1300	-	-	5	14	7
Beech 1900C	64	73	95	171	191
Bristol 175 Britannia	8	7	7	6	6
Canadair CL-44	15	14	15	13	11
CASA/Nurtanio C-212 Aviocar	97	103	112	104	109
Cessna 208 Caravan I	74	150	229	287	312
Cessna F406 Caravan II	-	-	14	19	21
Cessna 425/441 Conquest I/II	16	9	19	8	4
Convair 580/600/640	142	131	132	108	92
DHC-2 Turbo Beaver/Otter	3	3	3	4	4
DHC-5 Buffalo	2	2	1	1	1
DHC-6 Twin Otter	450	464	465	432	428
DHC-7 Dash 7	95	100	106	94	79
DHC-8 Dash 8	55	82	120	214	254
Dornier DO-228	59	79	90	113	96
Douglas DC-3T Turbo Express	1	1	-	-	1
Embraer EMB-110 Bandeirante	232	231	222	200	174
Embraer EMB-120 Brasilia	28	64	113	201	225
Fokker/Fairchild					
F-27/FH-227 Friendship	436	434	432	401	389
Fokker 50	-	13	45	101	121
GAF Nomad	23	16	14	9	8
Grumman G-21 Turbo Goose	-	-	-	-	1
Grumman G-73 Turbo Mallard	8	11	10	9	4
Grumman G-159 Gulfstream I	31	32	37	34	33
Handley Page Herald	21	15	17	17	17
Harbin Y-12 II	-	-	-	2	5
IAI Arava	-	3	4	3	1
Ilyushin IL-18	71	69	67	48	42
LET L-410	-	-	-	3	17
Lockheed L-188 Electra	77	79	83	74	67
Lockheed L-100/L-382 Hercules	56	52	58	56	54
Mitsubishi MU-2B	12	11	5	5	8
Nihon AMC YS-11	108	107	102	97	94
Pilatus PC-6 Turbo Porter	25	-	-	-	-
Pilatus Britten-Norman BN-2T					
Turbo Islander	5	3	3	2	3
Piper PA-31T/42 Cheyenne	18	28	35	29	25
Piper T-1040	8	9	15	15	12
GAF Nomad	23	16	14	9	8
Grumman G-21 Turbo Goose	-	-	-	-	1
Grumman G-73 Turbo Mallard	8	11	10	9	4
Grumman G-159 Gulfstream I	31	32	37	34	33
Handley Page Herald	21	15	17	17	17
Harbin Y-12 II	-	-	-	2	5
IAI Arava	-	3	4	3	1
Ilyushin IL-18	71	69	67	48	42
LET L-410	-	-	-	3	17
Lockheed L-188 Electra	77	79	83	74	67
Lockheed L-100/L-382 Hercules	56	52	58	56	54
Mitsubishi MU-2B	12	11	5	5	8
Nihon AMC YS-11	108	107	102	97	94
Pilatus PC-6 Turbo Porter	25	-	-	-	-
Pilatus Britten-Norman BN-2T					
Turbo Islander	5	3	3	2	3

(Continued on next page)

TURBINE-ENGINED AIRCRAFT IN THE WORLD AIRLINE FLEET (Continued)
(By Model, 1987 - 1991)

	1987	1988	1989	1990	1991
Turboprops (Continued)					
Piper PA-31T/42 Cheyenne	18	28	35	29	25
Piper T-1040	8	9	15	15	12
Rockwell Turbo Commander	9	11	16	14	15
Saab SF-340A/B	67	105	136	206	265
Saunders ST-27	11	9	2	-	-
Shorts SC-5 Belfast	5	5	5	5	5
Shorts SC-7 Skyliner/Skyvan	15	14	15	16	25
Shorts 330	71	76	68	64	51
Shorts 360	106	130	142	150	139
Swearingen Merlin	52	45	46	41	36
Swearingen Metro	302	356	361	249	338
Transall C-160	8	8	8	8	8
Xian (Antonov) Y-7	10	20	31	31	67
TOTAL AIRCRAFT IN SERVICE	**11,711**	**12,575**	**13,514**	**14,651**	**15,181**
Number Manufactured in U.S.	7,638	8,133	8,617	9,306	9,017
Percent Manufactured in U.S.	65.2%	64.7%	63.8%	63.5%	59.4%
Turbojet Aircraft in Service	7,600	8,085	8,587	9,426	9,819
Number Manufactured in U.S.	6,313	6,693	7,029	7,737	7,950
Percent Manufactured in U.S.	83.1%	82.8%	81.9%	82.1%	81.0%
Turboprop Aircraft in Service	3,808	4,219	4,687	5,049	5,174
Number Manufactured in U.S.	1,184	1,332	1,497	1,519	1,017
Percent Manufactured in U.S.	31.1%	31.6%	31.9%	30.1%	19.7%
Turbine-Powered Helicopters In Service	303	271	240	176	188
Number Manufactured in U.S.	141	108	91	51	50
Percent Manufactured in U.S.	46.5%	39.9%	37.9%	28.4%	26.6%
Turbine-Powered Helicopters--TOTAL	**303**	**271**	**240**	**176**	**188**
Aerospatiale SA-315 Lama	3	3	-	-	-
Aerospatiale SA-316 Alouette III	11	9	8	4	4
Aerospatiale SA-318 Alouette II	4	4	4	3	3
Aerospatiale SA-319 Alouette III Astazou	4	4	4	4	4
Aerospatiale SA-341 Gazelle	-	-	-	-	1
Aerospatiale (Nurtanio) SA-330 Puma	23	23	22	16	18
Aerospatiale AS-332 Super Puma	6	5	5	5	5
Aerospatiale AS-335	-	-	-	-	-
Aerospatiale AS-350 Ecureuil/AStar	5	6	7	10	10
Aerospatiale AS-355 Ecureuil 2/Twinstar	2	2	3	4	4
Aerospatiale SA-365 Dauphin II	9	9	12	10	10
Agusta A109	-	-	-	-	3
Bell (Agusta/Fuji) 204	6	6	5	6	5
Bell 205	2	2	2	2	2
Bell 206 Jetranger/Longranger	53	52	39	26	33
Bell 212	27	29	27	15	15
Bell (Fuji) 214/214ST	5	1	-	-	-
Bell 222 UT	5	4	1	-	-
Bell 412	5	5	2	3	4

(Continued on next page)

TURBINE-ENGINED AIRCRAFT IN THE WORLD AIRLINE FLEET (Continued)
(By Model, 1987 - 1991)

	1987	1988	1989	1990	1991
Turbine-Powered Helicopters (Continued)					
Boeing-Vertol 234 Chinook	3	3	-	-	-
Hughes (Kawasaki) 500/369D	10	1	1	1	1
MBB/Kawasaki BK 117	1	1	1	-	-
MBB/Nurtanio Bo.105	34	34	34	33	33
Sikorsky S-55T	5	5	5	5	5
Sikorsky S-58T	13	7	5	5	4
Sikorsky S-61	41	34	32	10	10
Sikorsky S-76	27	19	18	11	11
Westland 30	-	3	3	3	3

Source: Exxon International Company, "Air World Survey," compiled by Aviation Data Service, Inc. (Annually).

NOTE: The "Air World Survey" covers the world's airlines with the exception of Aeroflot, the USSR national airline, and covers aircraft in service as of March 31. Excludes air taxi operators.

TRAFFIC STATISTICS
WORLD AIRLINE SCHEDULED SERVICE
Calendar Years 1970 - 1991

Year	Passengers Carried (Millions)	Freight Tons Carried	Passenger-Miles Performed (Billions)	Seat-Miles Available	Passenger Load Factor (Percent)	Ton-Miles Performed Freight (Millions)	Ton-Miles Performed Mail (Millions)	TOTAL (Passengers & Baggage, Freight, Mail)
1970	383	6.7	286	522	55%	8,180	2,150	38,810
1971	411	7.4	307	568	54	9,060	1,990	41,420
1972	450	8.0	348	609	57	10,290	1,900	46,690
1973	489	9.0	384	667	58	12,010	1,970	51,910
1974	515	9.5	408	688	59	13,030	1,980	55,270
1975	534	9.6	433	733	59	13,270	1,990	58,080
1976	576	10.3	475	789	60	14,750	2,080	63,880
1977	610	11.1	508	837	61	16,190	2,180	68,790
1978	679	11.7	582	902	65	17,770	2,240	77,770
1979	754	12.1	659	999	66	19,190	2,350	86,900
1980	748	12.2	677	1,071	63	20,120	2,520	89,710
1981	752	12.0	695	1,091	64	21,150	2,600	92,800
1982	766	12.8	710	1,115	64	21,600	2,650	94,830
1983	798	13.5	739	1,151	64	24,050	2,740	100,270
1984	848	14.8	794	1,226	65	27,170	2,950	109,040
1985	899	15.1	850	1,293	66	27,290	3,010	114,860
1986	960	16.2	902	1,389	65	29,580	3,110	122,470
1987	1,028	17.7	987	1,471	67	33,100	3,210	134,540
1988	1,082	19.0	1,060	1,571	67	36,550	3,310	145,330
1989	1,118	20.0	1,107	1,630	68	39,220	3,460	153,330
1990	1,164	20.2	1,176	1,739	68	40,320	3,640	161,080
1991	1,125	19.1	1,128	1,709	66	38,700	3,510	154,110

Source: International Civil Aviation Organization (ICAO).
Includes international and domestic traffic on scheduled service performed by the airlines of the 164 states which were members of ICAO in 1991.

TRAFFIC STATISTICS
U.S. AIR CARRIER SCHEDULED SERVICE
Calendar Years 1964 - 1991

Year	Revenue Ton-Miles (Millions)			Total Available Ton-Miles (Millions)	Total Revenue Load Factor	Aircraft Revenue Miles (Millions)	Average Overall Flight Stage Length (Miles)	Average Available Seats per Aircraft Mile
	Passenger	Cargo	Total					
1964	5,630	1,803	7,434	15,514	47.9%	1,189	301	93
1965	6,629	2,356	8,986	18,408	48.8	1,354	322	96
1966	7,736	2,949	10,686	20,939	51.0	1,482	339	98
1967	9,561	3,475	13,036	26,968	48.3	1,834	371	101
1968	11,023	4,226	15,249	33,221	45.9	2,146	401	107
1969	12,197	4,701	16,898	38,664	43.7	2,385	443	112
1970	13,171	4,994	18,166	41,693	43.6	2,426	473	117
1971	13,565	5,120	18,685	44,139	42.3	2,378	476	125
1972	15,241	5,506	20,746	45,583	45.5	2,376	471	129
1973	16,196	6,046	22,242	49,019	45.4	2,448	477	135
1974	16,292	6,133	22,425	46,848	47.9	2,258	478	140
1975	16,281	5,905	22,186	47,254	46.9	2,241	476	143
1976	17,899	6,222	24,121	49,325	48.9	2,320	480	146
1977	19,322	6,587	25,909	52,284	49.6	2,419	490	149
1978	22,678	7,001	29,679	54,765	54.2	2,520	502	152
1979	26,202	7,189	33,390	60,844	54.9	2,791	517	154
1980	25,519	7,084	32,603	62,983	51.8	2,816	526	158
1981	24,889	7,060	31,949	61,186	52.2	2,703	519	161
1982	25,964	6,886	32,850	62,401	52.6	2,699	544	167
1983	28,183	7,573	35,756	65,385	54.7	2,809	558	169
1984	30,512	8,185	38,697	72,223	53.6	3,134	575	168
1985	33,640	7,689	41,329	76,059	54.3	3,320	569	168
1986	36,655	9,026	45,681	85,140	53.7	3,725	580	168
1987	40,453	10,016	50,469	92,209	54.7	3,988	606	167
1988	42,330	11,469	53,800	97,899	55.0	4,141	618	169
1989	43,271	12,187	55,458	100,082	55.4	4,193	633	169
1990	45,793	12,549	58,342	107,559	54.2	4,491	649	170
1991	44,780	12,109	56,889	105,534	53.9	4,414	651	188

Source: Department of Transportation, Office of Aviation Statistics, "Air Carrier Traffic Statistics Monthly" (Monthly).

NOTE: Detail may not add to totals because of rounding. Includes international and domestic operations.
Includes freight, air express, U.S. and foreign mail.

PASSENGER STATISTICS
U.S. AIR CARRIER SCHEDULED SERVICE
DOMESTIC AND INTERNATIONAL OPERATIONS
Calendar Years 1977 - 1991

Year	Revenue Passenger Enplanements (Thousands)	Average Passenger Trip-Length (Miles)	Revenue Passenger Miles (Millions)	Available Seat Miles (Millions)	Revenue Passenger Load Factor
DOMESTIC OPERATIONS					
1977	222,283	704	156,609	280,619	55.8%
1978	253,957	719	182,669	299,542	61.0
1979	292,700	714	208,891	332,796	62.8
1980	272,829	736	200,829	346,028	58.0
1981	265,304	749	198,715	346,172	57.4
1982	274,342	766	210,149	359,528	58.5
1983	296,721	765	226,909	379,150	59.8
1984	321,047	759	243,692	422,507	57.7
1985	357,109	758	270,584	445,826	60.7
1986	393,864	767	302,090	497,991	60.7
1987	416,831	779	324,637	526,958	61.6
1988	419,210	786	329,309	536,663	61.4
1989	416,331	793	329,975	530,079	62.3
1990	423,565	803	340,231	563,065	60.4
1991	412,269	806	332,407	543,413	61.2
INTERNATIONAL OPERATIONS					
1977	18,043	2,029	36,610	64,947	56.4%
1978	20,759	2,125	44,112	69,209	63.7
1979	24,163	2,199	53,132	83,330	63.8
1980	24,074	2,258	54,363	86,507	62.8
1981	20,672	2,427	50,173	78,725	63.7
1982	19,760	2,505	49,495	80,591	61.4
1983	21,917	2,506	54,920	85,388	64.3
1984	23,636	2,599	61,424	92,817	66.2
1985	24,913	2,642	65,819	101,963	64.6
1986	25,082	2,570	64,456	109,445	58.9
1987	30,847	2,588	79,834	121,763	65.6
1988	35,404	2,655	93,992	140,140	67.1
1989	37,361	2,750	102,739	154,297	66.6
1990	41,995	2,803	117,695	170,310	69.1
1991	39,941	2,889	115,389	171,561	67.3

Source: Department of Transportation, Office of Aviation Statistics, "Air Carrier Traffic Statistics Monthly" (Monthly).
Revenue passenger miles as a percent of available seat miles.

MARKET SHARE, U.S. AIRPORTS, 6 MONTHS 1992
Numbers in ( ) indicate ranking

	Enplaned Passengers	% of Total Passengers
Atlanta (3)		
Delta	7,990,860	84.08
Commuters	469,955	4.94
American	220,575	2.32
USAir	176,111	1.85
Continental	168,297	1.77
TWA	151,807	1.60
United	129,111	1.36
Northwest	124,539	1.31
America West	57,630	0.61
Midwest Express	15,193	0.16
Total	**9,504,078**	
Baltimore (30)		
USAir	1,106,576	53.09
Commuters	259,398	12.44
American	206,457	9.90
Delta	158,030	7.58
United	109,538	5.26
Northwest	74,688	3.58
TWA	70,656	3.39
Continental	51,307	2.46
America West	40,334	1.94
Others	7,406	0.36
Total	**2,084,390**	
Boston (12)		
Delta	976,187	20.87
USAir	826,531	17.67
Northwest	649,609	13.89
American	565,022	12.08
Commuters	449,959	9.62
United	399,747	8.55
Continental	334,137	7.14
Trump	201,534	4.31
TWA	194,851	4.17
America West	66,621	1.42
Others	12,525	0.27
Total	**4,676,723**	
Charlotte (15)		
USAir	3,910,284	87.89
Commuters	370,746	8.33
Delta	67,643	1.52
American	49,627	1.12
United	27,789	0.62
TWA	23,210	0.52
Total	**4,449,299**	
Chicago O'Hare (1)		
United	5,898,221	45.36
American	4,603,045	35.40
Simmons	534,535	4.11
Air Wis	458,767	3.53
Delta	383,252	2.95
Northwest	346,965	2.67
Continental	249,373	1.92
TWA	189,171	1.45
USAir	185,449	1.43
America West	122,851	0.94
Commuters	32,081	0.25
Total	**13,003,710**	

(Continued on next page)

MARKET SHARE, U.S. AIRPORTS, 6 MONTHS 1992 (CONTINUED)
Numbers in ( ) indicate ranking

	Enplaned Passengers	% of Total Passengers
Cincinnati (27)		
Delta	1,948,208	74.81
Commuters	389,242	14.95
USAir	82,666	3.17
American	58,119	2.23
United	41,115	1.58
TWA	37,361	1.43
Northwest	32,513	1.25
Continental	14,943	0.57
Total	**2,604,167**	
Dallas/Fort Worth (2)		
American	7,250,499	60.98
Delta	3,427,099	28.82
Commuters	562,815	4.73
Continental	143,087	1.20
United	125,649	1.06
Northwest	108,675	0.91
USAir	94,177	0.79
TWA	89,655	0.75
America West	63,266	0.53
Others	24,972	0.21
Total	**11,889,894**	
Denver (5)		
United	2,886,576	42.94
Continental	2,348,242	34.93
Commuters	460,495	6.85
Delta	282,576	4.20
American	250,421	3.72
Northwest	131,743	1.96
TWA	97,127	1.44
USAir	89,908	1.34
America West	87,375	1.30
Air Wis	73,305	1.09
Others	15,260	0.23
Total	**6,723,028**	
Detroit Metro Wayne (11)		
Northwest	3,446,413	72.69
American	251,800	5.31
Delta	205,764	4.34
Commuters	203,867	4.30
USAir	156,092	3.29
United	151,503	3.20
Continental	121,250	2.56
TWA	106,353	2.24
Southwest	75,267	1.59
Others	22,760	0.48
Total	**4,741,069**	
Honolulu (17)		
Hawaiian	1,058,268	24.35
Aloha	1,020,047	23.47
United	711,990	16.38
Northwest	445,179	10.24
Continental	306,710	7.06
Delta	281,977	6.49
American	226,852	5.22
TWA	97,373	2.24
Commuters	83,516	1.92
America West	79,389	1.83
American Trans Air	35,660	0.82
Total	**4,346,961**	

(Continued on next page)

MARKET SHARE, U.S. AIRPORTS, 6 MONTHS 1992 (CONTINUED)
Numbers in ( ) indicate ranking

	Enplaned Passengers	% of Total Passengers
Houston Intercontinental (20)		
Continental	3,225,904	78.65
American	240,511	5.86
Delta	188,592	4.60
Commuters	152,522	3.72
USAir	98,365	2.40
Southwest	71,505	1.74
United	62,531	1.52
America West	60,609	1.48
Others	818	0.02
Total	**4,101,357**	
Las Vegas (21)		
America West	1,350,865	32.97
Southwest	909,060	22.18
Delta	420,505	10.26
American	373,855	9.12
United	224,377	5.48
Northwest	224,297	5.47
Continental	192,382	4.69
USAir	188,898	4.61
TWA	98,980	2.42
Commuters	96,604	2.36
Hawaiian	17,849	0.44
Total	**4,097,672**	
Los Angeles (4)		
United	1,799,772	20.35
Delta	1,716,892	19.41
American	1,245,145	14.08
Southwest	726,459	8.21
USAir	675,518	7.64
Northwest	603,830	6.83
Continental	511,460	5.78
TWA	406,675	4.60
Commuters	374,521	4.23
America West	339,892	3.84
Alaska	243,538	2.75
WestAir	88,947	1.01
Hawaiian	67,613	0.76
Others	43,679	0.49
Total	**8,843,941**	
Miami (10)		
American	2,379,738	50.17
Delta	509,195	10.73
USAir	468,109	9.87
Commuters	398,466	8.40
Continental	287,162	6.05
United	268,008	5.65
Northwest	195,936	4.13
TWA	170,669	3.60
Others	66,477	1.40
Total	**4,743,760**	

(Continued on next page)

MARKET SHARE, U.S. AIRPORTS, 6 MONTHS 1992 (CONTINUED)
Numbers in ( ) indicate ranking

	Enplaned Passengers	% of Total Passengers
Minneapolis/St. Paul (14)		
Northwest	3,529,084	77.97
Commuters	233,770	5.16
United	178,615	3.95
American	132,688	2.93
Delta	130,864	2.89
Continental	115,582	2.55
TWA	86,568	1.91
USAir	70,288	1.55
America West	48,206	1.06
Others	770	0.02
Total	**4,526,435**	
New York Kennedy (19)		
TWA	1,473,869	35.63
American	1,095,859	26.49
Delta	556,890	13.46
Commuters	367,200	8.88
United	294,601	7.12
America West	115,933	2.80
USAir	62,806	1.52
Northwest	55,996	1.35
Others	113,627	2.75
Total	**4,136,781**	
New York LaGuardia (13)		
Delta	1,136,542	24.99
USAir	820,479	18.04
American	785,678	17.28
United	371,786	8.18
Continental	350,462	7.71
Trump	343,758	7.56
TWA	300,408	6.61
Northwest	270,027	5.94
Commuters	148,034	3.26
Others	20,156	0.44
Total	**4,547,330**	
Newark (9)		
Continental	2,632,856	50.69
United	556,266	10.71
USAir	487,442	9.38
American	474,027	9.13
Commuters	337,524	6.50
Delta	334,420	6.44
Northwest	164,473	3.17
TWA	100,350	1.93
America West	74,842	1.44
Others	32,217	0.62
Total	**5,194,417**	
Orlando (18)		
Delta	1,521,222	35.95
USAir	665,090	15.72
United	655,084	15.48
American	369,166	8.72
Continental	308,096	7.28
Northwest	300,066	7.09
TWA	190,477	4.50
Commuters	148,289	3.50
America West	43,384	1.03
American Trans Air	27,112	0.64
Others	4,096	0.10
Total	**4,232,082**	

(Continued on next page)

MARKET SHARE, U.S. AIRPORTS, 6 MONTHS 1992 (CONTINUED)
Numbers in ( ) indicate ranking

	Enplaned Passengers	% of Total Passengers
Philadelphia (24)		
USAir	1,831,008	52.00
Commuters	372,183	10.57
American	368,339	10.46
United	282,053	8.01
Delta	253,911	7.21
Northwest	153,943	4.37
TWA	121,823	3.46
Continental	117,142	3.33
Others	20,866	0.59
Total	**3,521,268**	
Phoenix (7)		
America West	2,292,184	41.90
Southwest	1,473,310	26.93
Delta	338,309	6.18
American	283,480	5.18
United	250,114	4.57
Northwest	239,658	4.38
Continental	175,193	3.20
USAir	154,302	2.82
TWA	120,922	2.21
Alaska	86,392	1.58
Commuters	57,083	1.04
Total	**5,470,947**	
Pittsburgh (16)		
USAir	3,642,448	83.50
Commuters	321,273	7.36
Delta	89,142	2.04
TWA	78,030	1.79
American	77,127	1.77
United	53,025	1.22
Northwest	48,105	1.10
Continental	47,227	1.08
Others	5,994	0.14
Total	**4,362,371**	
St. Louis (8)		
TWA	3,621,119	69.65
Southwest	467,014	8.98
Trans States Airlines	304,776	5.86
American	195,086	3.75
Northwest	138,008	2.65
United	128,943	2.48
Delta	117,609	2.26
USAir	90,644	1.74
America West	61,727	1.19
Continental	54,659	1.05
Commuters	19,589	0.38
Total	**5,199,174**	
Salt Lake City (26)		
Delta	2,152,582	79.73
Commuters	112,075	4.15
United	108,996	4.04
American	97,809	3.62
Continental	78,678	2.91
America West	60,300	2.23
TWA	46,685	1.73
Northwest	35,986	1.33
Horizon	6,785	0.25
Total	**2,699,896**	

(Continued on next page)

MARKET SHARE, U.S. AIRPORTS, 6 MONTHS 1992 (CONTINUED)
Numbers in ( ) indicate ranking

	Enplaned Passengers	% of Total Passengers
San Diego (25)		
Southwest	759,123	27.30
American	409,020	14.71
United	311,521	11.20
Delta	267,277	9.61
USAir	241,072	8.67
America West	203,408	7.32
Continental	160,432	5.77
Northwest	107,305	3.86
Commuters	105,567	3.80
Alaska	93,420	3.36
TWA	83,850	3.02
WestAir	29,803	1.07
Others	8,568	0.31
Total	**2,780,366**	
San Francisco (6)		
United	3,184,869	48.58
USAir	531,051	8.10
Delta	529,278	8.07
American	442,333	6.75
Continental	313,357	4.78
Northwest	305,015	4.65
WestAir	277,155	4.23
TWA	241,299	3.68
Southwest	236,928	3.61
America West	189,413	2.89
Alaska	157,300	2.40
Hawaiian	55,834	0.85
Commuters	44,035	0.67
Others	47,854	0.73
Total	**6,555,721**	
Seattle/Tacoma (22)		
United	820,375	22.29
Alaska	785,855	21.35
Northwest	436,765	11.87
Horizon	333,924	9.07
Delta	320,277	8.70
American	283,855	7.71
Continental	159,629	4.34
America West	124,372	3.38
USAir	116,176	3.16
TWA	102,529	2.79
WestAir	69,181	1.88
Markair	61,270	1.66
Hawaiian	56,349	1.53
Commuters	7,333	0.20
Others	3,037	0.08
Total	**3,680,927**	
Tampa (28)		
USAir	639,663	23.69
Delta	507,086	18.78
American	258,159	9.56
Continental	245,225	9.08
Northwest	241,030	8.93
United	139,167	5.15
TWA	128,568	4.76
Commuters	126,245	4.68
America West	32,311	1.20
Others	5,289	0.20
Total	**2,322,743**	

(Continued on next page)

MARKET SHARE, U.S. AIRPORTS, 6 MONTHS 1992 (CONTINUED)
Numbers in ( ) indicate ranking

	Enplaned Passengers	% of Total Passengers
Washington Dulles (29)		
United	1,435,747	63.86
Commuters	186,391	8.29
American	156,383	6.96
Delta	124,575	5.54
Air Wis	109,432	4.87
TWA	66,319	2.95
USAir	61,263	2.72
Continental	60,677	2.70
Northwest	47,510	2.11
Total	**2,248,297**	
Washington National (23)		
USAir	764,457	21.63
Delta	641,410	18.15
Northwest	504,325	14.27
American	472,264	13.36
Continental	286,399	8.10
United	233,102	6.60
Commuters	227,108	6.43
TWA	188,053	5.32
Trump	163,706	4.63
America West	36,554	1.03
Midwest Express	16,386	0.46
Total	**3,533,764**	

U.S. AIR CARRIERS
TOTAL ASSETS AND INVESTMENT IN EQUIPMENT
Calendar Years 1969 - 1991 (Millions of Dollars)

Year	TOTAL Assets	Value of Flight Equipment	Value of Ground Property & Equipment, & Other	Less: Reserves for Depreciation & Overhaul	Equals: Net Value of Owned Operating Property & Equipment	Investment in Operating Property and Equipment as a Percent of Total Assets
1969	$12,069	$ 9,943	$ 1,516	$ 3,560	$ 7,899	65.4%
1970	12,913	10,950	1,951	4,120	8,782	68.0
1971	12,998	11,221	2,028	4,649	8,600	66.2
1972	13,635	11,918	2,225	5,115	9,028	66.2
1973	14,464	12,908	2,424	5,693	9,639	66.6
1974	15,200	13,538	2,539	6,252	9,826	64.6
1975	15,064	14,035	2,635	6,823	9,847	65.4
1976	15,454	14,399	2,792	7,585	9,605	62.2
1977	16,869	14,822	2,997	8,141	9,679	57.4
1978	20,745	16,127	3,367	8,799	10,696	51.6
1979	24,907	18,561	3,985	9,746	12,800	51.4
1980	28,900	20,859	4,682	10,309	15,233	52.7
1981	30,513	22,375	5,175	11,028	16,521	54.1
1982	31,525	23,786	5,424	11,405	17,804	56.5
1983	35,213	26,588	6,191	12,910	19,868	56.4
1984	36,769	28,509	6,061	14,043	20,527	55.8
1985	40,978	30,402	6,772	15,467	21,707	53.0
1986	47,105	31,750	8,468	14,764	25,454	54.0
1987	51,436	33,177	9,223	15,580	26,820	52.1
1988	56,047	35,781	10,248	17,450	28,579	51.0
1989	62,454	38,812	11,903	19,018	31,697	50.8
1990	67,769	40,215	13,523	20,593	33,144	48.9
1991 p	69,396	42,103	14,255	21,873	34,485	49.7

Source: Department of Transportation, Office of Aviation Statistics, "Air Carrier Financial Statistics Quarterly" (Quarterly).
Includes land and construction in progress.
p Preliminary.

AIRLINES FILING BANKRUPTCY BETWEEN 1979-1992

	Airline	Bankruptcy Filing Date	Type of Filing	Successfully Reorganized
1.	**New York Airways**	5/18/79	Chapter 11	No
2.	Aeroamerica	11/19/79	Chapter 11	No
3.	Florida Airlines	1/24/80	Chapter 11	No
4.	Indiana Airways	3/3/80	Chapter 11	No
5.	Air Bahia	12/15/80	Chapter 11	No
6.	Tejas Airlines	12/31/80	Chapter 11	No
7.	Mountain West	3/6/81	Chapter 11	No
8.	LANICA	3/16/81	Chapter 11	No
9.	Coral Air	7/13/81	Chapter 11	No
10.	Pacific Coast	9/11/81	Chapter 11	No
11.	Swift Aire Line	9/18/81	Chapter 11	No
12.	Golden Gate	10/9/81	Chapter 11	No
13.	Pinehurst Airlines	1/26/82	Chapter 11	No
14.	Silver State Airlines	3/3/82	Chapter 11	No
15.	Air Pennsylvania	3/26/82	Chapter 11	No
16.	Air South	4/2/82	Chapter 11	No
17.	Cochise Airlines	4/16/82	Chapter 11	No
18.	**Braniff International**	5/13/82	Chapter 11	No
19.	Astec Air East	7/8/82	Chapter 11	No
20.	Will's Air	8/19/82	Chapter 11	No
21.	Aero Sun International	10/5/82	Chapter 11	No
22.	Aero Virgin Islands	10/19/82	Chapter 11	No
23.	Altair	11/9/82	Chapter 11	No
24.	North American	12/9/82	Chapter 11	No
25.	Inland Empire	2/1/83	Chapter 11	No
26.	State Airlines	2/14/83	Chapter 11	No
27.	Golden West	4/22/83	Chapter 11	No
28.	**Continental Air Lines**	9/24/83	Chapter 11	Yes
29.	National Florida	12/2/83	Chapter 7	No
30.	Air Vermont	1/30/84	Chapter 11	No
31.	**Pacific Express**	2/2/84	Chapter 11	No
32.	Dolphin	2/8/84	Chapter 11	No
33.	Combs Airways	4/9/84	Chapter 11	No
34.	New York Helicopter	5/1/84	Chapter 11	No
35.	**Air Florida**	7/3/84	Chapter 11	No
36.	Excellair	7/17/84	Chapter 7	No
37.	**American International**	7/19/84	Chapter 11	No
38.	**Emerald**	8/21/84	Chapter 11	No
39.	Hammonds Commuter	8/29/84	Chapter 11	No
40.	Air North	9/4/84	Chapter 11	No
41.	Wright Air Lines	9/27/84	Chapter 11	No
42.	Oceanaire Lines	10/2/84	Chapter 7	No
43.	Atlantic Gulf	10/10/84	Chapter 11	No
44.	Connectaire	10/10/84	Chapter 7	No
45.	**Air One**	10/26/84	Chapter 11	No
46.	Capitol Air	11/23/84	Chapter 11	No
47.	**Wien Air Alaska**	11/28/84	Chapter 11	No
48.	Northeastern International	1/8/85	Chapter 11	No
49.	Pompano Airways	1/22/85	Chapter 11	No
50.	Far West Airlines	2/22/85	Chapter 11	No
51.	American Central	3/8/85	Chapter 11	No
52.	Provincetown Boston	3/13/85	Chapter 11	No
53.	Sun West Airlines	3/19/85	Chapter 11	No
54.	Wise Airlines	5/1/85	Chapter 11	No
55.	Cascade Airways	8/19/85	Chapter 11	No
56.	Wheeler Airlines	10/7/85	Chapter 11	No
57.	Pride Air	12/2/85	Chapter 11	No
58.	Southern Express	1/21/86	Chapter 11	No
59.	Imperial Airlines	1/30/86	Chapter 11	No
60.	**Arrow Airways**	2/11/86	Chapter 11	No
61.	Sea Airmotive	4/9/86	Chapter 11	No
62.	SFO Helicopter	4/18/86	Chapter 11	No
63.	Trans Air	8/19/86	Chapter 11	No
64.	**Frontier Airlines**	8/28/86	Chapter 11	No

65.	Chicago Airlines	2/19/87	Chapter 11	No
66.	**McClain Airlines**	2/23/87	Chapter 11	No
67.	Rio Airways	2/27/87	Chapter 11	No
68.	Air Puerto Rico	3/6/87	Chapter 11	No
69.	Gull Air	3/10/87	Chapter 11	No
70.	Royal West Airlines	3/12/87	Chapter 11	No
71.	**Air Atlanta**	4/3/87	Chapter 11	No
72.	Air South, Inc.	6/17/87	Chapter 11	No
73.	Royale Airlines	9/9/87	Chapter 11	No
74.	Sun Coast Airlines	1/5/88	Chapter 11	No
75.	Air New Orleans	1/14/88	Chapter 11	No
76.	Air Virginia	1/15/88	Chapter 11	No
77.	Mid Pacific Airlines	1/19/88	Chapter 11	No
78.	Exec Express	3/4/88	Chapter 11	No
79.	Caribbean Express	5/6/88	Chapter 11	No
80.	Pocono Airlines, Inc.	5/25/88	Chapter 11	Yes
81.	Virgin Island Seaplane	6/20/88	Chapter 11	No
82.	Princeton Air Link Corp.	8/11/88	Chapter 7	No
83.	Qwest Air	9/14/88	Chapter 11	No
84.	Southern Jersey Airways	9/27/88	Chapter 11	No
85.	Eastern Air Lines	3/9/89	Chapter 11	No
86.	Big Sky Airlines	3/14/89	Chapter 11	No
87.	Air Kentucky	7/19/89	Chapter 7	No
88.	Braniff, Inc.	9/28/89	Chapter 11	No
89.	Presidential	10/26/89	Chapter 11	No
90.	Resort Commuter	11/17/89	Chapter 11	No
91.	Pocono Airlines, Inc.	1/23/90	Chapter 11	No
92.	SMB Stage Lines	5/10/90	Chapter 11	No
93.	CC Air	7/5/90	Chapter 11	No
94.	Continental Airlines	12/3/90	Chapter 11	No
95.	Britt Airways	12/3/90	Chapter 11	No
96.	Rocky Mountain Airways	12/3/90	Chapter 11	No
97.	Pan Am World Airways	1/8/90	Chapter 11	No
98.	Pan Am Express	1/8/91	Chapter 11	No
99.	L' Express	1/9/91	Chapter 11	No
100.	Eastern Air Lines	1/18/91	Chapter 7	No
101.	Bar Harbor Airlines	1/20/91	Chapter 11	No
102.	Northcoast Executive	1/29/91	Chapter 7	No
103.	Midway Airlines	3/25/91	Chapter 11	No
104.	Midway Airlines	11/7/91	Chapter 7	No
105.	Grand Airways	3/26/91	Chapter 11	No
106.	Metro Airlines	4/1/91	Chapter 11	No
107.	Jet Express	5/20/91	Chapter 11	No
108.	Metro Airlines Northeast	5/30/91	Chapter 11	No
109.	America West	6/27/91	Chapter 11	No
110.	Mohawk Airlines	8/12/91	Chapter 11	No
111.	Flagship Express	12/31/91	Chapter 11	No
112.	Virgin Island Seaplane	1/22/92	Chapter 11	No
113.	Trans World Airlines	1/31/92	Chapter 11	No
114.	L' Express	2/28/92	Chapter 7	No
115.	Markair	6/8/92	Chapter 11	No
116.	Hermans/Markair Express	6/8/92	Chapter 11	No
117.	States West Airlines, Inc.	12/15/92	Chapter 11	No

Note: Airlines listed in bold type are air carriers operating under FAR 121 and 14CFR401 authority from the Department of Transportation.

Source: Aviation Forecasting & Economics

STATUS OF AIRLINE LABOR CONTRACTS
December 29, 1992

Airlines	Pilots	Engineers	Flight Attendants	Dispatchers	Mechanics/ Related Personnel	Clerical/ Agents
ABX	IBT-7/31/95	-	-	-	-	-
Air Wisconsin	ALPA 9/30/93	-	AFA (6/30/92) In negotiation	TWU 10/1/94	IAM (12/31/90) In negotiation	IAM (12/31/90) In negotiation
Alaska	ALPA 5/1/93	-	AFA (10/1/90) In mediation	TWU 4/24/96	IAM 9/1/97	IAM (9/30/92) In negotiation
Aloha	ALPA 4/30/94	-	AFA 8/31/93	TWU 9/1/93	IAM 10/31/93	IAM 10/31/93
American	APA 8/31/94	FEIA 8/31/94	APFA 12/31/92 In negotiation	TWU 3/1/95	TWU 3/1/95	-
American Trans Air	-	-	AFA Initial Contract In negotiation	-	-	-
Continental	-	-	IAM 6/22/96	TWU 1/1/94	-	-
Delta	ALPA 1/1/95	-	-	PAFCA 1/1/94	-	-
DHL	ALPA Initial Contract	In mediation	-	-	-	-
Hawaiian	ALPA 10/31/94	-	AFA 10/31/94	TWU 10/31/94	IAM 10/31/94	IAM 10/31/94
Northwest	ALPA 3/1/94	-	IBT 1/1/93 In negotiation	TWU 12/31/94	IAM 5/1/92 In negotiation	IAM 5/1/92 In negotiation
Southwest	SAPA 8/31/94	-	TWU 6/1/96	SAEA 12/1/96	IBT 8/17/95	IAM 3/1/94 ROPA 12/31/94
TWA	ALPA 9/30/95	-	IFFA 9/1/97	TWU 10/8/95	IAM 9/10/95	IAM 9/10/95
United	ALPA 11/30/94	-	AFA 3/1/96	IAM 12/1/94	IAM 12/1/94	-
UPS	IPA 1/1/96	-	-	-	IBT 7/31/95	-
USAir	ALPA 4/30/96	-	AFA (8/31/89)	TWU Initial contract In mediation	IAM 9/30/95 In negotiation	-
USAir Shuttle	ALPA 7/2/93	-	AFA (12/31/89)	TWU Initial contract In negotiation	-	-
World	IBT (6/30/92) In mediation	IBT 6/30/92 In mediation	IBT 6/30/92 In negotiation	TWU 7/1/93	-	-

( ) denotes expired contract

AFA	-	**Association of Flight Attendants**	**IFFA**	-	**Independent Federation of Flight Attendants**
ALPA	-	**Air Line Pilots Association**	**IPA**	-	**Independent Pilots Association**
APA	-	**Allied Pilots Association**	**PAFCA**	-	**Professional Airline Flight Control Association**
APFA	-	**Association of Professional Flight Attendants**	**ROPA**	-	**Ramp Operations Provisioning Association**
FEIA	-	**Flight Engineers International Association**	**SAEA**	-	**Southwest Airline Employee Association**
IAM	-	**International Association of Machinists**	**SAPA**	-	**Southwest Airline Pilots Association**
IBT	-	**International Brotherhood of Teamsters**	**TWU**	-	**Transport Workers Union**

Source: J. Glass & Associates
Aviation Daily Chart

IMPORTANT PHONE NUMBERS FOR AIRLINE LABOR:

Air Line Pilots Association
1625 Massachusetts Avenue, N.W.
Washington, D.C. 20036
(p)703-689-2270
(f)202-797-4052
President: Randolph Babbitt
First VP: Duane Woerth
Secretary:Jerome Mugerditchian
Treasurer:John Donnelly
(41,000 members, 44 carriers)

Allied Pilots Association
Box 5524
Arlington, Texas 76005-5524
(214)988-3188
President: Richard LaVoy

Association Of Flight Attendants
1625 Massachusetts Avenue, N.W.
Washington, D.C. 20036
(p)202-328-5400
(f)202-328-5424
President: Dee Maki
VP: Frederick Casey
33,000 members at 20 U.S. airlines

Association of Professional Flight Attendants
1004 W. Euless Bvld., Euless, Tex.
(p)817-540-2077
President: Cheryle Leon
VP: Sheri Cappello

Flight Engineers International Association
Suite 202, 1926 Pacific Coast Highway
Redondo Beach, Calif. 90277-6145
(p)213-316-4094
President: Jerry Austin
VP: Donald Thielke

Independent Federation of Flight Attendants
President: Vicci Frankovich

Independent Pilots Association

International Association of Machinists & Aerospace Workers
9000 Machinists Pl., Upper Malboro, Md.
20772-2687
(p)301-967-4500
President George Kourpias

(Continued on next page)

IMPORTANT PHONE NUMBERS FOR AIRLINE LABOR: (Continued)

General VP: John Peterpaul
Airline Coordinator: William Sheri
Public Relations: James Connelly

International Brotherhood of Teamsters
President: Ron Carey
(Director Airline Division: Marv Griswald) check this!!!

National Mediation Board
1301 K. Street N.W.
Suite 250 East
Washington D.C. 20572
(202)523-5290
(202)523-4856
Board Members: Kimbery Madigan, Joshua Javits, Patrick Cleary
Pubilc Relations: Lew Townsend 523-5335

Professional Airline Flight Control Association

Ramp Operations Provisioning Association
8008 Cedar Springs
Lock Box #20
Love Field, Dallas, Tex 75235
(214)358-5405
President: Charles Cerf

Southwest Airline Employee Association
(dispatchers)

Southwest Airline Pilots Association
8036 Aviation Place
Lock Box #7
Dallas, Tex. 75235
(p)214-350-9237
President: Gary Kerans
VP John Kramer
1,400 members

Transport Workers Union
80 West End Ave, New York, N.Y. 10023
(212)873-6000
President: George Leitz
International VP and Director: John Kerrigan

AIRLINE CODES

Code	Airline
AA	American Airlines, Inc.
AC	Air Canada
AD	Lone Star Airlines
AF	Air France
AG	Provincial Airlines
AM	Aeromexico-Aerovias de Mexico S.A. de C.V.
AQ	Aloha Airlines, Inc.
AR	Aerolineas Argentinas
AS	Alaska Airlines
AT	Royal Air Maroc
AV	Avianca
AY	Finnair
AZ	Alitalia
BA	British Airways
BF	Markair, Inc.
BK	Paradise Island Airlines
BR	Eva Airways Corporation
BW	BWIA International
BZ	Keystone Air Service Ltd.
CA	Air China
CD	Trans-Provincial Airlines Ltd.
CH	Bemidji Airlines
CI	China Airlines
CM	Copa
CO	Continental Airlines
CP	Canadian Airlines International Ltd.
CQ	Air Alpha, Inc.
CU	Cubana Airlines
CX	Cathay Pacific Airways Ltd.
C2	Air Caribbean Ltd.
DH	Atlantic Coast Airlines
DL	Delta Air Lines, Inc.
DO	Dominicana de Aviacion
DQ	Coastal Air Transport
DV	Nantucket Airlines
DW	Helicopter Shuttle
DZ	Air Metro North, S.A.
ED	CCAir Inc.
EI	Aer Lingus P.L.C.
EJ	New England Airlines, Inc.
EM	Empire Airlines
EU	Empresa Ecuatoriana de Aviacion
EV	Atlantic Southeast Airlines, Inc.
FF	Tower Air, Inc.
FK	Flamenco Airways, Inc.
FN	F.S. Air Service, Inc.
FQ	Air Aruba
FY	Metroflight Airlines
GA	Garuda Indonesia
GD	TAESA-Transportes Aeros Ejecitivos, S.A. de C.V.
GQ	Big Sky Airlines
GU	Aviateca S.A.
GY	Guyana Airways
G2	Greenbrier Airlines
HA	Hawaiian Airlines
HD	New York Helicopter Corporation
HF	Hageland Aviation Services, Inc.
HG	Harbor Airlines, Inc.
HI	Papillon Airways, Inc.
HO	Airways International, Inc.
HP	America West Airlines, Inc.
HQ	Business Express
HW	North-Wright Air Ltd.

Code	Airline
IB	Iberia
IH	Loken Aviation, Inc.
IK	Road Air Lines
IW	AOM French Airlines
JB	Helijet Airways
JC	Rocky Mountain Airways
JF	L.A.B. Flying Service, Inc.
JI	Jet Express
JL	Japan Airlines
JM	Air Jamaica Limited
JQ	Trans-Jamaican Airlines Ltd.
JR	Aero California
JU	Yugoslav Airlines (JAT)
JV	Bearskin Lake Air Service Limited
JX	Jes Air
KE	Korean Air
KL	KLM-Royal Dutch Airlines
KN	Morris Air Service
KP	KIWI International Air Lines, Inc.
KS	Peninsula Airways, Inc.
KW	Carnival Air Lines
KX	Cayman Airways, Ltd.
LA	Lan Chile S.A.
LH	Lufthansa German Airlines
LI	Liat Ltd.
LM	ALM-Antillean Airlines
LR	LACSA
LS	LLiamna Air Taxi, Inc.
LT	LTU International Airways
LV	AEROPOSTAL
LW	Air Nevada
LY	El Al Israel Airlines
MH	Malaysia Airlines
MO	Calm Air International Ltd.
MQ	Simmons Airlines
MS	Egyptair
MU	China Eastern Airlines
MX	Mexicana de Aviacion
NA	Executive Air Charter
NJ	Northeast Express Regional Airlines
NK	Spirit Airlines
NN	C.A.A.A.- Air Martinique
NR	Norontair
NT	Norcanair
NV	Northwest Territorial Airways Ltd.
NW	Northwest Airlines, Inc.
NX	Nationair
NZ	Air New Zealand
N2	Business Helicopters
OA	Olympic Airways
OE	Westair Airlines
OG	Air Guadeloupe
OH	Comair, Inc.
OJ	Air St-Barthelemy
OK	Czechoslovak Airlines
OO	Sky West Airlines
OP	Chalk's International Airlines
OQ	Arizona Pacific Airline, Inc.
OT	Evergreen Alaska
OY	Sunaire
OZ	Asiana Airlines
PD	Pem Air Limited
PK	Pakistan International Airlines
PQ	Pacific Coast Airlines
PR	Philippine Airlines, Inc.

(Continued on next page)

AIRLINE CODES

Code	Airline
PY	Surinam Airways Ltd.
QA	Aero Caribe
QD	Grand Airways, Inc.
QF	Qantas Airways Ltd.
QK	Air Nova Inc.
QQ	Reno Air, Inc.
QR	Air Satellite, Inc.
QW	Turks and Caicos Airways, Ltd.
QX	Horizon Air
RG	Varig, S.A.
RJ	Royal Jordanian
RL	Ultr Air
RM	Wings West Airlines, Inc.
RP	Precision Airlines
RV	Reeve Aleutian Airways, Inc.
SE	Wings of Alaska
SK	SAS-Scandinavian Airlines System
SN	Sabena Belgian World Airlines
SQ	Singapore Airlines
SR	Swissair
SU	Aeroflot-Russian International Airlines
SV	Saudi Arabian Airlines
SY	Sun Country Airlines, Inc.
TA	Taca International Airlines, S.A.
TO	Alkan Air Ltd.
TP	TAP Air Portugal
TR	Transbrasil S/A Linhas Aereas
TV	Haiti Trans Air S.A.
TW	Trans World Airlines, Inc.
TZ	American Trans Air
T7	Translift Airways
UA	United Airlines
UC	LADECO Airlines
UE	Air LA
UF	SARO-Servicios Aereos Rutas Oriente S.A. de C.V.
UO	Direct Air, Inc.
UP	Bahamasair
US	USAir
UW	Perimeter Airlines (Inland) Ltd.
VA	Viasa
VI	Vieques Air Link, Inc.
VP	Vasp
VS	Virgin Atlantic Airways Ltd.
VW	Aeromar Airlines
WJ	Labrador Airways, Ltd.
WM	Windward Island Airways International N.V.
WN	Southwest Airlines
WP	Aloha Islandair, Inc.
WS	Westates Airlines
WU	Air West Airlines, Inc.
XJ	Mesaba Aviation
XP	Casino Express Airlines
XQ	Action Airlines
XW	Walker's International
XY	Ryan Air, Inc.
YE	Grand Canyon Airlines, Inc.
YH	Air Baffin Ltd.
YI	Air Sunshine Inc.
YL	Long Island Airlines
YN	Air Creebec Inc.
YR	Scenic Airlines Inc.
YU	Aerolineas Dominicanas, S.A. - Dominair

Code	Airline
YV	Mesa Airlines
YW	Stateswest Airlines, Inc.
YX	Midwest Express Airlines, Inc.
ZD	Ross Aviation, Inc.
ZK	Great Lakes Aviation, Ltd.
ZN	Key Airlines, Inc.
ZO	Mohawk Airlines
ZP	Virgin Air, Inc.
ZS	Hispaniola Airways C. por A.
ZV	Air Midwest
ZW	Air Wisconsin
ZX	AirBC, Ltd.
2D	Southeast Airlines
2E	Markair Express, Inc.
2F	Frontier Flying Service
2P	Prairie Flying Service (1976) Ltd.
2S	Island Express
2U	Air Caraibes (Dominica) Limited
2Y	Koyukon Air, Inc.
2Z	Aerolitoral - Servicios Aeros Litoral S.A. de C.V.
3C	Camai Air
3E	Northwestern Air Lease
3H	Air Inuit (1985) Ltd.
3J	Air Alliance
3K	Tatonduk Flying Service
3L	West Isle Air, Inc.
3M	Gulfstream International Airlines, Inc.
3N	AirVantage Inc.
3S	Shuswap Air
3T	Contact Air
3V	Waglisla Air Inc.
4A	Canadian Eagle Airlines Ltd.
4B	Olson Air Service, Inc.
4E	Tanana Air Service
4G	Alliance Air, Inc.
4H	Hanna's Air Saltspring
4K	Kenn Borek Air Ltd.
4L	Air Alma, Inc.
4N	Air North
4T	Turquoise Airways Ltd.
4V	Voyageur Airways
4W	Warbelow's Air Ventures, Inc.
4X	L'Express Airlines
4Y	Yute Air Alaska, Inc.
5A	Alpine Aviation, Inc.
5B	Bellair Inc.
5C	Conquest Airlines Corp.
5D	Trans Dominican Airways (TRADO)
5F	Arctic Circle Air
5K	Kenmore Air
5N	Connectair Charters Ltd.
5P	Ptarmigan Airways, Ltd.
5S	Airspeed Aviation Inc.
5T	Aviacion del Noroeste
5U	Skagway Air Service, Inc.
6A	Aviacsa - Consorcio Aviaxsa, S.A. de C.V.
6B	Baxter Aviation
6C	Cape Smythe Air Service, Inc.
6D	Alaska Island Air, Inc.
6G	Las Vegas Airlines
6L	Aklak Air Ltd.
6M	40-Mile Air, Ltd.
6Q	Barrow Air, Inc.

(Continued on next page)

AIRLINE CODES

Code	Airline
6S	Ketchikan Air Service, Inc.
6T	Tyee Airways, Limited
6V	Air Vegas, Inc.
6W	Wilderness Airline (1975) Ltd.
7A	Haines Airways, Inc.
7B	Simpson Air Ltd.
7D	Air Molokai, Inc.
7F	First Air
7H	ERA Aviation
7K	Larry's Flying Service, Inc.
7M	Aeromonterrey, S.A. de C.V.
7N	Air Manitoba
7R	Redwing Airways, Inc.
7T	Trans Cote Inc.
7V	Alpha Air
7W	Air Sask Aviation 1991
7Z	Laker Airways (Bahamas) Ltd.
8B	Baker Aviation, Inc.
8D	Awood Air Ltd.
8E	Bering Air, Inc.
8F	Haiti National Airlines (HANAIR)
8G	GP Express Airlines, Inc.
8H	Harbor Air Service
8L	Servicio Aereo Leo Lopez
8M	Skymaster
8N	Flagship Airlines, Inc.
8P	Pacific Coastal Airlines Limited
8R	WRA, Inc.
8U	Atlantic Air BVI Ltd.
8V	Wright Air Service, Inc.
9A	Air Atlantic Ltd.
9E	Express Airlines I, Inc.
9F	Skycraft Air Transport Inc.
9K	Cape Air
9L	COLGAN AIR
9M	Central Mountain Air Ltd.
9N	Trans States Airlines, Inc.
9Q	Taquan Air Service, Inc.
9S	Sabourin Lake Airways, Ltd.
9T	Athabaska Airways, Ltd.
9V	Air Laurentian
9Y	Yutana Airlines
IS	Island Airlines, Inc.
KO	Rolling Hills Aviation, Inc.
V2	Arizona Air
3L	West Isle Air, Inc.
4Q	Trans North Aviation, Ltd.
MC	Military Airlift Command (MAC)

Source: Official Airline Guides

Section II.

Regional Airlines

TOP 50 REGIONAL AIRLINE COMPANIES IN 1991

Individual Group	Carrier Group	Carrier Group/ Individual Carrier and Codes Shared (1991)	Individual Carrier	Carrier Group
	1	AMR Eagle (AA*)		
3		Flagship Airlines [1]	2,364,839	6,412,630
6		Simmons Airlines	2,082,085	
16		Wings West Airlines	1,021,680	
18		Executive Air	944,026	
	2	USAir Express (US*)		
8		Henson Aviation	1,948,679	4,437,781
17		Allegheny Commuter Airlines	965,111	
19		Pennsylvania Commuter Airlines	889,237	
23		Jetstream International Airlines	634,754	
	3	Continental Express (CO*)		
1		Continental Express	3,294,811	
2	4	Air Wisconsin (UA*) [2]	2,490,842	
4	5	Atlantic Southeast Airlines (DL*)	2,251,036	
5	6	WestAir Airlines (UA*) [3]	2,246,779	
7	7	Comair (DL*)	2,022,032	
9	8	Metroflight (AA*)	1,610,344	
10	9	Horizon Air Industries (AS*/NW*)	1,422,557	
11	10	Business Express Airlines (DL*)	1,343,108	
	11	Mesa Airlines		
20		Mesa Airlines (UA*/US*/YX*) [4]	888,988	1,312,005
27		Air Midwest (TW*)	423,017	
12	12	Express Airlines, I (NW*)	1,268,880	
13	13	SkyWest Airlines (DL*)	1,245,851	
14	14	TransStates Airlines (TW*)	1,121,291	
15	15	Mesaba Airlines (NW*)	1,041,984	
21	16	CCAir (US*)	753,370	
22	17	Trans World Express (PA/TW*) [5]	721,168	
24	18	Atlantic Coast Airlines (UA*)	612,000	
25	19	Midway Commuter Airlines (ML*) [6]	538,554	
	20	Northeast Express Regional Airlines (NW*)		
32		Precision Airlines	296,800	523,031
37		Northeast Express Airlines	226,231	
26	21	Chautauqua Airlines (US*)	481,052	
28	22	Aloha IslandAir (AQ*)	358,800	
29	23	Crown Airways (US*)	349,614	
30	24	Sunaire Express	348,336	
31	25	StatesWest Airlines (NW*/US*)	316,512	
33	26	Paradise Island Airways	289,772	
34	27	Scenic Airlines	287,697	
35	28	Commutair (US*)	272,124	
36	29	Great Lakes Aviation	249,251	
38	30	ERA Aviation (AS*)	208,090	
39	31	Markair Express (BF*)	199,155	
40	33	Vieques Air Link	127,482	
41	34	Nantucket Airlines	110,163	
42	35	Conquest Airlines	90,402	
43	36	Peninsula Airways	81,265	
44	38	Mohawk Airlines	78,325	
45	39	L'Express Airlines	77,858	
46	40	SouthCentral Air	72,629	
47	41	Air Nevada	61,270	
48	42	GP Express Airlines	60,005	
49	43	Airways International	49,428	
50	44	Lone Star Airlines	48,681	
Total Enplanements - Top 50 Airlines			**40,888,045**	**97.5%**
Total Industry Enplanements			**41,956,662**	
Total Revenue Passenger Miles - Top 50 Airlines (000's)			**7,712,043**	**98.9%**
Total Industry Revenue Passenger Miles (000's)			**7,801,722**	

Notes:
1 Includes Nashville Eagle and Command Airways.
2 Includes Aspen Airways.
3 Excludes Atlantic Coast Airlines.
4 Inclusive of Skyway and FloridaGulf.
5 Inclusive of Pan Am Express.
6 Carrier ceased operations during 1991.

Source: AvStat Associates for the Regional Airline Association based on U.S.DOT filings.

TOP 50 INDIVIDUAL REGIONAL AIRLINES IN 1991

1991 Rank	Carrier	Codes Shared (1991)	1991 Passengers
1	Continental Express	CO	3,294,811
2	Air Wisconsin 1/	UA	2,490,842
3	Flagship Airlines 2/	AA	2,364,839
4	Atlantic Southeast Airlines	DL	2,251,036
5	WestAir Airlines 3/	UA	2,246,779
6	Simmons Airlines	AA	2,082,085
7	Comair	DL	2,022,032
8	Henson Aviation	US	1,948,679
9	Metroflight	AA	1,610,344
10	Horizon Air Industries	AS/NW	1,422,557
11	Business Express Airlines	DL	1,343,188
12	Express Airlines, I	NW	1,268,880
13	SkyWest Airlines	DL	1,245,851
14	Trans States Airlines	TW	1,121,291
15	Mesaba Aviation	NW	1,041,984
16	Wings West Airlines	AA	1,021,680
17	Allegheny Commuter Airlines	US	965,111
18	Executive Air	AA	944,026
19	Pennsylvania Commuter Airlines	US	889,237
20	Mesa Airlines 4/	UA/US/YX	888,988
21	CCAir	US	753,370
22	Trans World Express 5/	PA/TW	721,168
23	Jetstream International Airlines	US	634,754
24	Atlantic Coast Airlines	UA	612,000
25	Midway Commuter Airlines 6/	ML	538,554
26	Chautauqua Airlines	US	481,052
27	Air Midwest	TW	423,017
28	Aloha IslandAir	AQ	358,800
29	Crown Airways	US	349,614
30	Sunaire Express		348,336
31	StatesWest Airlines	NW/US	316,512
32	Precision Airlines	NW	296,800
33	Paradise Island Airways		289,772
34	Scenic Airlines		287,697
35	Commutair	US	272,124
36	Great Lakes Aviation	UA	249,251
37	Northeast Express Airlines	NW	226,231
38	ERA Aviation	AS	208,090
39	Markair Express	BF	199,155
40	Vieques Air Link		127,482
41	Nantucket Airlines		110,163
42	Conquest Airlines		90,402
43	Peninsula Airways		81,265
44	Mohawk Airlines		78,325
45	L'Express Airlines		77,858
46	SouthCentral Air		72,629
47	Air Nevada		61,270
48	GP-Express Airlines		60,005
49	Airways International		49,428
50	Lone Star Airlines		48,681

Total Enplanements - Top 50 Airlines	**40,888,045**
Total Industry Enplanements	**41,956,662**
Total Revenue Passenger Miles Top 50 Airlines (000's)	**7,712,043**
Total Industry Revenue Passenger Miles (000's)	**7,801,722**

Notes:
1. Includes Aspen Airways.
2. Includes Nashville Eagle and Command Airways.
3. Excludes Atlantic Coast Airlines.
4. Inclusive of Skyway and FloridaGulf.
5. Inclusive of Pan Am Express.
6. Carrier ceased operations during 1991.

Source: AvStat Associates for the Regional Airline Association based on U.S.DOT filings

HISTORICAL GROWTH - REGIONAL TRAFFIC AND AIRCRAFT, 1970-1991

Year	Passengers Enplaned	Percent Change	Passenger Aircraft Operated	Percent Change
1970	4,270,000		687	
1971	4,698,000	10.0	782	13.8
1972	5,262,000	12.0	791	1.2
1973	5,688,000	8.1	885	11.9
1974	6,842,000	20.3	997	12.7
1975	7,243,000	5.9	1,073	7.6
1976	7,914,000	9.3	1,009	-6.0
1977	9,185,000	16.1	1,119	10.9
1978	11,026,000	20.0	1,047	-6.4
1979	13,972,000	26.7	1,265	20.8
1980	14,810,000	6.0	1,339	5.8
1981	15,400,000	4.0	1,463	9.3
1982	18,550,000	20.5	1,573	7.5
1983	21,820,000	17.6	1,545	-1.8
1984	26,140,000	19.8	1,747	13.1
1985	26,000,000	-0.5	1,745	-0.1
1986	28,360,000	9.1	1,806	3.5
1987	31,787,539	12.1	1,841	1.9
1988	35,188,223	10.7	1,801	-2.2
1989	37,359,656	6.2	1,907	5.9
1990	42,099,429	12.7	1,917	0.5
1991	41,974,242	-0.3	1,992	3.9

Source: AvStat Associates for the Regional Airline Association.

TOP REGIONAL AIRCRAFT 1991

Rank	Manufacturer	Model	Total Aircraft in Airline Service[1]	Total Seats	Fleet Seating Capacity	1991 Fleet Flying Hours
1	Embraer	Brasilia	173	5,190	11.4	508,203
2	Saab	340	157	5,024	11.1	344,401
3	Aerospatiale	ATR-42	92	4,512	9.9	197,415
4	British Aerospace	Jetstream	222	4,218	9.3	473,770
5	Beech	1900	188	3,572	7.9	502,717
6	Fairchild	Metro	187	3,553	7.8	370,251
7	Shorts	360	84	3,024	6.7	139,782
8	deHavilland	Dash 8[2]	75	2,910	6.4	204,646
9	British Aerospace	146	17	1,670	3.7	47,022
10	deHavilland	Dash 7	27	1,350	3.0	55,294
11	deHavilland	Twin Otter	71	1,349	3.0	114,638
12	Fairchild/Fokker	FH227/F27	27	1,288	2.8	57,282
Total - Top Aircraft [3]			**1,320** (65.3%)	**37,660**	83.0	**3,015,421** (78.4%)
All Others in Service			672 (34.7%)	7,701	17.0	655,197 (21.6%)
Industry Total			1,992 (100%)	45,361	100.0	3,670,618 (100%)

Notes: 1 Aircraft in service as of 1/1/92.
  2 Includes both the -100 and -300 models.
  3 Aircraft accounting for 2.0% or more of 1991 passenger capacity.
Source: AvStat Associates for the Regional Airline Association.

AIRCRAFT IN REGIONAL AIRLINE PASSENGER SERVICE

| | Passenger Seating Capacity | | | | | | |
	1-9	10-19	20-30	31-40	41 & Up	Other[1]	Total
Aircraft in Active Fleet							
1991	470	796	190	311	184	41	**1,992**
1978	667	270	36	36	22	16	**1,047**
Percent Change	(29.5)	194.8	427.8	763.9	736.4	156.3	**90.3**
Hours Flown (000's)							
1991	387	1,633	529	676	385	61	**3,671**
1978	566	474	34	23	23	11	**1,131**
Percent Change	(31.7)	244.5	1,456.1	2,837.7	1,576.0	451.4	**224.5**
Revenue Passenger Miles							
1991	2.1%	31.7%	16.7%	24.7%	19.9%	4.8%	**7.80 bil.**
1978	22.0%	58.0%	8.5%	4.5%	6.0%	1.0%	**1.28 bil.**
Percent Change							**509.4%**

Notes: 1 Other inclusive of helicopters and jets.

Source: AvStat Associates for the Regional Airline Association.

SUMMARY OF PASSENGER AIRCRAFT IN REGIONAL AIRLINE USE - 1991

Manufacturer	Piston Single Engine	Piston Multi Engine	Turboprop	Jet	Helicopter	Total
Aerospatiale/Mohawk			106		4	**110**
Beech	1	3	225			**229**
Bell					14	**14**
British Aerospace			231	17		**248**
Britten Norman		28				**28**
CASA			10			**10**
Cessna	166	101				**267**
Convair			13			**13**
deHavilland	34		186			**220**
Dornier			16			**16**
Embraer			192			**192**
Fairchild			190			**190**
Fokker			24	3		**27**
G.A.F.			4			**4**
Grumman		5	5			**10**
Mitsubishi			1			**1**
Piper	7	149	9			**165**
Saab			157			**157**
Shorts			88			**88**
Sikorsky					3	**3**
Total 1991	**208**	**286**	**1,457**	**20**	**21**	**1,992**
Total 1990	**198**	**294**	**1,380**	**24**	**21**	**1,917**
Percent Change	**5.1**	**(2.7)**	**5.6**	**(16.7)**	**0.0**	**3.9**

Source: Avstat Associates for the Regional Airline Association

SUMMARY OF ALL-CARGO AIRCRAFT IN REGIONAL AIRLINE USE - 1991

Manufacturer	Piston Single Engine	Piston Multi Engine	Turboprop	Jet	Helicopter	Total
Beech	6	51	44			**101**
Bell					1	**1**
Britten Norman		1				**1**
CASA			2			**2**
Cessna	227	51				**278**
Convair		17	12			**29**
Dassault				3		**3**
deHavilland			15			**15**
Dornier			4			**4**
Douglas		24	1	1		**26**
Embraer			2			**2**
Fairchild			11			**11**
Fokker			14			**14**
G.A.F.			1			**1**
Lear				8		**8**
Mitsubishi			20			**20**
NAMC			4			**4**
Piper		114	2			**116**
Rockwell		25	2	2		**29**
Shorts			18			**18**
Total	**233**	**283**	**152**	**14**	**1**	**683**

Source: Avstat Associates for the Regional Airline Association

Section III.

General Aviation and Helicopters

ACTIVE GENERAL AVIATION AIRCRAFT
December 31, 1963 - 1990

Year	TOTAL	Carrier	TOTAL	Multi-Engine	Single-Engine 4-place & over	3-place & over	Rotor-craft	Other
1963	87,167	2,079	85,088	9,695	42,647	30,977	1,171	588
1964	90,799	2,057	88,742	10,644	45,777	30,367	1,306	648
1965	97,567	2,125	95,442	11,977	49,789	31,364	1,503	809
1966	106,978	2,272	104,706	13,548	52,972	35,687	1,622	877
1967	116,638	2,452	114,186	14,651	56,865	39,675	1,899	1,096
1968	126,823	2,586	124,237	16,760	60,977	42,830	2,350	1,320
1969	133,496	2,690	130,806	18,111	63,703	45,001	2,557	1,434
1970	134,422	2,679	131,743	18,291	64,759	44,884	2,255	1,554
1971	133,790	2,642	131,148	17,855	64,464	44,792	2,352	1,685
1972	147,593	2,583	145,010	19,849	70,998	49,448	2,787	1,928
1973	156,139	2,599	153,540	21,929	74,831	51,386	3,143	2,251
1974	163,974	2,472	161,502	23,418	78,924	53,008	3,610	2,542
1975	170,970	2,495	168,475	24,559	82,621	54,390	4,073	2,832
1976	180,796	2,492	178,304	25,684	88,211	56,730	4,505	3,174
1977	186,767	2,473	184,294	26,652	91,960	57,340	4,726	3,616
1978	201,323	2,545	198,778	28,782	101,466	59,185	5,315	4,028
1979	213,948	3,609	210,339	31,311	106,028	62,362	5,864	4,770
1980	214,853	3,808	211,045	31,664	107,930	60,505	6,001	4,945
1981	217,199	3,973	213,226	33,301	107,983	59,914	6,974	5,049
1982	213,851	4,027	209,779	34,204	106,503	57,670	6,169	6,209
1983	217,496	4,203	213,293	34,404	107,228	59,199	6,540	5,233
1984	225,313	4,370	220,943	35,649	109,933	61,989	7,096	6,275
1985	215,332	4,678	210,654	33,589	105,555	58,829	6,418	6,263
1986	224,953	4,909	220,044	34,313	109,351	62,427	6,943	7,010
1987	222,436	5,253	217,183	33,032	107,502	63,533	6,333	6,783
1988	215,926	5,660	210,266	32,243	105,207	59,553	6,406	6,857
1989	225,515	5,778	219,737	34,171	107,752	62,618	7,475	7,721
1990	218,312	6,083	212,229	32,727	104,566	60,507	7,397	7,032

Source: Federal Aviation Administration, "FAA Statistical Handbook of Aviation" (Annually).
NOTE: Detail may not add to totals because of estimating procedures.
"Active aircraft" must have a current U.S. registration and have flown during the calendar year. Prior to 1971, only a current U.S. registration was necessary.
Effective 1978, includes certificated route air carriers, supplemental air carriers (charters), multi-engine aircraft in commuter passenger service, and all aircraft over 12,500 pounds operated by air taxis, commercial operators, and travel clubs.
Includes autogiros; excludes air carrier helicopters.
Includes gliders, dirigibles, and balloons.

U.S. GENERAL AVIATION AIRCRAFT SHIPMENTS
By Selected Manufacturers
Calendar Years 1987 - 1991

	1987	1988	1989	1990	1991
NUMBER OF AIRCRAFT SHIPPED	**1,085**	**1,143**	**1,535**	**1,144**	**1,021**
Single-Engine, Piston	613	628	1,023	608	564
Multi-Engine, Piston	87	67	87	87	49
Turboprop	263	291	268	281	222
Turbojet	122	157	157	168	186
VALUE OF SHIPMENTS					
(Millions of Dollars)	**$1,364**	**$1,918**	**$1,804**	**$2,008**	**$1,968**
Single-Engine, Piston	$ 80	$ 66	$ 104	$ 68	$ 93
Multi-Engine, Piston	18	12	24	24	(b)
Turboprop	477	596	524	644	527
Turbojet	789	1,242	1,149	1,272	1,348
Number of Aircraft By Selected Manufacturer					
American General	NA	NA	NA	10	82
Aviat	NA	NA	NA	NA	71
Beech	314	372	371	433	402
Bellanca	NA	NA	7	4	1
Cessna	187	161	183	171	176
Christen	NA	NA	75	68	-
Classic	NA	NA	NA	8	8
Fairchild	36	29	12	14	10
Gates Learjet	16	23	25	25	25
Gulfstream	30	51	40	34	29
Lake	23	28	23	17	11
Maule	54	55	35	28	66
Mooney	143	142	143	147	88
Piper	282	282	621	178	41
Taylorcraft	NA	NA	NA	7	11

Source: General Aviation Manufacturers' Association.
Manufacturers' net billing price.
(b) "Multi-Engine, Piston" combined with "Single-Engine, Piston"
NA Not available.

U.S. GENERAL AVIATION
ACTIVE AIRCRAFT AND HOURS FLOWN
BY PRIMARY USE
Calendar Years 1986 - 1990

Primary Use	1986	1987	1988	1989	1990
ACTIVE AIRCRAFT AS OF DECEMBER 31					
TOTAL	**220,044**	**217,183**	**210,266**	**219,737**	**212,229**
Executive	12,075	11,960	10,882	12,285	10,906
Business	43,780	39,943	34,918	37,507	35,496
Commuter	1,721	1,014	973	1,444	1,242
Air Taxi	7,568	6,228	6,518	7,115	6,186
Instructional	15,812	15,727	16,674	17,780	19,889
Personal	120,308	123,487	122,557	124,786	120,636
Aerial Application	7,068	6,516	7,042	7,093	6,687
Aerial Observation	4,716	4,858	4,759	5,784	5,302
Other Work	1,274	1,577	1,841	2,139	1,525
Other	5,707	5,873	4,081	3,802	4,358
THOUSANDS OF HOURS FLOWN					
TOTAL	**34,416**	**33,443**	**33,593**	**35,012**	**34,767**
Executive	3,781	3,403	3,748	3,739	3,155
Business	5,896	5,713	4,960	4,689	4,784
Commuter$Sb	2,185	1,359	1,118	1,508	1,444
Air Taxi$Sb	2,913	2,877	2,842	3,270	2,436
Instructional	4,677	4,904	5,309	6,489	7,847
Personal	10,097	10,787	10,813	10,328	10,048
Aerial Application	1,985	1,666	1,989	2,023	2,028
Aerial Observation	1,620	1,412	1,412	1,861	1,891
Other Work	323	379	567	560	619
Other	939	943	835	549	514

Source: Federal Aviation Administration, "FAA Statistical Handbook of Aviation" (Annually).
NOTE: Detail may not add to totals because of rounding and estimating procedures.
Definitions of "primary use" categories available in Glossary of "FAA Statistical Handbook."
Limited to single-engine commuters or air taxis under 12,500 pounds.

U.S. GENERAL AVIATION
TYPE OF AIRCRAFT AND HOURS FLOWN
Calendar Years 1986 - 1990

	1986	1987	1988	1989	1990
Number of Active Aircraft by Type					
All Aircraft -- TOTAL	**220,044**	**217,183**	**210,266**	**219,737**	**212,229**
Fixed Wing: Piston:					
Single Engine: 1-3 Seats	62,427	63,533	59,553	62,618	60,507
4 + Seats	109,351	107,502	105,207	107,752	104,566
Twin Engine: 1-6 Seats	16,166	15,741	15,143	15,927	15,186
7 + Seats	7,555	7,566	7,554	7,432	7,421
Other	148	112	99	86	94
Turboprop:					
Twin Engine: 1-12 Seats	4,809	4,337	4,231	4,888	4,320
13 + Seats	970	723	826	1,206	937
Other	185	214	202	230	395
Turbojet: Twin Engine	4,037	3,900	3,821	4,004	3,950
Other	444	438	367	398	425
Rotorcraft: Piston	2,921	2,813	2,584	3,244	3,459
Turbine	4,022	3,520	3,822	4,232	3,938
Balloons, Dirigibles, and Gliders	7,010	6,783	6,857	7,721	7,032
Thousands of Hours Flown by Type of Aircraft					
All Aircraft -- TOTAL	**34,416**	**33,443**	**33,593**	**35,012**	**34,767**
Fixed Wing: Piston	26,861	27,039	26,226	26,971	27,973
Turboprop	2,882	2,177	2,370	3,132	2,521
Turbojet	1,654	1,528	1,678	1,654	1,512
Rotorcraft: Piston	804	652	576	749	775
Turbine	1,821	1,631	2,131	2,077	1,617
Balloons, Dirigibles, and Gliders	394	416	613	429	369
Average Hours Flown per Year per Aircraft by Type					
All Aircraft -- TOTAL	**149**	**148**	**154**	**155**	**159**
Fixed Wing: Piston:					
Single Engine: 1-3 Seats	125	134	132	132	149
4 + Seats	130	126	134	131	141
Twin Engine: 1-6 Seats	172	165	150	169	163
7 + Seats	280	289	256	259	231
Other	111	140	225	133	623
Turboprop:					
Twin Engine: 1-12 Seats	335	337	373	341	334
13 + Seats	1,013	652	895	1,044	1,024
Other	499	839	392	569	446
Turbojet: Twin Engine	385	371	412	385	359
Other	154	229	347	275	293
Rotorcraft: Piston	273	229	228	236	216
Turbine	460	485	577	497	425
Balloons, Dirigibles, and Gliders	56	62	95	56	52

Source: General Aviation Manufacturers Association, "General Aviation Statistical Databook" (Annually), based on data from the Federal Aviation Administration, "FAA Statistical Handbook of Aviation" and the Federal Aviation Administration, Office of Management Systems.

NOTE: Detail may not add to totals because of rounding and/or estimating procedures.

SPECIFICATIONS OF U.S. CIVIL HELICOPTERS
In Production as of 1991

Company	Commercial Model	Number of Places	Useful Load (Lbs.)	Range with Useful Load (N.Miles)	External Cargo Payload (Lbs.)
Bell Helicopter Textron	212	15	5,228	231	5,000
	214 Series	20	7,889	457	7,900
	412	15	5,305	402	11,900
Enstrom Helicopter	F-28 Series	3	1,030	228	1,000
	280 Series	3	1,015	260	1,000
McDonnell Douglas Helicopter	500 Series	5	1,559	367	2,000
	520 Series	5	1,806	239	2,306
	530 Series	5	1,536	275	2,000
Robinson Helicopter	R22	2	546	209	-
Schweizer Aircraft	300C	3	950	201	1,050
Sikorsky Aircraft	S-76B	14	5,091	357	3,300

Source: Helicopter Association International, "1991 Helicopter Annual" (Annually).

CIVIL HELICOPTER SHIPMENTS
Calendar Years 1987 - 1991

Company and Model	1987	1988	1989	1990	1991
CIVIL SHIPMENTS	**358**	**383**	**515**	**603**	**571**
Value (Millions of Dollars)	$277	$334	$251	$254	$211
Bell--TOTAL	**127**	**62**	**22**	**16**	**4**
206 series	74	-	-	-	-
212	11	13	3	1	-
214 series	13	18	2	1	-
222	12	11	-	-	-
412	17	20	17	14	4
Enstrom--TOTAL	**12**	**17**	**24**	**27**	**17**
F-28 series	7	7	6	12	8
280 series	5	10	18	15	9
McDonnell Douglas--TOTAL	**41**	**44**	**73**	**77**	**50**
500 series	37	39	64	65	42
520N series		-	-	-	3
530 series	4	5	9	12	5
Robinson--TOTAL	**127**	**204**	**310**	**384**	**402**
R22	127	204	310	384	402
Rogerson--TOTAL	**-**	**-**	**-**	**-**	**2**
UH12E	-	-	-	-	2
Schweizer--TOTAL	**37**	**45**	**69**	**83**	**78**
300C	37	45	69	83	78
Sikorsky--TOTAL	**14**	**11**	**17**	**16**	**18**
S-76	13	11	17	16	18
S-70C series	1	-	-	-	-

Source: Aerospace Industries Association, based on company reports.
NOTE: All data exclude production by foreign licensees.
Domestic and export helicopter shipments for non-military use. Helicopters in military configuration exported to foreign governments and purchased under commercial contract are reported elsewhere. Models which may be shipped in either a civil or a military configuration appear in both tables.
Bell Helicopter moved production of its 206 series helicopters to its Canadian facility in 1987.

ACTIVE U.S. AIRMAN CERTIFICATES HELD
As of December 31, 1987 - 1991

	1987	1988	1989	1990	1991
Pilots -- TOTAL	**699,653**	**694,016**	**700,010**	**702,659**	**692,095**
Students	146,016	136,913	142,544	128,663	120,203
Private	300,949	299,786	293,179	299,111	293,306
Commercial	143,645	143,030	144,540	149,666	148,365
Airline Transport	91,287	96,968	102,087	107,732	112,167
Helicopter (only)	8,702	8,608	8,863	9,567	9,860
Glider (only)	7,901	7,600	7,708	7,833	8,033
Lighter-Than-Air	1,153	1,111	1,089	-	-
Recreational	-	-	-	87	161
Non-Pilots -- TOTAL	**427,962**	**448,710**	**468,405**	**492,237**	**517,462**
Mechanics	297,178	312,419	326,243	344,282	366,392
Parachute Rigger	9,659	9,770	9,879	10,094	7,916
Ground Instructor	60,861	62,582	64,503	66,882	70,086
Dispatcher	9,491	10,020	10,455	11,002	11,607
Flight Navigator	1,445	1,400	1,357	1,290	1,225
Flight Engineer	49,328	52,519	55,968	58,687	60,236
Flight Instructor Certificates	60,316	61,798	61,472	63,775	69,209
Instrument Ratings	266,122	273,804	282,804	297,073	303,193

Source: Federal Aviation Administration, "FAA Statistical Handbook of Aviation" (Annually)
Glider and lighter-than-air pilots are not required to have a medical examination;
however, the totals above are the pilots who received a medical.
Lighter-than-air type ratings are no longer being issued.
Special ratings shown on pilot certificates represented above, not additional certificates.

Section IV.

Airports

AIRPORT RANKING BY TOTAL AIRCRAFT MOVEMENTS - 1991

RANK	CITY	AIRPORT	TOTAL A/C MOVE-MENTS	% CHG 1991/ 1990
1	CHICAGO, ILLINOIS	O'HARE INT'L	813,896	0.4
2	DALLAS, TEXAS	DALLAS/FT WORTH INT'L	736,127	0.7
3	LOS ANGELES, CALIFORNIA	LOS ANGELES INT'L	657,436	-3.4
4	ATLANTA, GEORGIA	HARTSFIELD ATLANTA INT'L	589,470	-25.4
5	COSTA MESA, CALIFORNIA	JOHN WAYNE	569,241	8.9
6	PHOENIX, ARIZONA	SKY HARBOR INT'L	496,244	-0.5
7	DENVER, COLORADO	STAPLETON INT'L	482,446	2.6
8	CHARLOTTE, NORTH CAROLINA	CHARLOTTE/DOUGLAS INT'L	446,370	-0.9
9	LONG BEACH, CALIFORNIA	LONG BEACH	439,325	-11.7
10	BOSTON, MASSACHUSETTS	BOSTON LOGAN INT'L	430,411	1.4
11	SAN FRANCISCO, CALIFORNIA	SAN FRANCISCO INT'L	418,632	-2.7
12	ST LOUIS, MISSOURI	LAMBERT-ST LOUIS INT'L	413,212	-5.9
13	OAKLAND, CALIFORNIA	OAKLAND INT'L	412,836	2.6
14	MIAMI, FLORIDA	MIAMI INT'L	407,303	1
15	HONOLULU, HAWAII	HONOLULU INT'L	403,566	-0.9
16	LAS VEGAS, NEVADA	MCCARRAN INT'L	398,143	-0.6
17	DETROIT, MICHIGAN	DETROIT METRO WAYNE COUNTY	396,278	2.2
18	PITTSBURGH, PENNSYLVANIA	GREATER PITTSBURGH INT'L	387,095	0.3
19	SEATTLE, WASHINGTON	BOEING FIELD/KING COUNTY INT'L	382,863	-5
20	MINNEAPOLIS-ST PAUL, MINNESOTA	MINNEAPOLIS-ST PAUL INT'L		
21	LONDON	HEATHROW	381,726	-2.2
22	NEWARK, NEW JERSEY	NEWARK INT'L	378,787	-0.2
23	PHILADELPHIA, PENNSYLVANIA	PHILADELPHIA INT'L	371,516	-8.8
24	DAYTONA BEACH, FLORIDA	DAYTONA BEACH REGIONAL	341,868	1.4
25	SEATTLE, WASHINGTON	SEATTLE TACOMA INT'L	338,607	-4.6
26	SAN JOSE, CALIFORNIA	SAN JOSE INT'L	329,913	2
27	NEW YORK, NEW YORK	LA GUARDIA	326,776	-8.3
28	MEMPHIS, TENNESSEE	MEMPHIS INT'L	323,774	-1
29	FRANKFURT	FRANKFURT	319,825	-1.4
30	HOUSTON, TEXAS	HOUSTON INTERCONTINENTAL	303,996	-0.3
31	SALT LAKE CITY, UTAH	SALT LAKE CITY INT'L	301,657	-0.6
32	CINCINNATI, OHIO	CINCINNATI/NO KENTUCKY INT'L	294,033	2.1
33	WASHINGTON, D.C.	WASHINGTON NATIONAL	292,926	-6.6
34	TULSA, OKLAHOMA	RICHARD LLOYD JONES, JR	286,208	-7.3
35	CHICAGO, ILLINOIS	MIDWAY	281,110	-12.2
36	NASHVILLE, TENNESSEE	NASHVILLE INT'L	280,777	7
37	NEW YORK, NEW YORK	J.F. KENNEDY INT'L	276,984	-8.8
38	BALTIMORE, MARYLAND	BALTIMORE/WASHINGTON, DC INT'L	275,903	-8.7
39	ORLANDO, FLORIDA	ORLANDO INT'L	272,755	-3.3
40	RALEIGH-DURHAM, NORTH CAROLINA	RALEIGH-DURHAM INT'L	272,281	-3.8
41	WASHINGTON, D.C.	WASHINGTON DULLES INT'L	264,597	9.2
42	MEXICO CITY	JUAREZ INT'L	259,785	19.1
43	PARIS	CHARLES DE GAULLE	258,162	6.9
44	PORTLAND, OREGON	PORTLAND INT'L	257,023	-2.2
45	HOUSTON, TEXAS	WILLIAM P. HOBBY	248,101	-3.4
46	AMSTERDAM	SCHIPHOL	247,065	0.2
47	TUCSON, ARIZONA	TUCSON INT'L	238,852	5.7
48	CLEVELAND, OHIO	CLEVELAND HOPKINS INT'L	234,356	-14.6
49	INDIANAPOLIS, INDIANA	INDIANAPOLIS INT'L	233,442	2.2
50	MELBOURNE, FLORIDA	MELBOURNE REGIONAL	230,704	-16.1
51	ANCHORAGE, ALASKA	ANCHORAGE INT'L	225,733	2.8
52	TAMPA, FLORIDA	TAMPA INT'L	224,085	-6.1
53	BURBANK, CALIFORNIA	BURBANK-GLENDALE-PASADENA	224,033	-6.1
54	STOCKHOLM	ARLANDA	224,008	-13
55	NEWPORT NEWS, VIRGINIA	NEWPORT NEWS/WILLIAMSBRG INT'L	222,955	-12.4
56	ZURICH	ZURICH	221,395	0.7
57	WEST PALM BEACH, FLORIDA	PALM BEACH INT'L	217,604	-9.3
58	ALBUQUERQUE, NEW MEXICO	ALBUQUERQUE INT'L	210,230	-6.9
59	DALLAS, TEXAS	LOVE FIELD	207,402	-3.1
60	COPENHAGEN	COPENHAGEN	204,144	-4
61	FT. LAUDERDALE/HOLLYWOOD, FLORIDA	FT. LAUDERDALE/HOLLYWOOD	203,318	-9.5
62	SAN JUAN	LUIS MUNOZ MARIN	203,055	-0.9

(Continued on next page)

AIRPORT RANKING BY TOTAL AIRCRAFT MOVEMENTS - 1991
(Continued)

RANK	CITY	AIRPORT	TOTAL A/C MOVE-MENTS	% CHG 1991/ 1990
63	MIWAUKEE, WISCONSIN	GENERAL MITCHEL INT'L	202,000	-2.3
64	PARIS	ORLY	200,813	-0.5
65	COLORADO SPRINGS, COLORADO	COLORADO SPRINGS	198,445	13.8
66	TOKYO	TOKYO INT'L (HANEDA)	193,318	5.1
67	COLUMBUS, OHIO	PORT COLUMBUS INT'L	192,423	-7.2
68	DAYTON, OHIO	DAYTON INT'L	185,922	-1.6
69	MUNICH	MUNICH	183,932	-4.1
70	BOISE, IDAHO	BOISE AIR TERMINAL	183,757	-8
71	ROCHESTER, NEW YORK	ROCHESTER	183,653	0
72	TULSA, OKLAHOMA	TULSA INT'L	182,760	-6.5
73	AUSTIN, TEXAS	ROBERT MUELLER MUNICIPAL	182,079	-5.6
74	KAHULUI, HAWAII	KAHULUI	180,857	-1
75	ROME	FIUMICINO	175,196	-0.7
76	LONDON	GATWICK	174,300	-14.2
77	WICHITA, KANSAS	WICHITA MID-CONTINENT	173,113	-1.1
78	SYRACUSE, NEW YORK	HANCOCK INT'L	172,112	-3
79	EAST FAMINGDALE, NEW YORK	REPUBLIC	171,754	-0.9
80	KANSAS CITY, MISSOURI	KANSAS CITY INT'L	170,098	4.3
81	HARTFORD/SPRINGFIELD, CONNECTICUT	BRADLEY INT'L	168,383	-6.5
82	OMAHA, NEBRASKA	EPPLEY AIRFIELD	166,009	6.8
83	MADRID	BARAJAS	163,781	3.5
84	EL PASO, TEXAS	EL PASO INT'L	162,869	-8.6
85	RENO, NEVADA	RENO CANNON INT'L	159,164	-0.1
86	SAN ANTONIO, TEXAS	SAN ANTONIO INT'L	156,352	5.2
87	ONTARIO, CALIFORNIA	ONTARIO INT'L	156,003	1.4
88	SACRAMENTO, CALIFORNIA	SACRAMENTO METRO	154,609	-4.2
89	DUSSELDORF	DUSSELDORF	153,068	-1.3
90	KANSAS CITY, MISSOURI	KANSAS CITY DOWNTOWN	152,227	-7.3
91	MANCHESTER	MANCHESTER	151,648	0.3
92	OKLAHOMA CITY, OKLAHOMA	WILL ROGERS WORLD	150,244	0.6
93	JACKSONVILLE, FLORIDA	JACKSONVILLE INT'L	147,910	-3.2
94	GENEVA	COINTRIN	147,294	-1.8
95	KNOXVILLE, TENNESSEE	MCGHEE TYSON	146,429	-12.3
96	MONTPELLIER	FREJORGUES	145,640	3.4
97	BUFFALO, NEW YORK	GREATER BUFFALO INT'L	144,696	-5.2
98	BERLIN	TEGEL & TEMPELHOF	144,317	36.1
99	DES MOINES, IOWA	DES MOINES INT'L	142,571	-3.6
100	LANSING, MICHIGAN	CAPITAL CITY	141,325	-5.7
101	HAMBURG	HAMBURG-FULHSBUTTEL	141,265	0.2
102	AKRON/CANTON, OHIO	AKRON-CANTON REGIONAL	140,085	-6.4
103	MANCHESTER, NEW HAMPSHIRE	MANCHESTER	137,395	1.3
104	RICHMOND, VIRGINIA	RICHMOND INT'L	136,992	-17
105	ALLENTOWN, PENNSYLVANIA	ALLENTOWN-BETHLEHEM-EASTON INT'L	136,973	-3
106	NORFOLK, VIRGINIA	NORFOLK INT'L	136,594	-11.7
107	BANGKOK	BANGKOK INT'L	135,916	8.6
108	STUTTGART	STUTTGART	134,418	-1.3
109	GREENSBORO, NORTH CAROLINA	GREENSBORO	133,813	-12.2
110	MILAN	LINATE	133,069	-1.7
111	CHARLESTON, SOUTH CAROLINA	CHARLESTON INT'L	132,432	-1.2
112	NICE	NICE-COTE D'AZUR	132,170	-7.2
113	HONG KONG	HONG KONG INT'L	130,799	6.4
114	HELSINKI	HELSINKI VANTAA	129,788	-2.4
115	VIENNA	VIENNA INT'L	124,416	13.1
116	SINGAPORE	CHANGI	123,234	13.5
117	BARCELONA	BARCELONA	123,049	4.5
118	DUBLIN	DUBLIN	123,008	-0.3
119	BIRMINGHAM, ALABAMA	BIRMINGHAM	122,684	-11.3
120	SEOUL	KIMPO INT'L	122,610	8.3
121	BANGOR, MAINE	BANGOR INT'L	122,081	-6.6
122	SAO PAULO	GUARULHOS	120,500	14.2
123	COLOGNE	COLOGNE BONN	118,283	-0.5
124	ABERDEEN	ABERDEEN	117,953	3.1

(Continued on next page)

AIRPORT RANKING BY TOTAL AIRCRAFT MOVEMENTS - 1991
(Continued)

RANK	CITY	AIRPORT	TOTAL A/C MOVE-MENTS	% CHG 1991/ 1990
125	BATON ROUGE, LOUISIANA	BATON ROUGE METROPOLITAN	117,579	-10.2
126	CHRISTCHURCH	CHRISTCHURCH INT'L	117,058	-14.4
127	SPRINGFIELD, ILLINOIS	CAPITAL	115,902	-5
128	ROANOKE, VIRGINIA	ROANOKE REGIONAL	114,186	-11.5
129	KAUIA ISLAND, HAWAII	LIIHUE	112,679	-1
130	MANILA	B. AQUINA INT'L	112,135	-9.8
131	BAKERSFIELD, CALIFORNIA	MEADOWS FIELD	110,390	-7
132	MARSEILLE	MARSEILLE PROVENCE	106,323	-4.4
133	KARACHI	KARACHI INT'L	105,800	-1.7
134	GLASGOW	GLASGOW	105,638	-4.7
135	MONTEREY, CALIFORNIA	MONTEREY PENINSULA AIRPORT	104,543	-9.7
136	READING, PENNSYLVANIA	READING REGIONAL	104,337	-11.1
137	EDINBURGH	EDINBURGH	104,229	-0.4
138	KALAMAZOO, MICHIGAN	KALAMAZOO/BATTLE CREEK INT'L	104,087	1.9
139	FORT MYERS, FLORIDA	PAGE FIELD	103,997	-9.1
140	YUMA, ARIZONA	YUMA INT'L	103,958	1.2
141	SPRINGFIELD, MISSOURI	SPRINGFIELD REGIONAL	103,243	2
142	GLASGOW	PRESTWICK	102,887	-9.3
143	SAVANNAH, GEORGIA	SAVANNAH INT'L	102,805	-7.4
144	GAINESVILLE, FLORIDA	GAINESVILLE REGIONAL	100,250	-6.8
145	JAKARTA	SOEKARNO HATTA INT'L	99,991	9.1
146	GUADALAJARA	MIGUEL HIDALGO	99,067	18.4
147	AUCKLAND	AUCKLAND INT'L	98,637	16
148	MIDDLETOWN, PENNSYLVANIA	HARRISBURG INT'L	97,362	-24.3
149	RIO DE JANEIRO	RIO DE JANEIRO	96,768	4.5
150	MAASTRICHT	MAASTRICHT	95,706	10.4
151	LAFAYETTE, LOUISIANA	LAFAYETTE REGIONAL	95,127	-13.6
152	BELFAST	BELFAST INT'L	94,674	-1.7
153	HANOVER	HANOVER	94,663	-0.9
154	BASEL-MULHOUSE	BASEL-MULHOUSE	94,454	-1
155	NEWBURGH, NEW YORK	STEWART INT'L	93,062	4.3
156	PALMA DE MALLORCA	PALMA DE MALLORCA	92,695	6.9
157	PEORIA, ILLINOIS	GREATER PEORIA REGIONAL	91,279	-14
158	EGELSBACH	EGELSBACH	90,245	9.4
159	BIRMINGHAM	BIRMINGHAM INT'L	89,857	-3.7
160	BRASILIA	BRASILIA	88,988	8.3
161	CLEVELAND, OHIO	CLEVELAND BURKE LAKEFRONT	88,335	0.2
162	HILO, HAWAII	HILO INT'L	88,206	-10.7
163	SIOUX FALLS, SOUTH DAKOTA	JOE FOSS FIELD	87,173	-1.1
164	MOLINE, ILLINOIS	QUAD CITY	85,677	4.2
165	MOSCOW	SHREMETYEVO	85,437	----
166	JOHANNESBURG	JAN SMUTS AIRPORT	82,441	-15.5
167	ISTANBUL	ATATURK	82,206	-9.2
168	LYON	BRON	81,643	-6.4
169	WILMINGTON, NORTH CAROLINA	NEW HANOVER INT'L	79,408	5.2
170	LIVERPOOL	LIVERPOOL	78,807	-14.3
171	TARBES-LOURDES	TARBES-LOURDES	76,026	23.2
172	PASCO, WASHINGTON	TRI-CITIES	75,659	-9.2
173	NUREMBERG	NUREMBERG	75,327	5
174	DUBAI	DUBAI INT'L	75,305	-2.4
175	GUAYAQUIL	SIMON BOLIVAR	74,950	40.6
176	ASHEVILLE, NORTH CAROLINA	ASHEVILLE REGIONAL	72,245	-5.4
177	LYON	SATOLAS	71,915	-3.7
178	PAU PYRENEES	UZEIN	71,574	17.2
179	GRAN CANARIA	GRAN CANARIA	71,043	11.5
180	LISBON	LISBON	70,352	3.5
181	BORDEAUX	MERIGNAC	69,497	-2.8
182	ERIE, PENNSYLVANIA	ERIE INT'L	69,113	-0.5
183	LUXEMBOURG	LUXEMBOURG	68,654	9
184	VALENCE	CHABEUIL	6,541	-0.3
185	NANTES	NANTES ATLANTIQUE	66,293	----
186	OSAKA	OSAKA INT'L	65,362	0.1

(Continued on next page)

AIRPORT RANKING BY TOTAL AIRCRAFT MOVEMENTS - 1991
(Continued)

RANK	CITY	AIRPORT	TOTAL A/C MOVE-MENTS	% CHG 1991/ 1990
187	BOMBAY	BOMBAY	65,316	-2.7
188	JACKSON, MISSISSIPPI	JACKSON INT'L	65,112	-7.9
189	BRISTOL	BRISTOL	64,769	-12.2
190	CORK	CORK	64,549	0.6
191	SIOUX CITY, IOWA	SIOUX GATEWAY	64,239	4.1
192	NEWCASTLE	NEWCASTLE	63,667	2.7
193	TOKYO	NEW TOKYO (NARITA) INT'L	62,204	-0.6
194	MISSOULA, MONTANA	MISSOULA COUTNY	62,169	5.3
195	TAIPEI	CHIANG KAI SHEK INT'L	62,080	9.8
196	SURABAYA	JUANDA	61,661	4.9
197	MIDLAND, TEXAS	MIDLAND INT'L	61,280	0.6
198	PORTO ALEGRE	SALGADO FILHO	61,264	46.8
199	JAKARTA	HALIM PERDANKUSUMA	60,937	5.6
200	SAN JUAN	ISLA GRANDE	60,297	-33.9
201	CLERMONT-FERRAND	AULNAT	59,872	-5.3
202	RIYADH	KIND KHALED INT'L	59,514	-13
203	JEDDAH	KING ABDULAZIZ INT'L	59,221	-14.4
204	GREER, SOUTH CAROLINA	GREENVILLE-SPARTANBURG	58,867	-15.9
205	EAST MIDLANDS	EAST MIDLANDS INT'L	56,841	-17.4
206	NEW DELHI	INDIRA GANDHI INT'L	56,414	-2.3
207	KAILUA-KONA, HAWAII	KE-AHOLE	56,140	-6.6
208	TEESIDE	TEESIDE INT'L	55,195	-14.2
209	CASPER, WYOMING	NATRONA COUNTY INT'L	54,278	-1.5
210	SOUTHAMPTON	SOUTHAMPTON	54,164	----
211	AUGUSTA, GEORGIA	BUSH FIELD	53,793	-21
212	MUENSTER	MUENSTER-OSNABRUECK	53,762	-4.8
213	SHANNON	SHANNON	53,300	-10
214	CARDIFF WALES	CARDIFF WALES	53,109	-17
215	LONDON	STANSTED	52,799	12
216	SALVADOR	DOIS JULHO	52,422	5.8
217	BREMEN	BREMEN	51,929	-2.1
218	LEEDS BRADFORD	LEEDS	51,800	-19
219	NEW ORLEANS, LOUISIANA	NEW ORLEANS INT'L	50,936	-8.7
220	GOTHENBURG	LANDVETTER	50,682	-9
221	HARLINGEN, TEXAS	VALLEY INT'L	50,669	-10.9
222	BREST-GUIPAVAS	BREST-GUIPAVAS	50,579	7.4
223	DURANGO, COLORADO	DURANGO-LA PLATA COUNTY	50,238	-5.3
224	CANCUN	CANCUN	49,654	9.3
225	FAJARDO	FAJARDO	4,390	80.2
226	FREELAND, MICHIGAN	TRI-CITY INT'L	48,178	-2.9
227	TURIN	CITTA DI TORINO	47,444	-0.3
228	TENERIFE SUR	TENERIFE SUR	46,922	12
229	RECIFE	GUARARAPES	46,160	11.9
230	BEAUVAIS	BEAUVAIS-TILLE	46,068	1.7
231	MALAGA	MALAGA	46,038	3.6
232	MONACO	HELIPORT DE MONACO	45,573	-5.1
233	LILLE	LILLE LESQUIN	45,264	6
234	FUKUOKA	FUKUOKA INT'L	45,136	4
235	TOPEKA, KANSAS	FORBES FIELD	45,103	-6.7
236	MONTERREY	MARIANO ESCOBEDO	44,261	37.1
237	CHAMBERY	CHAMBERY/AIX	43,791	-11.2
238	BALIKPAPAN	SEPINGGAN	43,515	-1
239	QUITO	MARISCAL SUCRE	43,241	29.7
240	PADUCAH, KENTUCKY	BARKLEY REGIONAL	42,568	-4.7
241	CARACAS	SIMON BOLIVAR EN MAIQUETIA	42,446	0.7
242	ST ETIENNE	BOUTHEON	39,841	6
243	LIMOGES	LIMOGES BELLEGARDE	39,373	----
244	ALGIERS	HOUARI BOUMEDIENE	39,157	-10.3
245	WARSAW	OKECIE	38,980	-17
246	FORT MYERS, FLORIDA	SOUTHWEST FLORIDA REGIONAL	38,716	-11.3
247	KANSAS CITY, MISSOURI	RICHARDS-GEBAUR	38,353	3
248	LIBREVILLE	INT'L LEON M'BA LIBREVILLE	37,479	5.8

(Continued on next page)

AIRPORT RANKING BY TOTAL AIRCRAFT MOVEMENTS - 1991
(Continued)

RANK	CITY	AIRPORT	TOTAL A/C MOVE-MENTS	% CHG 1991/ 1990
249	ROME	CIAMPINO	37,408	-0.6
250	SHARJAH	SHARJAH INT'L	37,054	-11.9
251	VIEQUES	VIEQUES	37,002	47
252	MANAUS	EDUARDO GOMES	36,526	7.9
253	NADI	NADI INT'L	36,038	18.7
254	TIJUANA RODRIGUEZ		35,991	21.4
255	MILAN	MALPENSA	35,896	8.4
256	BERGERAC	ROUMANIERES	35,519	-1.7
257	DENPANSAR-BALI	NGURAH RAI	35,321	14.5
258	ARUBA	REINA BEATRIX	34,721	25.6
259	MAPUTO	MAPUTO INT'L	34,509	----
260	FARO	FARO	34,075	24
261	ACAPULCO	ACAPULCO	34,071	1.3
262	ABU DHABI	ABU DHABI INT'L	33,660	-14
263	SANTIAGO	A. MERINO BENITEZ	33,417	0.8
264	POINTE A PITRE	LE RAIZET	33,179	6.7
265	PUERTO VALLARTA	ORDAZ	32,968	14.9
266	VICHY-CHARMEIL	VICHY-CHARMEIL	32,394	-1.1
267	PORTO	PORTO	32,008	3.8
268	PORT ELIZABETH	H.F. VERWOERD AIRPORT	31,829	-9.9
269	BELEM	VAL DE CAES	30,917	13.4
270	DURBAN	LOUIS BOTHA AIRPORT	30,701	0.1
271	STRASBOURG	ENTZHEIM	29,242	-5.3
272	MAZATLAN	GEN. RAFAEL BUELNA	28.782	7.9
273	GENOA	CRISTOFORO COLOMBO	28,468	----
274	DHARAN	DHARAN INT'L	28,112	-11.9
275	ANKARA	ESENBOGA	27,881	0.3
276	ANGOULEME	ANGOULEME	27,568	-19
277	MERIDA	REJON	27,127	16.3
278	MADRAS	MADRAS	27,075	5.2
279	DOLE TAVAUX	DOLE TAVAUX	2,102	-15.9
280	TEL AVIV	BEN GURION INT'L	26,061	-1.7
281	DHAKA	ZIA INTERNATIONAL	26,028	-10.2
282	CULEBRA	CULEBRA	25,436	59
283	BELO HORIZONTE	CONFINS INT'L	25,283	15.9
284	VALENCIA	VALENCIA	25,029	20.8
285	IBIZA	IBIZA	25,019	20.4
286	LANZAROTE	LANZAROTE	24,251	19.3
287	TRIESTE	DEI LEGIONARI	24,102	5
288	TENERIFE NORTE	TENERIFE NORTE	23,953	28.1
289	CALCUTTA	CALCUTTA	23,547	-10.3
290	UJUNG PANDANG	HASANNUDDIN	23,254	7.5
291	MAYAGUEZ	EUGENIO M. DE HOSTOS	23,216	-20.9
292	ALICANTE	ALICANTE	23,083	4.9
293	MONTEVIDEO	AEROPUERTO INT'L DE CARRASCO	23,007	-6.1
294	PALERMO	PUNTA RAISI	22,222	3.7
295	KINSHASA	N'DJILI	21,862	-17.5
296	SEVILLA	SEVILLA	21,374	20.1
297	MEDAN	POLONIA	21,215	9.3
298	AGUADILLA	RAFAEL HERNANDEZ (BORINQUEN)	21,180	-17.3
299	ATLANTIC CITY, NEW JERSEY	ATLANTIC CITY INT'L	21,172	0
300	ROUEN	ROUEN	20,882	-47.2
301	GEORGE	P.W. BOTHA AIRPORT	20,772	54.1
302	HUMACAO	HUMACAO	20,284	4.3
303	SOFIA	SOFIA	19,637	-54.5
304	METZ-FRESCATY	METZ-FRESCATY	19,052	----
305	ADNAN	MENDERES	19,012	-16.9
306	LUSAKA	LUSAKA INT'L	18,559	-31.5
307	EINDHOVEN	EINDHOVEN	18,281	-4
308	COLOMBO	COLOMBO KATUNAYAKE	18,125	1.5
309	ANTALYA	ANTALYA	18,112	-15.9
310	BOA VISTA	BOA VISTA	17,612	-20

(Continued on next page)

AIRPORT RANKING BY TOTAL AIRCRAFT MOVEMENTS - 1991
(Continued)

RANK	CITY	AIRPORT	TOTAL A/C MOVE-MENTS	% CHG 1991/ 1990
311	SAN DIEGO, CALIFORNIA	SAN DIEGO INT'L-LINDBERGH FIELD	17,409	2.1
312	NOUMEA	LA TONTOUTA	17,108	43
313	CAMPO GRANDE	CAMPO GRANDE	16,918	8
314	NATAL	AUGUSTO SEVERO	16,619	14.6
315	CURACAO	AEROPUERTO HATO	16,259	0.8
316	KETCHIKAN, ALASKA	KETCHIKAN INT'L	15,770	-13.9
317	BILBAO	BILBAO	15,066	15.6
318	ST. GEORGE'S	POINT SALINE	14,904	20.5
319	MENORCA	MENORCA	14,267	1.8
320	FUERTEVENTURA	FUERTEVENTURA	14,126	17.4
321	SAN JOSE DEL CABO	LOS CABOS	13,689	5.1
322	MCALLEN, TEXAS	MCALLEN-MILLER INT'L	13,084	17.8
323	FOZ DE IGUACU	CATARATAS	12,704	9.5
324	RIO BRANCO	PRES. MEDICE	12,110	-13.1
325	PERIGUEUX	PERIGUEUX-BASSILLAC	11,616	-10
326	SANTIAGO DE COMPOSTELA		10,871	13.7
327	BONAIRE	FLAMINGO	10,529	7
328	BOLOGNA	GUGLIEMO MARCONI	10,383	9
329	KIMBERLEY	B.J. VORSTER AIRPORT	10,148	4.3
330	NIMES	GARONS	10,129	-26.3
331	BANGUI	BANGUI-M'POKO	9,935	-27.4
332	LONDON	LONDON CITY	9,728	-27
333	SARASOTA, FLORIDA	SARASOTA-BRADENTON	9,566	-11.8
334	BIAK	FRANS KAISIEPO	9,431	12.7
335	CAMPINAS	VIRACOPOS	9,385	37.7
336	LJUBLJANA	BRNIK	8,794	-45.9
337	MACAPA	MACAPA	8,584	0.7
338	ZARAGOZA	ZARAGOZA	8,145	11.1
339	LA PALMA	LA PALMA	8,070	19
340	UPINGTON	PIERRE VAN RYNEVELD	8,057	-19.2
341	MALE'	MALE' INT'L	7,810	9.3
342	DALAMAN	DALAMAN	7,806	-22.3
343	ZAGREB	ZAGREB	7,649	-53.8
344	PUSAN	KIMHAE	7,433	12.4
345	PONTA DELGADA	NORDELA	7,314	-9.5
346	MANADO	SAM RATULANGI	6,774	-2.7
347	BIARRITZ	BAYONNE-ANGLET	6,748	8.3
348	FUJAIRAH	FUJAIRAH INT'L	6,608	141.2
349	KEFLAVIK	KEFLAVIK	6,441	2.9
350	CORUMBA	CORUMBA	6,385	-0.4
351	ILHA DO SAL	AMILCAR CABRAL INT'L	6,380	----
352	ADANA	ADANA	6,026	-1.2
353	ALMERIA	ALMERIA	6,019	23.9
354	MELILLA	MELILLA	5,906	7.6
355	TRIVANDRUM	TRIVANDRUM	5,716	31
356	ASTURIAS	ASTURIAS	5,292	7.7
357	PONTA PORA	PONTA PORA	4,740	31.5
358	JEREZ	JEREZ	4,604	-11.5
359	VIGO	VIGO	4,473	6.3
360	ARECIBO	ARECIBO	4,342	-34.5
361	GRANADA	GRANADA	3,911	-4.1
362	VITORIA	VITORIA	3,069	-10.3
363	PAMPLONA	PAMPLONA	2,943	6.6
364	SAN SEBASTIAN	SAN SEBASTIAN	2,922	-7.4
365	SANTANDER	SANTANDER	2,791	-1.7
366	LA CORUNA	LA CORUNA	2,658	90.7
367	MURCIA	MURCIA	2,522	1
368	GERONA	GERONA	2,364	-29.6
369	CHEJU	CHEJU	2,257	34.3
370	HIERRO	HIERRO	2,189	-8
371	URUGUAIANA	RUBERN BERTA	2,138	2.2
372	TABATINGA	TABATINGA	2,110	166.1

(Continued on next page)

AIRPORT RANKING BY TOTAL AIRCRAFT MOVEMENTS - 1991
(Continued)

RANK	CITY	AIRPORT	TOTAL A/C MOVE-MENTS	% CHG 1991/ 1990
373	CRUZEIRO DO SOL	CAMPO INT'L	1,864	-3.9
374	PATILLAS	PATILLAS	1,826	737.6
375	VALLADOLID	VALLADOLID	1,652	-23.3
376	DUBROVNIK	DUBROVNIK	1,266	----
377	REUS	REUS	975	22.5
378	RAS AL KHAIMAH	RAS AL KHAIMAH INT'L	262	-42.8

Source: Airports Council International

AIRPORT RANKINGS - TOTAL PASSENGERS - 1991

RANK	CITY	AIRPORT	TOTAL PASSENGERS	% CHG 1991/ 1992
1	CHICAGO, ILLINOIS	O'HARE INT'L	59,852,330	-0.3
2	DALLAS, TEXAS	DALLAS/FT WORTH INT'L	48,198,208	-0.7
3	LOS ANGELES, CALIFORNIA	LOS ANGELES INT'L	45,668,204	-0.3
4	TOKYO	TOKYO INT'L (HANEDA)	42,015,096	4.5
5	LONDON	HEATHROW	40,495,508	-5.7
6	ATLANTA, GEORGIA	HARTSFIELD ATLANTA INT'L	37,915,024	-21.1
7	SAN FRANCISCO, CALIFORNIA	SAN FRANCISCO INT'L	31,774,845	2.3
8	DENVER, COLORADO	STAPLETON INT'L	28,285,189	3.1
9	FRANKFURT	FRANKFURT	27,978,403	-4.7
10	NEW YORK, NEW YORK	J.F. KENNEDY INT'L	27,441,937	-7.9
11	MIAMI, FLORIDA	MIAMI INT'L	26,591,415	2.9
12	OSAKA	OSAKA INT'L	23,483,451	-0.1
13	PARIS	ORLY	23,320,456	-4.2
14	NEWARK, NEW JERSEY	NEWARK INT'L	23,055,537	3.6
15	HONOLULU, HAWAII	HONOLULU INT'L	22,224,595	-4.9
16	PHOENIX, ARIZONA	SKY HARBOR INT'L	22,140,437	1.9
17	PARIS	CHARLES DE GAULLE	21,975,428	-2.4
18	BOSTON, MASSACHUSETTS	BOSTON LOGAN INT'L	21,547,026	-6.1
19	DETROIT, MICHIGAN	DETROIT METRO WAYNE COUNTY	21,309,046	-2.2
20	TOKYO	NEW TOKYO (NARITA) INT'L	20,710,083	-4.4
21	MINNEAPOLIS-ST PAUL, MINNESOTA	MINNEAPOLIS-ST PAUL INT'L	20,601,177	1.1
22	NEW YORK, NEW YORK	LA GUARDIA	20,545,060	-9.7
23	LAS VEGAS, NEVADA	MCCARRAN INT'L	20,171,557	5.7
24	HONG KONG	HONG KONG INT'L	19,747,543	2
25	ST LOUIS, MISSOURI	LAMBERT-ST LOUIS INT'L	19,151,278	-4.6
26	LONDON	GATWICK	18,821,442	-11.2
27	SEOUL	KIMPO INT'L	18,467,328	9.8
28	ORLANDO, FLORIDA	ORLANDO INT'L	18,411,945	0.1
29	HOUSTON, TEXAS	HOUSTON INTERCONTINENTAL	18,117,587	3.4
30	CHARLOTTE, NORTH CAROLINA	CHARLOTTE/DOUGLAS INT'L	16,876,779	8.1
31	PITTSBURGH, PENNSYLVANIA	GREATER PITTSBURGH INT'L	16,735,015	-2.4
32	AMSTERDAM	SCHIPHOL	16,541,928	0.4
33	ROME	FIUMICINO	16,491,732	-6.9
34	SEATTLE, WASHINGTON	SEATTLE TACOMA INT'L	16,313,090	0.4
35	SINGAPORE	CHANGI	16,285,039	4.3
36	MADRID	BARAJAS	16,019,369	-1.3
37	BANGKOK	BANGKOK INT'L	15,239,037	-4.3
38	PHILADELPHIA, PENNSYLVANIA	PHILADELPHIA INT'L	15,041,936	-7.8
39	WASHINGTON, D.C.	WASHINGTON NATIONAL	14,863,063	-4.5
40	FUKUOKA	FUKUOKA INT'L	13,443,416	1.6
41	MEXICO CITY	JUAREZ INT'L	13,001,967	7
42	SALT LAKE CITY, UTAH	SALT LAKE CITY INT'L	12,477,926	4.1
43	ZURICH	ZURICH	12,150,453	-4.3
44	STOCKHOLM	ARLANDA	11,922,376	-14.7
45	PALMA DE MALLORCA	PALMA DE MALLORCA	11,754,752	3.8
46	SAN DIEGO, CALIFORNIA	SAN DIEGO INT'L-LINDBERGH FIELD	11,423,067	1.9
47	COPENHAGEN	COPENHAGEN	11,400,351	-6
48	DUSSELDORF	DUSSELDORF	11,310,364	-5.2
49	MANCHESTER	MANCHESTER	10,865,114	0.5
50	MUNICH	MUNICH	10,797,984	-5.5
51	WASHINGTON, D.C.	WASHINGTON DULLES INT'L	10,790,217	5.4
52	TAIPEI	CHIANG KAI SHEK INT'L	10,613,850	5.1
53	CINCINNATI, OHIO	CINCINNATI/NO KENTUCKY INT'L	10,126,819	10.1
54	BALTIMORE, MARYLAND	BALTIMORE/WASHINGTON, DC INT'L	9,885,615	-3.2
55	TAMPA, FLORIDA	TAMPA INT'L	9,488,137	-10.4
56	RALEIGH-DURHAM, NORTH CAROLINA	RALEIGH-DURHAM INT'L	9,381,586	1.2
57	BARCELONA	BARCELONA	8,961,272	-0.9
58	MILAN	LINATE	8,882,391	-5.6
59	NASHVILLE, TENNESSEE	NASHVILLE INT'L	8,844,392	18.9
60	JAKARTA	SOEKARNO HATTA INT'L	8,366,109	5.2
61	MEMPHIS, TENNESSEE	MEMPHIS INT'L	8,176,294	-7.1
62	SAN JUAN	LUIS MUNOZ MARIN	8,156,734	-6.5

(Continued on next page)

AIRPORT RANKINGS - TOTAL PASSENGERS - 1991
(Continued)

RANK	CITY	AIRPORT	TOTAL PASSEN-GERS	% CHG 1991/ 1992
63	CLEVELAND, OHIO	CLEVELAND HOPKINS INT'L	8,142,644	-6.7
64	FT. LAUDERDALE/HOLLYWOOD, FLORIDA	FT. LAUDERDALE/HOLLYWOOD	8,045,712	-11.6
65	BOMBAY	BOMBAY	8,031,595	-5
66	HOUSTON, TEXAS	WILLIAM P. HOBBY	7,650,005	-4.8
67	MANILA	B. AQUINA INT'L	7,569,910	1.3
68	SAO PAULO	GUARULHOS	7,548,743	6.8
69	CARACAS	SIMON BOLIVAR EN MAIQUETIA	7,345,863	-0.4
70	MOSCOW	SHREMETYEVO	7,278,038	----
71	CHICAGO, ILLINOIS	MIDWAY	7,245,709	-17.6
72	KANSAS CITY, MISSOURI	KANSAS CITY INT'L	7,108,081	-0.7
73	SAN JOSE, CALIFORNIA	SAN JOSE INT'L	7,044,942	4
74	HELSINKI	HELSINKI VANTAA	7,000,244	-7.5
75	JEDDAH	KING ABDULAZIZ INT'L	6,922,763	-10.2
76	BERLIN	TEGEL & TEMPELHOF	6,873,917	2.3
77	NEW ORLEANS, LOUISIANA	NEW ORLEANS INT'L	6,589,904	-5.3
78	HAMBURG	HAMBURG-FUHLSBUTTEL	6,484,492	-5.5
79	GRAN CANARIA	GRAN CANARIA	6,436,143	5.2
80	PORTLAND, OREGON	PORTLAND INT'L	6,364,934	-0.2
81	RIYADH	KIND KHALED INT'L	6,360,769	-8.8
82	OAKLAND, CALIFORNIA	OAKLAND INT'L	6,181,251	12.1
83	TENERIFE SUR	TENERIFE SUR	6,122,006	9.7
84	VIENNA	VIENNA INT'L	5,845,558	2.3
85	ONTARIO, CALIFORNIA	ONTARIO INT'L	6,181,251	12.1
86	INDIANAPOLIS, INDIANA	INDIANAPOLIS INT'L	5,696,213	-0.2
87	DALLAS, TEXAS	LOVE FIELD	5,582,533	-2.9
88	NICE	NICE-COTE D'AZUR	5,537,629	-3.2
89	GENEVA	COINTRIN	5,531,306	-7.2
90	NEW DELHI	INDIRA GANDHI INT'L	5,497,076	-6
91	JOHANNESBURG	JAN SMUTS AIRPORT	5,476,104	-5.8
92	RIO DE JANEIRO	RIO DE JANEIRO	5,375,713	-30.3
93	COSTA MESA, CALIFORNIA	JOHN WAYNE	5,345,284	16.5
94	LISBON	LISBON	5,307,703	0.5
95	DUBLIN	DUBLIN	5,278,534	-4
96	AUCKLAND	AUCKLAND INT'L	5,249,798	-2.3
97	ISTANBUL	ATATURK	5,243,285	-19
98	SAN ANTONIO, TEXAS	SAN ANTONIO INT'L	5,233,589	-3.5
99	WEST PALM BEACH, FLORIDA	PALM BEACH INT'L	4,944,075	-10.3
100	ALBUQUERQUE, NEW MEXICO	ALBUQUERQUE INT'L	4,938,431	-1
101	KARACHI	KARACHI INT'L	4,800,664	-3.9
102	KAHULUI, HAWAII	KAHULUI	4,741,901	-3.9
103	MALAGA	MALAGA	4,630,852	-2.3
104	HARTFORD/SPRINGFIELD, CONNECTICUT	BRADLEY INT'L	4,613,857	-8.6
105	GUADALAJARA	MIGUEL HIDALGO	4,562,304	10.6
106	MARSEILLE	MARSEILLE PROVENCE	4,449,898	-10.7
107	DUBAI	DUBAI INT'L	4,396,230	-12.4
108	SACRAMENTO, CALIFORNIA	SACRAMENTO METRO	4,351,964	19.8
109	GLASGOW	GLASGOW	4,261,399	-3.3
110	STUTTGART	STUTTGART	4,237,728	-4.2
111	MILWAUKEE, WISCONSIN	GENERAL MITCHEL INT'L	4,114,051	-8.3
112	AUSTIN, TEXAS	ROBERT MUELLER MUNICIPAL	4,105,517	-4
113	MONTEVIDEO	AEROPUERTO INT'L DE CARRASCO	4,005,270	504.2
114	DAYTON, OHIO	DAYTON INT'L	3,994,086	-4.8
115	BURBANK, CALIFORNIA	BURBANK-GLENDALE-PASADENA	3,711,975	6
116	CANCUN	CANCUN	3,634,674	14.4
117	LYON	SATOLAS	3,560,487	-7.3
118	ANCHORAGE, ALASKA	ANCHORAGE INT'L	3,472,157	10.3
119	COLUMBUS, OHIO	PORT COLUMBUS INT'L	3,438,026	-6.1
120	FORT MYERS, FLORIDA	SOUTHWEST FLORIDA REGIONAL	3,438,026	-6.1
121	TEL AVIV	BEN GURION INT'L	3,428,994	-5.5
122	RENO, NEVADA	RENO CANNON INT'L	3,410,429	10.7
123	BIRMINGHAM	BIRMINGHAM INT'L	3,400,030	-6
124	FARO	FARO	3,323,986	20.5

(Continued on next page)

AIRPORT RANKINGS - TOTAL PASSENGERS - 1991
(Continued)

RANK	CITY	AIRPORT	TOTAL PASSENGERS	% CHG 1991/ 1992
125	ALGIERS	HOUARI BOUMEDIENE	3,299,812	-19.2
126	BRASILIA	BRASILIA	3,292,240	4
127	EL PASO, TEXAS	EL PASO INT'L	3,287,728	0
128	BUFFALO, NEW YORK	GREATER BUFFALO INT'L	3,172,753	-6.9
129	COLOGNE	COLOGNE BONN	3,042,993	-1.4
130	OKLAHOMA CITY, OKLAHOMA	WILL ROGERS WORLD	2,993,854	-3.6
131	HANOVER	HANOVER	2,894,763	2.9
132	TULSA, OKLAHOMA	TULSA INT'L	2,842,135	-5
133	LANZAROTE	LANZAROTE	2,799,407	12.9
134	DENPANSAR-BALI	NGURAH RAI	2,776,852	9.3
135	CHRISTCHURCH	CHRISTCHURCH INT'L	2,736,558	-8
136	KAUIA ISLAND, HAWAII	LIIHUE	2,696,749	-5.3
137	ALICANTE	ALICANTE	2,628,984	-1.4
138	GOTHENBURG	LANDVETTER	2,585,423	-13
139	DHARAN	DHARAN INT'L	2,577,737	-13.5
140	JACKSONVILLE, FLORIDA	JACKSONVILLE INT'L	2,577,148	-6.4
141	MILAN	MALPENSA	2,567,956	7.7
142	NORFOLK, VIRGINIA	NORFOLK INT'L	2,532,566	-6.1
143	IBIZA	IBIZA	2,516,255	3.2
144	TUCSON, ARIZONA	TUCSON INT'L	2,449,206	-8.4
145	EDINBURGH	EDINBURGH	2,434,760	-6.5
146	SYRACUSE, NEW YORK	HANCOCK INT'L	2,361,668	-10.5
147	OMAHA, NEBRASKA	EPPLEY AIRFIELD	2,303,312	970.4
148	ROCHESTER, NEW YORK	ROCHESTER	2,256,481	-8.4
149	CALCUTTA	CALCUTTA	2,234,338	-4.2
150	BORDEAUX	MERIGNAC	2,214,571	-13.8
151	BELFAST	BELFAST INT'L	2,182,756	-5.3
152	SURABAYA	JUANDA	2,133,798	5.9
153	KAILUA-KONA, HAWAII	KE-AHOLE	2,118,777	-2.8
154	SALVADOR	DOIS JULHO	2,105,785	2.7
155	RECIFE	GUARARAPES	2,065,951	0.8
156	ABERDEEN	ABERDEEN	2,037,408	3.3
157	WARSAW	OKECIE	1,981,119	-26
158	ABU DHABI	ABU DHABI INT'L	1,972,261	-15.9
159	MADRAS	MADRAS	1,958,384	7.8
160	SANTIAGO	A. MERINO BENITEZ	1,947,657	11.3
161	BIRMINGHAM, ALABAMA	BIRMINGHAM	1,934,305	-7.1
162	MONTERREY	MARIANO ESCOBEDO	1,862,903	14.4
163	DURBAN	LOUIS BOTHA AIRPORT	1,858,295	-7
164	BASEL-MULHOUSE	BASEL-MULHOUSE	1,843,993	0
165	RICHMOND, VIRGINIA	RICHMOND INT'L	1,765,333	-5.4
166	PALERMO	PUNTA RAISI	1,758,530	-1.6
167	QUITO	MARISCAL SUCRE	1,734,965	12.9
168	LONDON	STANSTED	1,723,036	46.4
169	TIJUANA	RODRIGUEZ	1,716,086	-1.4
170	GREENSBORO, NORTH CAROLINA	GREENSBORO	1,705,264	-7.7
171	ANKARA	ESENBOGA	1,694,413	-15.7
172	PUERTO VALLARTA	ORDAZ	1,690,257	1.1
173	SEVILLA	SEVILLA	1,667,749	0.2
174	PORTO ALEGRE	SALDAGO FILHO	1,613,683	9.5
175	NEWCASTLE	NEWCASTLE	1,592,126	-2.7
176	ANTALYA	ANTALYA	1,564,733	-25.7
177	SHANNON	SHANNON	1,542,899	-5
178	COLOMBO	COLOMBO KATUNAYAKE	1,521,472	3.4
179	POINTE A PITRE	LE RAIZET	1,510,603	3
180	ACAPULCO	ACAPULCO	1,503,409	-2.2
181	ARUBA	REINA BEATRIX	1,499,757	15.7
182	DES MOINES, IOWA	DES MOINES INT'L	1,490,829	4.4
183	MENORCA	MENORCA	1,487,828	1.4
184	DHAKA	ZIA INTERNATIONAL	1,480,455	3.1
185	VALENCIA	VALENCIA	1,479,603	7.5
186	PORTO	PORTO	1,454,466	7.3

(Continued on next page)

AIRPORT RANKINGS - TOTAL PASSENGERS - 1991
(Continued)

RANK	CITY	AIRPORT	TOTAL PASSEN-GERS	% CHG 1991/ 1992
187	HILO, HAWAII	HILO INT'L	1,439,811	-1.4
188	STRASBOURG	ENTZHEIM	1,427,518	-8.1
189	NUREMBERG	NUREMBERG	1,427,230	-3.1
190	FUERTEVENTURA	FUERTEVENTURA	1,377,164	23
191	TURIN	CITTA DI TORINO	1,362,167	-4.5
192	LONG BEACH, CALIFORNIA	LONG BEACH	1,353,087	-7.1
193	BANGOR, MAINE	BANGOR INT'L	1,338,129	6.2
194	GUAYAQUIL	SIMON BOLIVAR	1,335,041	0.7
195	ADNAN MENDERES	IZMIR	1,304,832	-25.8
196	BOLOGNA	GUGLIEMO MARCONI	1,266,307	-2
197	MANAUS	EDUARDO GOMES	1,262,572	-7.8
198	COLORADO SPRINGS, COLORADO	COLORADO SPRINGS	1,255,307	5.4
199	CHARLESTON, SOUTH CAROLINA	CHARLESTON INT'L	1,242,698	-10.5
200	TENERIFE NORTE	TENERIFE NORTE	1,203,801	20.7
201	BELEM	VAL DE CAES	1,191,931	-8.2
202	MEDAN	POLONIA	1,189,864	4.2
203	MIDDLETOWN, PENNSYLVANIA	HARRISBURG INT'L	1,187,581	-4.3
204	BOISE, IDAHO	BOISE AIR TERMINAL	1,180,568	-3.5
205	BILBAO	BILBAO	1,178,477	9.8
206	WICHITA, KANSAS	WICHITA MID-CONTINENT	1,177,036	-3.7
207	BELO HORIZONTE	CONFINS INT'L	1,165,639	-3.4
208	KNOXVILLE, TENNESSEE	MCGHEE TYSON	1,148,396	-1.4
209	EAST MIDLANDS	EAST MIDLANDS INT'L	1,147,551	-11
210	MAZATLAN	GEN. RAFAEL BUELNA	1,145,040	-2.9
211	MONTPELLIER	FREJORGUES	1,092,939	-1.3
212	GREER, SOUTH CAROLINA	GREENVILLE-SPARTANBURG	1,055,823	-10.9
213	CURACAO	AEROPUERTO HATO	1,047,645	-5.6
214	MIDLAND, TEXAS	MIDLAND INT'L	1,032,705	-10.4
215	BREMEN	BREMEN	1,021,770	-7.5
216	PUSAN	KIMHAE	1,021,095	7.8
217	LUXEMBOURG	LUXEMBOURG	1,009,386	-5
218	SAVANNAH, GEORGIA	SAVANNAH INT'L	987,997	-9.1
219	NANTES	NANTES ATLANTIQUE	945,045	----
220	UJUNG PANDANG	HASANNUDDIN	937,340	7.7
221	HARLINGEN, TEXAS	VALLEY INT'L	932,100	-12.1
222	SOFIA	SOFIA	921,141	-61.4
223	DAYTONA BEACH, FLORIDA	DAYTONA BEACH REGIONAL	899,529	-13.4
224	ALLENTOWN, PENNSYLVANIA	ALLENTOWN-BETHLEHEM-EASTON INT'L	866,271	-1
225	MERIDA	REJON	864,805	11.5
226	ATLANTIC CITY, NEW JERSEY	ATLANTIC CITY INT'L	860,281	5.4
227	SANTIAGO DE COMPOSTELA	SANTIAGO DE COMPOSTELA	854,109	-4.4
228	NADI	NADI INT'L	846,644	-7.4
229	MANCHESTER, NEW HAMPSHIRE	MANCHESTER	825,317	6.2
230	BRISTOL	BRISTOL	820,998	0.3
231	NEWBURGH, NEW YORK	STEWART INT'L	805,142	110.7
232	BALIKPAPAN	SEPINGGAN	802,451	8.3
233	JACKSON, MISSISSIPPI	JACKSON INT'L	783,948	-2.6
234	BATON ROUGE, LOUISIANA	BATON ROUGE METROPOLITAN	762,829	-1
235	LILLE	LILLE LESQUIN	752,292	-6.2
236	GENOA	CRISTOFORO COLOMBO	749,847	----
237	PORT ELIZABETH	H.F. VERWOERD AIRPORT	714,142	-6.3
238	SAN JOSE DEL CABO	LOS CABOS	688,858	17.8
239	KEFLAVIK	KEFLAVIK	687,215	-1.8
240	JAKARTA	HALIM PERDANKUSUMA	685,898	8.2
241	SHARJAH	SHARJAH INT'L	684,104	-16.4
242	MELBOURNE, FLORIDA	MELOURNE REGIONAL	661,591	-12.3
243	LEEDS BRADFORD	LEEDS BRADFORD	657,016	-21
244	ROANOKE, VIRGINIA	ROANOKE REGIONAL	655,964	-9.3
245	LIBREVILLE	INT'L LEON M'BA LIBREVILLE	646,015	13.1
246	CORK	CORK	644,896	-9
247	ZAGREB	ZAGREB	644,487	-60.5
248	DALAMAN	DALAMAN	643,777	-10.2

(Continued on next page)

AIRPORT RANKINGS - TOTAL PASSENGERS - 1991
(Continued)

RANK	CITY	AIRPORT	TOTAL PASSEN-GERS	% CHG 1991/1992
249	MOLINE, ILLINOIS	QUAD CITY	606,582	1
250	TRIVANDRUM	TRIVANDRUM	593,583	4.9
251	NATAL	AUGUSTO SEVERO	590,694	3
252	AKRON/CANTON, OHIO	AKRON-CANTON REGIONAL	590,016	-11.4
253	CARDIFF WALES	CARDIFF WALES	576,189	-12.1
254	MALE'	MALE' INT'L	566,232	2.2
255	FOZ DE IGUACU	CATARATAS	559,955	-2.3
256	PAU PYRENEES	UZEIN	549,866	1.7
257	LA PALMA	LA PALMA	534,948	9.7
258	BIARRITZ	BAYONNE-ANGLET	527,065	-4
259	ASHEVILLE, NORTH CAROLINA	ASHEVILLE REGIONAL	520,745	-3.2
260	KINSHASA	N'DJILI	514,983	-18.8
261	MCALLEN, TEXAS	MCALLEN-MILLER INT'L	512,937	15.1
262	LANSING, MICHIGAN	CAPITAL CITY	512,844	-3.1
263	KALAMAZOO, MICHIGAN	KALAMAZOO/BATTLE CREEK INT'L	506,837	2.7
264	LIVERPOOL	LIVERPOOL	488,208	-5.5
265	ROME	CIAMPINO	476,927	-10.8
266	ALMERIA	ALMERIA	471,619	-5.9
267	MONTEREY, CALIFORNIA	MONTEREY PENINSULA AIRPORT	468,858	-6.4
268	LUSAKA	LUSAKA INT'L	466,032	-21
269	FREELAND, MICHIGAN	TRI-CITY INT'L	465,622	-0.5
270	BREST-GUIPAVAS	BREST-GUIPAVAS	464,211	-7.2
271	AUGUSTA, GEORGIA	BUSH FIELD	464,011	2.5
272	SIOUX FALLS, SOUTH DAKOTA	JOE FOSS FIELD	458,355	-1.3
273	PEORIA, ILLINOIS	GREATER PEORIA REGIONAL	443,794	8.3
274	SPRINGFIELD, MISSOURI	SPRINGFIELD REGIONAL	434,004	-5.2
275	ASTURIAS	ASTURIAS	427,581	13.5
276	SOUTHAMPTON	SOUTHAMPTON	425,681	----
277	TRIESTE	DEI LEGIONARI	423,111	2
278	TARBES-LOURDES	TARBES-LOURDES	421,477	-4.9
279	MAPUTO	MAPUTO INT'L	409,200	----
280	PONTA DELGADA	NORDELA	398,793	2.6
281	CAMPO GRANDE	CAMPO GRANDE	373,162	-2.6
282	ILHA DO SAL	AMILCAR CABRAL INT'L	363,178	----
283	NIMES	GARONS	356,306	-1.9
284	VIGO	VIGO	353,268	4.1
285	GAINESVILLE, FLORIDA	GAINESVILLE REGIONAL	349,850	-20
286	LJUBLJANA	BRNIK	347,583	-54.6
287	WILMINGTON, NORTH CAROLINA	NEW HANOVER INT'L	346,806	4.5
288	ST. GEORGE'S	POINT SALINE	344,377	10.2
289	TEESIDE	TEESIDE INT'L	336,532	-18.8
290	JEREZ	JEREZ	335,659	7
291	ADANA	ADANA	331,141	-21.2
292	PASCO, WASHINGTON	TRI-CITIES	321,882	5.9
293	ERIE, PENNSYLVANIA	ERIE INT'L	319,547	-3.1
294	NEWPORT NEWS, VIRGINIA	NEWPORT NEWS/WILLIAMSBURG INT'L	310,263	3.3
295	MUENSTER	MUENSTER-OSNABRUECK	307,311	0.1
296	CHEJU	CHEJU	305,486	33.5
297	GRANADA	GRANADA	303,514	0.7
298	MISSOULA, MONTANA	MISSOULA COUNTY	284,224	3.7
299	GERONA	GERONA	283,588	-29.4
300	NOUMEA	LA TONTOUTA	283,374	-5.5
301	VITORIA	VITORIA	268,145	-7.4
302	BAKERSFIELD, CALIFORNIA	MEADOWS FIELD	265,927	-2.2
303	LAFAYETTE, LOUISIANA	LAFAYETTE REGIONAL	261,130	5.2
304	BONAIRE	FLAMINGO	250,735	17.5
305	CLERMONT-FERRAND	AULNAT	249,687	-7
306	GEORGE	P.W. BOTHA AIRPORT	248,364	10.2
307	BIAK	FRANS KAISIEPO	239,727	3.3
308	DUBROVNIK	DUBROVNIK	238,699	----
309	SPRINGFIELD, ILLINOIS	CAPITAL	234,785	-5.2
310	MACAPA	MACAPA	217,772	-5.6

(Continued on next page)

AIRPORT RANKINGS - TOTAL PASSENGERS - 1991
(Continued)

RANK	CITY	AIRPORT	TOTAL PASSEN-GERS	% CHG 1991/1992
311	MELILLA	MELILLA	208,723	12.6
312	MANADO	SAM RATULANGI	205,719	-0.7
313	FAJARDO	FAJARDO	199,409	57.3
314	DURANGO, COLORADO	DURANGO-LA PLATA COUNTY	190,657	-1.6
315	ZARAGOZA	ZARAGOZA	189,886	7.7
316	SANTANDER	SANTANDER	186,287	10.6
317	EGELSBACH	EGELSBACH	184,724	8
318	VIEQUES	VIEQUES	183,800	26.9
319	SIOUX CITY, IOWA	SIOUX GATEWAY	182,934	-19.1
320	LA CORUNA	LA CORUNA	177,602	96.1
321	LONDON	LONDON CITY	171,216	-26
322	MAASTRICHT	MAASTRICHT	169,737	-44.6
323	EINDHOVEN	EINDHOVEN	169,031	-11
324	PAMPLONA	PAMPLONA	166,609	4.7
325	RIO BRANCO	PRES. MEDICE	163,223	-17.1
326	SARASOTA, FLORIDA	SARASOTA-BRADENTON	160,575	-6.6
327	SAN JUAN	ISLA GRANDE	159,541	12.7
328	KETCHIKAN, ALASKA	KETCHIKAN INT'L	150,259	-3.4
329	LIMOGES	LIMOGES BELLEGARDE	141,981	----
330	CASPER, WYOMING	NATRONA COUNTY INT'L	139,849	1.1
331	MONACO	HELIPORT DE MONACO	121,359	-3.2
332	READING, PENNSYLVANIA	READING REGIONAL	121,307	-4.6
333	CAMPINAS	VIRACOPOS	121,236	2.3
334	YUMA, ARIZONA	YUMA INT'L	120,592	-2.4
335	BOA VISTA	BOA VISTA	119,906	-9.7
336	MAYAGUEZ	EUGENIO M. DE HOSTOS	119,103	-26.1
337	MURCIA	MURCIA	118,533	-11.8
338	AGUADILLA	RAFAEL HERNANDEZ (BORINQUEN)	107,341	231.7
339	VALLADOLID	VALLADOLID	106,625	45.9
340	BEAUVAIS	BEAUVAIS-TILLE	105,320	-23.7
341	KIMBERLEY	B.J. VORSTER AIRPORT	103,639	-11.3
342	SAN SEBASTIAN	SAN SEBASTIAN	100,023	-9.9
343	CULEBRA	CULEBRA	96,427	101.5
344	METZ-FRESCATY	METZ-FRESCATY	94,203	----
345	REUS	REUS	91,730	3
346	HIERRO	HIERRO	90,917	4.3
347	ST ETIENNE	BOUTHEON	85,176	-2
348	CHAMBERY	CHAMBERY/AIX	78,706	1.8
349	BANGUI	BANGUI-M'POKO	77,218	-1.8
350	PADUCAH, KENTUCKY	BARKLEY REGIONAL	61,143	-0
351	CORUMBA	CORUMBA	56,554	-29.3
352	BERGERAC	ROUMANIERES	46,180	-26.4
353	RAS AL KHAIMAH	RAS AL KHAIMAH INT'L	38,406	-33.1
354	UPINGTON	PIERRE VAN RYNEVELD	38,226	-4.2
355	PERIGUEUX	PERIGUEUX-BASSILLAC	37,937	-4.2
356	GLASGOW	PRESTWICK	35,600	-63.1
357	ROUEN	ROUEN	33,816	2.1
358	TABATINGA	TABATINGA	29,414	9.4
359	EAST FAMINGDALE, NEW YORK	REPUBLIC	25,848	-12.4
360	CRUZEIRO DO SUL	CAMPO INT'L	25,099	-17.4
361	VALENCE	CHABEUIL	23,337	39.9
362	ANGOULEME	ANGOULEME	19,160	-50
363	TOPEKA, KANSAS	FORBES FIELD	18,360	-21.8
364	HUMACAO	HUMACAO	17,700	9.4
365	FUJAIRAH	FUJAIRAH INT'L	14,395	-18.4
366	LYON	BRON	11,313	0.7
367	DOLE TAVAUX	DOLE TAVAUX	5,737	-44
368	URUGUAIANA	RUBERN BERTA	5,327	4.9
369	PONTA PORA	PONTA PORA	3,506	99.2
370	ARECIBO	ARECIBO	3,154	-58.7
371	VICHY-CHARMEIL	VICHY-CHARMEIL	2,881	-10
372	CLEVELAND, OHIO	CLEVELAND BURKE LAKEFRONT	1,262	-93.3
373	PATILLAS	PATILLAS	1,011	440.6

AIRPORT RANKING BY TOTAL CARGO (METRIC TONNES) - 1991

RANK	CITY	AIRPORT	TOTAL CARGO	% CHG 1991/ 1990
1	TOKYO	NEW TOKYO (NARITA) INT'L	1,383,599	-0.5
2	NEW YORK, NEW YORK	J.F. KENNEDY INT'L	1,257,069	-4.9
3	FRANKFURT	FRANKFURT	1,206,316	-2.8
4	LOS ANGELES, CALIFORNIA	LOS ANGELES INT'L	1,141,196	-2
5	CHICAGO, ILLINOIS	O'HARE INT'L	1,071,598	-1.5
6	MIAMI, FLORIDA	MIAMI INT'L	967,239	0.1
7	HONG KONG	HONG KONG INT'L	849,786	6
8	LONDON	HEATHROW	736,324	-5.5
9	SEOUL	KIMPO INT'L	704,334	0.1
10	AMSTERDAM	SCHIPHOL	654,195	3.8
11	SINGAPORE	CHANGI	652,922	3.2
12	TAIPEI	CHIANG KAI SHEK INT'L	632,311	6.3
13	PARIS	CHARLES DE GAULLE	615,699	-4.9
14	SAN FRANCISCO, CALIFORNIA	SAN FRANCISCO INT'L	606,008	6.8
15	ATLANTA, GEORGIA	HARTSFIELD ATLANTA INT'L	599,674	-1.8
16	ANCHORAGE, ALASKA	ANCHORAGE INT'L	587,817	8.5
17	DALLAS, TEXAS	DALLAS/FT WORTH INT'L	547,008	-1.7
18	OSAKA	OSAKA INT'L	499,604	1
19	TOKYO	TOKYO INT'L (HANEDA)	489,255	0.9
20	NEWARK, NEW JERSEY	NEWARK INT'L	483,622	-4.2
21	COLOMBO	COLOMBO KATUNAYAKE	474,600	1091.6
22	DAYTON, OHIO	DAYTON INT'L	441,418	-19.9
23	DUBROVNIK	DUBROVNIK	440,423	----
24	BANGKOK	BANGKOK INT'L	403,087	-0.7
25	INDIANAPOLIS, INDIANA	INDIANAPOLIS INT'L	394,162	28.8
26	HONOLULU, HAWAII	HONOLULU INT'L	382,167	1.9
27	PHILADELPHIA, PENNSYLVANIA	PHILADELPHIA INT'L	351,059	5.7
28	BOSTON, MASSACHUSETTS	BOSTON LOGAN INT'L	347,735	-4.5
29	SAO PAULO	GUARULHOS	326,867	13.4
30	MEMPHIS, TENNESSEE	MEMPHIS INT'L	312,016	215.9
31	PARIS	ORLY	296,421	3
32	DENVER, COLORADO	STAPLETON INT'L	292,625	4.5
33	ROME	FIUMICINO	278,550	-2.5
34	MINNEAPOLIS-ST PAUL, MINNESOTA	MINNEAPOLIS-ST PAUL INT'L	268,114	0.7
35	ZURICH	ZURICH	263,595	-2.8
36	ONTARIO, CALIFORNIA	ONTARIO INT'L	256,280	3.6
37	OAKLAND, CALIFORNIA	OAKLAND INT'L	252,855	18.9
38	SEATTLE, WASHINGTON	SEATTLE TACOMA INT'L	247,124	0.8
39	NIMES	GARONS	237,453	-20.6
40	HOUSTON, TEXAS	HOUSTON INTERCONTINENTAL	230,304	3.3
41	MANILA	B. AQUINO INT'L	228,695	-3.3
42	FUKUOKA	FUKUOKA INT'L	222,425	2.5
43	LONDON	GATWICK	212,905	-6.5
44	DETROIT, MICHIGAN	DETROIT METRO WAYNE COUNTY	210,785	9.9
45	TEL AVIV	BEN GURION INT'L	207,602	4
46	SAN JUAN	LUIS MUNOZ MARIN	206,471	1.7
47	COLOGNE	COLOGNE BONN	202,501	12.7
48	MADRID	BARAJAS	188,739	-16.7
49	COPENHAGEN	COPENHAGEN	176,327	2
50	BOMBAY	BOMBAY	174,073	-13.3
51	JAKARTA	SOEKARNO HATTA INT'L	172,180	-0.2
52	WASHINGTON, D.C.	WASHINGTON DULLES INT'L	163,823	-6.3
53	CINCINNATI, OHIO	CINCINNATI/NO KENTUCKY INT'L	162,553	7.7
54	PORTLAND, OREGON	PORTLAND INT'L	160,077	13.2
55	LUXEMBOURG	LUXEMBOURG	153,073	7
56	AUCKLAND	AUCKLAND INT'L	150,717	2.1
57	BALTIMORE, MARYLAND	BALTIMORE/WASHINGTON, DC INT'L	148,531	-1.1
58	ORLANDO, FLORIDA	ORLANDO INT'L	145,119	14.7
59	DUBAI	DUBAI INT'L	144,086	-2.2
60	JEDDAH	KING ABDULAZIZ INT'L	141,300	-18
61	CHARLOTTE, NORTH CAROLINA	CHARLOTTE/DOUGLAS INT'L	140,341	13
62	MEXICO CITY	JUAREZ INT'L	136,248	3.1
63	RIO DE JANEIRO	RIO DE JANEIRO	135,743	1.3

(Continued on next page)

AIRPORT RANKING BY TOTAL CARGO (METRIC TONNES) - 1991
(Continued)

RANK	CITY	AIRPORT	TOTAL CARGO	% CHG 1991/ 1990
64	JOHANNESBURG	JAN SMUTS AIRPORT	132,615	-6.7
65	SALT LAKE CITY, UTAH	SALT LAKE CITY INT'L	132,138	14.2
66	HARTFORD/SPRINGFIELD, CONNECTICUT	BRADLEY INT'L	129,483	11.9
67	NEW DELHI	INDIRA GANDI INT'L	126,425	-6.5
68	PITTSBURGH, PENNSYLVANIA	GREATER PITTSBURGH INT'L	126,243	-1.4
69	PHOENIX, ARIZONA	SKY HARBOR INT'L	118,846	4
70	SANTIAGO	A. MARINO BENITEZ	113,913	12.3
71	NEW YORK, NEW YORK	LA GUARDIA	108,777	-6.9
72	MANAUS	EDUARDO GOMES	107,172	-26.7
73	ST LOUIS, MISSOURI	LAMBERT-ST LOUIS INT'L	105,416	-2.4
74	STOCKHOLM	ARLANDA	97,489	3.5
75	MILWAUKEE, WISCONSIN	GENERAL MITCHEL INT'L	94,357	16.8
76	RIYADH	KIND KHALED INT'L	92,598	-16.9
77	TAMPA, FLORIDA	TAMPA INT'L	88,276	0.1
78	KANSAS CITY, MISSOURI	KANSAS CITY INT'L	84,065	-0.4
79	LISBON	LISBON	83,484	-10.6
80	FT. LAUDERDALE/HOLLYWOOD, FLORIDA	FT. LAUDERDALE/HOLLYWOOD	82,648	4.6
81	SAN JOSE, CALIFORNIA	SAN JOSE INT'L	77,229	-7.1
82	RALEIGH-DURHAM, NORTH CAROLINA	RALEIGH-DURHAM INT'L	77,197	13.7
83	MILAN	LINATE	76,432	-9
84	MUNICH	MUNICH	75,375	-3.4
85	CLEVELAND, OHIO	CLEVELAND HOPKINS INT'L	74,064	0.2
86	MILAN	LINATE	76,432	-9
87	MANCHESTER	MANCHESTER	73,324	-9.7
88	NEWBURGH, NEW YORK	STEWART INT'L	72,819	34.6
89	DES MOINES, IOWA	DES MOINES INT'L	72,022	1.8
90	VIENNA	VIENNA INT'L	69,656	7.9
91	MOSCOW	SHREMETYEVO	69,395	----
92	ISTANBUL	ATATURK	67,473	-20.4
93	NEW ORLEANS, LOUISIANA	NEW ORLEANS INT'L	66,179	5.4
94	HELSINKI	HELSINKI VANTAA	63,469	-8.3
95	BARCELONA	BARCELONA	62,663	-5.1
96	GENEVA	COINTRIN	59,676	-9.4
97	WASHINGTON, D.C.	WASHINGTON NATIONAL	59,432	-2.9
98	HAMBURG	HAMBURG-FUHLSBUTTEL	55,646	-4
99	OMAHA, NEBRASKA	EPPLEY AIRFIELD	54,981	9.3
100	AGUADILLA	RAFAEL HERNANDEZ (BORINQUEN)	54,600	13.7
101	KINSHASA	N'DJILI	53,415	-30.7
102	DUBLIN	DUBLIN	50,722	-4.5
103	CARACAS	SIMON BOLIVAR EN MAIQUETIA	49,604	-3.3
104	DUSSELDORF	DUSSELDORF	49,095	-4.3
105	SAN DIEGO, CALIFORNIA	SAN DIEGO INT'L-LINDBERGH FIELD	48,177	-8.8
106	MARSEILLE	MARSEILLE PROVENCE	47,554	-2.6
107	BRASILIA	BRASILIA	43,908	16.2
108	RICHMOND, VIRGINIA	RICHMOND INT'L	43,769	12.4
109	DHARAN	DHARAN INT'L	42,583	-22.7
110	GREENSBORO, NORTH CAROLINA	GREENSBORO	42,550	-12.1
111	MADRAS	MADRAS	41,448	-3.4
112	ALBUQUERQUE, NEW MEXICO	ALBUQUERQUE INT'L	40,295	13.4
113	NASHVILLE, TENNESSEE	NASHVILLE INT'L	39,541	-1.9
114	BUFFALO, NEW YORK	GREATER BUFFALO INT'L	38,523	9.3
115	EL PASO, TEXAS	EL PASO INT'L	38,312	-0.6
116	KAHULUI, HAWAII	KAHULUI	38,193	5.3
117	GRAN CANARIA	GRAN CANARIA	38,100	8.9
118	TULSA, OKLAHOMA	TULSA INT'L	37,826	-6.7
119	JACKSONVILLE, FLORIDA	JACKSONVILLE INT'L	37,169	-4.2
120	DHAKA	ZIA INTERNATIONAL	36,185	-10.4
121	CALCUTTA	CALCUTTA	35,612	-6.4
122	LONDON	STANSTED	35,512	4.9
123	COLUMBUS, OHIO	PORT COLUMBUS IN'L	35,299	35.7
124	BIRMINGHAM, ALABAMA	BIRMINGHAM	35,065	-18.6
125	OKLAHOMA CITY, OKLAHOMA	WILL ROGERS WORLD	35,015	5.1
126	LYON	SATOLAS	33,796	6.6

(Continued on next page)

AIRPORT RANKING BY TOTAL CARGO (METRIC TONNES) - 1991
(Continued)

RANK	CITY	AIRPORT	TOTAL CARGO	% CHG 1991/ 1990
127	NICE	NICE-COTE D'AZUR	33,757	6.6
128	ABU DHABI	ABU DHABI INT'L	32,665	-4.3
129	BELFAST	BELFAST INT'L	32,011	-0.2
130	AUSTIN, TEXAS	ROBERT MUELLER MUNCIPAL	31,676	-10.7
131	GUAYAQUIL	SIMON BOLIVAR	31,262	12.1
132	SACRAMENTO, CALIFORNIA	SACRAMENTO METRO	31,017	5
133	BERLIN	TEGEL & TEMPELHOF	29,807	-1.2
134	RECIFE	GUARARAPES	29,777	-3.1
135	PORTO ALEGRE	SALGADO FILHO	29,467	-13.4
136	CAMPINAS	VIRACOPOS	29,244	-33.5
137	STUTTGART	STUTTGART	28,415	-6.1
138	SHARJAH	SHARJAH INT'L	28,073	-2.6
139	QUITO	MARISCAL SUCRE	27,264	4.3
140	BASEL-MULHOUSE	BASEL-MULHOUSE	26,831	-7
141	BIRMINGHAM	BIRMINGHAM INT'L	26,768	19.7
142	SAN ANTONIO, TEXAS	SAN ANTONIO INT'L	26,757	6.3
143	LIVERPOOL	LIVERPOOL	26,247	-1.1
144	KNOXVILLE, TENNESSEE	MCGHEE TYSON	26,212	-8.1
145	MAASTRICHT	MAASTRICHT	26,167	3.1
146	DENPANSAR-BALI	NGURAH RAI	25,684	34.3
147	HILO, HAWAII	HILO INT'L	25,533	-0.7
148	HANOVER	HANOVER	25,077	13.5
149	GUADALAJARA	MIGUEL HIDALGO	24,995	19
150	PUSAN	KIMHAE	24,900	13
151	LONG BEACH, CALIFORNIA	LONG BEACH	24,886	37.1
152	SALVADOR	DOIS JULHO	24,466	-15.6
153	SEATTLE, WASHINGTON	BOEING FIELD/KING COUNTY INT'L	23,880	7.4
154	DURBAN	LOUIS BOTHA AIRPORT	23,456	5.3
155	SURABAYA	JUANDA	23,406	5.9
156	MEDAN	POLONIA	22,734	-1.3
157	BELEM	VAL DE CAES	22,073	-17.5
158	GLASGOW	GLASGOW	21,763	-16.6
159	CHRISTCHURCH	CHRISTCHURCH INT'L	21,620	8.1
160	CHICAGO, ILLINOIS	MIDWAY	21,607	2
161	UJUNG PANDANG	HASANNUDDIN	21,305	-4.2
162	NORFOLK, VIRGINIA	NORFOLK INT'L	21,256	-14.2
163	WICHITA, KANSAS	WICHITA MID-CONTINENT	20,795	10.9
164	WARSAW	OKECIE	19,627	-4.4
165	ALGIERS	HOUARI BOUMEDIENE	19,418	-15.3
166	NUREMBERG	NUREMBERG	19,348	10.6
167	EAST MIDLANDS	EAST MIDLANDS INT'L	19,105	-6.9
168	PORTO	PORTO	18,902	-3.1
169	KAILUA-KONA, HAWAII	KE-AHOLE	18,791	5.6
170	SYRACUSE, NEW YORK	HANCOCK INT'L	18,372	-5.1
171	TENERIFE SUR	TENERIFE SUR	17,733	-2
172	GOTHENBURG	LANDVETTER	17,699	-28.6
173	BURBANK, CALIFORNIA	BURBANK-GLENDALE-PASADENA	17,640	-11.8
174	MONTEVIDEO	AEROPUERTO INT'L DE CARRASCO	17,455	16.8
175	JACKSON, MISSISSIPPI	JACKSON INT'L	17,116	-12.4
176	PALMA DE MALLORCA	PALMA DE MALLORCA	16,837	-5.8
177	WEST PALM BEACH, FLORIDA	PALM BEACH INT'L	16,501	9.8
178	BELO HORIZONTE	CONFINS INT'L	16,318	1.4
179	LIBREVILLE	INT'L LEON M'BA LIBREVILLE	15,843	32.1
180	RENO, NEVADA	RENO CANNON INT'L	15,587	-1.2
181	TURIN	CITTA DI TORINO	15,537	-15.2
182	CLEVELAND, OHIO	CLEVELAND BURKE LAKEFRONT	15,516	-0.5
183	MERIDA	REJON	15,351	39.5
184	GLASGOW	PRESTWICK	15,286	-1.9
185	KEFLAVIK	KEFLAVIK	15,220	0.8
186	EDINBURGH	EDINBURGH	15,075	1.5
187	LANSING, MICHIGAN	CAPITAL CITY	14,849	22.4
188	POINTE A PITRE	LE RAIZET	14,733	-7.1
189	KAUIA ISLAND, HAWAII	LIIHUE	14,357	-6.1

(Continued on next page)

AIRPORT RANKING BY TOTAL CARGO (METRIC TONNES) - 1991
(Continued)

RANK	CITY	AIRPORT	TOTAL CARGO	% CHG 1991/ 1990
190	BOISE, IDAHO	BOISE AIR TERMINAL	13,823	3.4
191	MONTERREY	MARIANO ESCOBEDO	13,372	26.3
192	PEORIA, ILLINOIS	GREATER PEORIA REGIONAL	12,631	-7.9
193	NADI	NADI INT'L	12,318	-11
194	TUCSON, ARIZONA	TUCSON INT'L	12,051	-30.5
195	BORDEAUX	MERIGNAC	11,732	-15.4
196	FUJAIRAH	FUJAIRAH INT'L	11,721	394.9
197	ANKARA	ESENBOGA	11,245	-21.5
198	GENOA	CRISTOFORO COLOMBO	11,226	----
199	BALIKPAPAN	SEPINGGAN	10,987	15.8
200	MALE'	MALE' INT'L	10,317	43.7
201	LUSAKA	LUSAKA INT'L	10,299	-28.4
202	TRIVANDRUM	TRIVANDRUM	10,203	-9.7
203	CURACAO	AEROPUERTO HATO	9,331	-14.3
204	SOFIA	SOFIA	9,066	-52
205	ROANOKE, VIRGINIA	ROANOKE REGIONAL	9,018	-13
206	PORT ELIZABETH	H.F. VERWOERD AIRPORT	8.797	9.2
207	TIJUANA	RODRIGUEZ	8,412	30.4
208	FORT MYERS, FLORIDA	SOUTHWEST FLORIDA REGIONAL	8,087	4.6
209	MONTPELLIER	FREJORGUES	7,791	-14.5
210	ABERDEEN	ABERDEEN	7,558	-7.7
211	CANCUN	CANCUN	7,390	-9.5
212	CAMPO GRANDE	CAMPO GRANDE	7,342	4.5
213	VALENCIA	VALENCIA	7,288	-52.7
214	PONTA DELGADA	NORDELA	7,143	-4.7
215	BREMEN	BREMEN	7,087	9.2
216	SIOUX FALLS, SOUTH DAKOTA	JOE FOSS FIELD	6,996	-1.5
217	PALERMO	PUNTA RAISI	6,993	-14.4
218	STRASBOURG	ENTZHEIM	6,934	-7.8
219	ZARAGOZA	ZARAGOZA	6,748	14
220	ALLENTOWN, PENNSYLVANIA	ALLENTOWN-BETHLEHEM-EASTON INT'L	6,616	-3.7
221	ADNAN MENDERES	IZMIR	6,057	-35.9
222	ROME	CIAMPINO	5,983	12.9
223	LANZAROTE	LANZAROTE	5,917	49
224	MALAGA	MALAGA	5,847	-23.9
225	NOUMEA	LA TONTOUTA	5,843	-16.1
226	HOUSTON, TEXAS	WILLIAM P. HOBBY	5,795	3.6
227	CHARLESTON, SOUTH CAROLINA	CHARLESTON INT'L	5,670	-4.8
228	MANCHESTER, NEW HAMPSHIRE	MANCHESTER	5,515	-74.8
229	BOLOGNA	GUGLIEMO MARCONI	5,288	20.5
230	TENERIFE NORTE	TENERIFE NORTE	5,275	20.5
231	MAPUTO	MAPUTO INT'L	4,978	----
232	SPRINGFIELD, MISSOURI	SPRINGFIELD REGIONAL	4,905	27.7
233	MANADO	SAM RATULANGI	4,888	25.1
234	GREER, SOUTH CAROLINA	GREENVILLE-SPARTANBURG	4,857	-31.2
235	CLERMONT-FERRAND	AULNAT	4,842	-4.4
236	RIO BRANCO	PRES. MEDICE	4,813	24.8
237	ALICANTE	ALICANTE	4,719	-26.8
238	ZAGREB	ZAGREB	4,582	-67.5
239	NATAL	AUGUSTO SEVERO	4,481	4.6
240	MENORCA	MENORCA	4,462	-5.4
241	IBIZA	IBIZA	4,393	-6.1
242	BATON ROUGE, LOUISIANA	BATON ROUGE METROPOLITAN	4,388	1.2
243	BANGUI	BANGUI-M'POKO	4,131	-22.3
244	NANTES	NANTES ATLANTIQUE	4,099	----
245	MUENSTER	MUENSTER-OSNABRUECK	3,955	16.1
246	ROCHESTER, NEW YORK	ROCHESTER	3,823	23.3
247	MACAPA	MACAPA	3,772	-1.6
248	ADANA	ADANA	3,663	-33.4
249	LILLE	LILLE LESQUIN	3,567	-30.7
250	BRISTOL	BRISTOL	3,562	21.8
251	SAVANNAH, GEORGIA	SAVANNAH INT'L	3,454	-3
252	SANTIAGO DE COMPOSTELA	SANTIAGO DE COMPOSTELA	3,354	8.5

(Continued on next page)

AIRPORT RANKING BY TOTAL CARGO (METRIC TONNES) - 1991
(Continued)

RANK	CITY	AIRPORT	TOTAL CARGO	% CHG 1991/ 1990
253	MAZATLAN	GEN. RAFAEL BUELNA	3,308	-8.6
254	FOZ DE IGUACU	CATARATAS	3,255	6.5
255	SAN JOSE DEL CABO	LOS CABOS	3,246	49.7
256	LAS VEGAS, NEVADA	MCCARRAN INT'L	3,167	-8.4
257	MIDDLETOWN, PENNSYLVANIA	HARRISBURG INT'L	3,156	4.5
258	BIAK	FRANS KAISIEPO	3,123	-10
259	ACAPULCO	ACAPULCO	3,075	2.9
260	PAU PYRENEES	UZEIN	3,040	27.2
261	SEVILLA	SEVILLA	3,027	-21
262	NEWCASTLE	NEWCASTLE	3,007	-9.8
263	ILHA DO SAL	AMILCAR CABRAL INT'L	2,901	----
264	MIDLAND, TEXAS	MIDLAND INT'L	2,887	23.3
265	BREST-GUIPAVAS	BREST-GUIPAVAS	2,863	-27
266	BILBAO	BILBAO	2,816	0.4
267	PUERTO VALLARTA	ORDAZ	2,780	-13.5
268	FUERTEVENTURA	FUERTEVENTURA	2,657	-2.4
269	ANTALYA	ANTALYA	2,654	-32.5
270	LJUBLJANA	BRNIK	2,641	-41.9
271	COSTA MESA, CALIFORNIA	JOHN WAYNE	2,515	33.5
272	ST. GEORGE'S	POINT SALINE	2,413	17.7
273	HARLINGEN, TEXAS	VALLEY INT'L	2,333	-39.2
274	COLORADO SPRINGS, COLORADO	COLORADO SPRINGS	2,266	-27.4
275	JAKARTA	HALIM PERDANKUSUMA	2,208	-1.4
276	BOA VISTA	BOA VISTA	2,200	-58.3
277	LA PALMA	LA PALMA	2,023	-10.1
278	MOLINE, ILLINOIS	QUAD CITY	1,963	-22.8
279	FARO	FARO	1,940	3.3
280	CORUMBA	CORUMBA	1,767	-6.2
281	CARDIFF WALES	CARDIFF WALES	1,734	-47.6
282	MCALLEN, TEXAS	MCALLEN-MILLER INT'L	1,670	19.2
283	CRUZEIRO DO SUL	CAMPO INT'L	1,653	-22.1
284	WILMINGTON, NORTH CAROLINA	NEW HANOVER INT'L	1,649	11.9
285	GEORGE	P.W. BOTHA AIRPORT	1,587	0.3
286	KETCHIKAN, ALASKA	KETCHIKAN INT'L	1,392	-2.5
287	MISSOULA, MONTANA	MISSOULA COUNTY	1,386	19.6
288	BANGOR, MAINE	BANGOR INT'L	1,322	6.5
289	AKRON/CANTON, OHIO	AKRON-CANTON REGIONAL	1,298	-21
290	ERIE, PENNSYLVANIA	ERIE INT'L	1,155	24.9
291	SAN JUAN	ISLA GRANDE	1,039	-19.2
292	SOUTHAMPTON	SOUTHAMPTON	1,019	----
293	TRIESTE	DEI LEGIONARI	966	10
294	MELBOURNE, FLORIDA	MELBOURNE REGIONAL	961	85.4
295	KANSAS CITY, MISSOURI	KANSAS CITY DOWNTOWN	912	-1.9
296	PASCO, WASHINGTON	TRI-CITIES	892	0.7
297	ASHEVILLE, NORTH CAROLINA	ASHEVILLE REGIONAL	743	8.5
298	DAYTONA BEACH, FLORIDA	DAYTONA BEACH REGIONAL	661	-0.7
299	VITORIA	VITORIA	656	-26.5
300	BIARRITZ	BAYONNE-ANGLET	655	4.3
301	VIGO	VIGO	652	-10.6
302	AUGUSTA, GEORGIA	BUSH FIELD	630	2.9
303	MAYAGUEZ	EUGENIO M. DE HOSTOS	626	-8.2
304	FREELAND, MICHIGAN	TRI-CITY INT'L	574	14.4
305	CHEJU	CHEJU	522	54.4
306	KIMBERLEY	B.J. VORSTER AIRPORT	518	-32.8
307	MELILLA	MELILLA	514	-10
308	KALAMAZOO, MICHIGAN	KALAMAZOO/BATTLE CREEK INT'L	485	-3.5
309	JEREZ	JEREZ	469	-32.4
310	CASPER, WYOMING	NATRONA COUNTY INT'L	426	5.8
311	LEEDS BRADFORD	LEEDS BRADFORD	400	-34.6
312	ASTURIAS	ASTURIAS	376	-14.6
313	SAN SEBASTIAN	SAN SEBASTIAN	334	-8.7
314	MURCIA	MURCIA	331	-2.8
315	TABATINGA	TABATINGA	326	136.2

(Continued on next page)

AIRPORT RANKING BY TOTAL CARGO (METRIC TONNES) - 1991
(Continued)

RANK	CITY	AIRPORT	TOTAL CARGO	% CHG 1991/ 1990
316	HIERRO	HIERRO	307	7.2
317	VIEQUES	VIEQUES	303	8.6
318	BONAIRE	FLAMINGO	293	10.2
319	BEAUVAIS	BEAUVAIS-TILLE	285	15.4
320	ALMERIA	ALMERIA	276	-6.2
321	ARECIBO	ARECIBO	271	90.8
322	GAINESVILLE, FLORIDA	GAINESVILLE REGIONAL	258	2.6
323	SIOUX CITY, IOWA	SIOUX GATEWAY	256	-33.7
324	DURANGO, COLORADO	DURANGO-LA PLATA COUNTY	249	-3.7
325	LAFAYETTE, LOUISIANA	LAFAYETTE REGIONAL	248	4.6
326	LA CORUNA	LA CORUNA	208	23.7
327	SANTANDER	SANTANDER	198	-54.4
328	LONDON	LONDON CITY	195	52
329	GRANADA	GRANADA	195	-31.6
330	TARBES-LOURDES	TARBES-LOURDES	181	-33.7
331	UPINGTON	PIERRE VAN RYNEVELD	135	-39.9
332	FORT MYERS, FLORIDA	PAGE FIELD	131	78.6
333	PAMPLONA	PAMPLONA	118	16.4
334	LIMOGES	LIMOGES BELLEGARDE	114	----
335	RAS AL KHAIMAH	RAS AL KHAIMAH INT'L	98	-87.4
336	TEESIDE	TEESIDE INT'L	98	1
337	SPRINGFIELD, ILLINOIS	CAPITAL	91	-4
338	FAJARDO	FAJARDO	84	21.7
339	PONTA PORA	PONTA PORA	82	2.5
340	DALAMAN	DALAMAN	80	-1.2
341	SARASOTA, FLORIDA	SARASOTA-BRADENTON	79	-9.5
342	YUMA, ARIZONA	YUMA INT'L	68	----
343	VALLADOLID	VALLADOLID	49	-57.9
344	CULEBRA	CULEBRA	44	22.2
345	PADUCAH, KENTUCKY	BARKELY REGIONAL	38	-12
346	CHAMBERY	CHAMBERY/AIX	33	-2.1
347	HUMACAO	HUMACAO	32	-3
348	METZ-FRESCATY	METZ-FRESCATY	25	----
349	GERONA	GERONA	23	57432.5
350	ANGOULEME	ANGOULEME	20	-31
351	URUGUAIANA	RUBERN BERTA	12	-33.3
352	PERIGUEUX	PERIGUEUX-BASSILLAC	10	44.1
353	DOLE TAVAUX	DOLE TAVAUX	4	-73.3
354	LYON	BRON	3	-3.5
355	BERGERAC	ROUMANIERES	3	-22.5
356	VICHY-CHARMEIL	VICHY-CHARMEIL	2	----
357	REUS	REUS	1	-96.6
358	VALENCE	CHABEUIL	0	-74.9

Source: Airports Council International

(Continued on next page)

U.S. CIVIL AND JOINT-USE AIRCRAFT FACILITIES
BY TYPE AND STATE
As of December 31, 1991

State	TOTAL	Public	Paved	Lighted
Alabama	203	103	135	98
Alaska	545	419	61	145
Arizona	273	75	154	71
Arkansas	236	97	160	86
California	922	269	667	248
Colorado	373	84	160	84
Connecticut	134	27	83	27
Delaware	34	10	14	12
Dist. of Col.	16	2	14	4
Florida	698	130	307	145
Georgia	375	113	192	116
Hawaii	48	13	41	13
Idaho	211	122	78	47
Illinois	924	126	277	166
Indiana	576	116	163	120
Iowa	290	139	157	139
Kansas	394	148	135	133
Kentucky	144	69	92	58
Louisiana	413	88	240	75
Maine	159	78	49	33
Maryland	168	39	72	49
Massachusetts	197	51	114	43
Michigan	432	221	180	175
Minnesota	477	161	138	141
Mississippi	207	86	115	79
Missouri	462	145	221	141
Montana	227	126	97	87
Nebraska	307	98	106	93
Nevada	122	61	59	34
New Hampshire	79	27	46	19
New Jersey	329	57	143	50
New Mexico	170	72	79	50
New York	512	171	204	132
North Carolina	336	119	146	112
North Dakota	464	100	78	97
Ohio	714	194	279	190
Oklahoma	403	157	213	131
Oregon	378	103	154	76
Pennsylvania	752	153	309	141
Rhode Island	23	8	16	7
South Carolina	146	68	77	65
South Dakota	160	76	63	74
Tennessee	218	90	132	86
Texas	1,662	401	830	414
Utah	112	48	76	44
Vermont	70	17	17	11
Virginia	341	74	153	84
Washington	411	134	199	133
West Virginia	94	40	56	32
Wisconsin	472	144	178	140
Wyoming	96	41	48	37
50 States and DC --Total	**17,509**	**5,510**	**7,777**	**4,787**

(Continued on next page)

U.S. CIVIL AND JOINT-USE AIRCRAFT FACILITIES (Continued)
BY TYPE AND STATE
As of December 31, 1991

State	TOTAL	Public	Paved	Lighted
Puerto Rico	29	11	24	11
Virgin Islands	8	2	3	2
S. Pacific	35	28	18	11
TOTAL	**17,581**	**5,551**	**7,822**	4,811

FACILITIES BY CLASS

Class	Total	Public	Private
Airports	12,904	5,245	7,659
Heliports	4,199	98	4,101
Stolports	70	6	64
Seaplane Bases	408	202	206
Total Facilities	**17,581**	**5,551**	**12,030**

Source: Federal Aviation Administration, "FAA Statistical Handbook of Aviation" (Annually).
Included in these data are facilities having joint civil-military use.
"Public" refers to use, whether publicly or privately owned.

ACTIVE U.S. CIVIL AIRCRAFT
BY PRIMARY USE AND TYPE OF AIRCRAFT
As of December 31, 1990

Primary Use	TOTAL	Fixed Wing Turbojet	Turboprop	Piston	Rotorcraft	Other
TOTAL -- ALL AIRCRAFT	**218,312**	**8,522**	**7,247**	**188,102**	**7,408**	**7,032**
Air Carrier -- TOTAL	**6,083**	**4,148**	**1,595**	**329**	**11**	-
Large	4,665	4,145	438	82	-	-
Small	1,418	3	1,157	247	11	-
General Aviation -- TOTAL	**212,229**	**4,374**	**5,652**	**187,773**	**7,397**	**7,032**
Executive	10,906	3,204	2,861	3,933	863	45
Business	35,496	340	847	33,863	393	55
Commuter	1,242	-	466	643	126	7
Air Taxi	6,186	374	640	3,853	1,132	190
Instructional	19,889	4	38	18,603	877	367
Personal	120,636	115	262	113,429	1,369	5,459
Aerial Application	6,687	-	220	5,402	1,065	-
Aerial Observation	5,302	17	23	4,011	995	256
Other Work	1,525	-	16	1,041	224	245
Other	4,358	321	280	2,995	355	408

Source: Federal Aviation Administration, "FAA Statistical Handbook of Aviation" (Annually).

NOTE: Detail may not add to totals because of estimating procedures.
Definitions of "primary use" categories available in Glossary of "FAA Statistical Handbook."
Limited to single-engine commuters or Air taxis under 12,500 pounds. Otherwise, aircraft included in "Air Carrier."

HELIPORTS/HELIPADS IN THE UNITED STATES
By State
As of 1991

State	Total Helipads in state	Private Use Heliports & Helistops	Private Use Helipads at Airports	Public Use Heliports & Helistops	Public Use Helipads at Airports
Alabama	54	52	-	1	1
Alaska	32	15	3	6	8
Arizona	88	86	-	-	2
Arkansas	73	70	2	-	1
California	392	368	3	-	21
Colorado	165	159	1	1	4
Connecticut	73	68	1	3	1
Delaware	12	11	-	1	-
District of Columbia	16	16	-	-	-
Florida	190	189	-	-	1
Georgia	78	77	-	-	1
Hawaii	19	15	-	1	3
Idaho	28	26	-	1	1
Illinois	237	227	3	7	-
Indiana	102	96	2	3	1
Iowa	65	64	-	-	1
Kansas	31	27	-	-	4
Kentucky	30	30	-	-	-
Louisiana	201	193	2	5	1
Maine	17	15	-	2	-
Maryland	51	48	2	1	-
Massachusetts	109	105	-	2	2
Michigan	58	56	1	1	-
Minnesota	29	26	1	-	2
Mississippi	30	30	-	-	-
Missouri	103	95	1	4	3
Montana	19	18	-	1	-
Nebraska	23	22	1	-	-
Nevada	24	24	-	-	-
New Hampshire	32	30	-	1	1
New Jersey	199	195	-	4	-
New Mexico	19	17	-	2	-
New York	124	115	-	9	-
North Carolina	55	52	1	2	-
North Dakota	7	7	-	-	-
Ohio	196	173	1	17	5
Oklahoma	85	81	-	4	-
Oregon	84	80	2	2	-
Pennsylvania	273	264	1	8	-
Rhode Island	12	11	-	1	-
South Carolina	24	24	-	-	-
South Dakota	9	9	-	-	-
Tennessee	64	58	2	3	1
Texas	388	374	3	9	2
Utah	34	29	-	-	5
Vermont	17	17	-	-	-
Virginia	102	100	-	1	1
Washington	102	96	2	-	4
West Virginia	23	23	-	-	-
Wisconsin	63	63	-	-	-
Wyoming	13	12	-	-	1
Total U.S.	**4,239**	**4,058**	**35**	**103**	**78**

HELIPORTS/HELIPADS IN THE UNITED STATES (Continued)
By State
As of 1991

Source: Helicopter Association International, "1992 Helicopter Annual" (Annually).
NOTE: 96.6 percent of all U.S. helicopter landing areas are private, while 3.4 percent are public.
Excludes temporary heliports, offshore heliports, and infrequently used helicopter landing sites.
'Helipads at Airports' not updated since 1991/1992 edition.

Section V.

The Aerospace Industry

AEROSPACE INDUSTRY SALES BY CUSTOMER
Calendar Years 1977 - 1991 (Millions of Dollars)

Aerospace Products and Services

Year	TOTAL SALES	U.S. Government			Other Customers	Related Products and Services
		Total	Dept. of Defense	NASA and Other Agencies		
CURRENT DOLLARS						
1977	$ 32,199	$ 26,095	$14,368	$ 3,012	$ 8,715	$ 6,104
1978	37,702	30,889	15,533	3,151	12,205	6,813
1979	45,420	37,705	18,918	3,453	15,334	7,715
1980	54,697	45,878	22,795	4,106	18,977	8,819
1981	63,974	53,090	27,244	4,709	21,137	10,884
1982	67,756	56,366	34,016	4,899	17,451	11,390
1983	79,975	66,646	41,558	5,910	19,178	13,329
1984	83,486	69,572	45,969	6,063	17,540	13,914
1985	96,571	80,476	53,178	6,262	21,036	16,095
1986	106,183	88,486	59,161	6,236	23,089	17,697
1987	110,008	91,673	61,817	6,813	23,043	18,335
1988	114,562	95,468	61,327	7,899	26,242	19,094
1989	120,534	100,445	61,199	9,601	29,645	20,089
1990	134,375	111,979	60,502	11,097	40,379	22,396
1991	138,885	115,737	55,867	11,682	48,189	23,147
CONSTANT DOLLARS (1987 = 100)						
1977	$ 58,973	$ 47,793	$26,315	$ 5,516	$15,962	$11,179
1978	65,569	53,720	27,014	5,480	21,226	11,849
1979	71,528	59,378	29,792	5,438	24,148	12,150
1980	77,475	64,983	32,288	5,816	26,880	12,492
1981	80,470	66,780	34,269	5,923	26,587	13,691
1982	77,083	64,125	38,699	5,573	19,853	12,958
1983	86,741	72,284	45,074	6,410	20,800	14,457
1984	83,653	69,711	46,061	6,075	17,575	13,942
1985	97,843	81,536	53,878	6,344	21,313	16,307
1986	106,396	88,663	59,280	6,248	23,135	17,732
1987	110,008	91,673	61,817	6,813	23,043	18,335
1988	112,426	93,688	60,184	7,752	25,753	18,738
1989	113,604	94,670	57,680	9,049	27,941	18,934
1990	121,606	101,338	54,753	10,043	36,542	20,268
1991	121,191	100,992	48,750	10,194	42,050	20,198

Source: Aerospace Industries Association.

AEROSPACE INDUSTRY SALES BY PRODUCT GROUP
Calendar Years 1977 - 1991 (Millions of Dollars)

Year	TOTAL SALES	Aircraft Total	Civil	Military	Missiles	Space	Related Products & Services
CURRENT DOLLARS							
1977	$ 32,199	$16,988	$ 6,183	$10,805	$ 4,106	$ 5,001	$ 6,104
1978	37,702	21,074	8,222	12,852	4,098	5,717	6,813
1979	45,420	26,382	13,227	13,155	4,778	6,545	7,715
1980	54,697	31,464	16,285	15,179	6,469	7,945	8,819
1981	63,974	36,062	16,427	19,635	7,640	9,388	10,884
1982	67,756	35,484	10,982	24,502	10,368	10,514	11,390
1983	79,975	42,431	12,373	30,058	10,269	13,946	13,329
1984	83,486	41,905	10,690	31,215	11,335	16,332	13,914
1985	96,571	50,482	13,730	36,752	11,438	18,556	16,095
1986	106,183	56,405	15,718	40,687	11,964	20,117	17,697
1987	110,008	59,188	15,465	43,723	10,219	22,266	18,335
1988	114,562	60,886	19,019	41,867	10,270	24,312	19,094
1989	120,534	61,550	21,903	39,646	13,622	25,274	20,089
1990	134,375	71,353	31,362	40,091	14,180	26,446	22,396
1991	138,885	76,126	37,653	38,474	10,930	28,681	23,147
CONSTANT DOLLARS (1987 = 100)							
1977	$ 58,973	$31,114	$11,324	$19,789	$ 7,520	$ 9,159	$11,179
1978	65,569	36,650	14,299	22,351	7,127	9,943	11,849
1979	71,528	41,546	20,830	20,717	7,524	10,307	12,150
1980	77,475	44,567	23,067	21,500	9,163	11,254	12,492
1981	80,470	45,361	20,663	24,698	9,610	11,809	13,691
1982	77,083	40,369	12,494	27,875	11,795	11,961	12,958
1983	86,741	46,021	13,420	32,601	11,138	15,126	14,457
1984	83,653	41,989	10,711	31,278	11,358	16,365	13,942
1985	97,843	51,147	13,911	37,236	11,589	18,800	16,307
1986	106,396	56,518	15,749	40,769	11,988	20,157	17,732
1987	110,008	59,188	15,465	43,723	10,219	22,266	18,335
1988	112,426	59,751	18,664	41,086	10,079	23,859	18,738
1989	113,604	58,011	20,644	37,367	12,839	23,821	18,934
1990	121,606	64,573	28,382	36,281	12,833	23,933	20,268
1991	121,191	66,428	32,856	33,572	9,538	25,027	20,198

SALES OF MAJOR AEROSPACE COMPANIES
AS REPORTED BY THE BUREAU OF THE CENSUS
Calendar Years 1977 - 1991 (Millions of Dollars)

Year	GRAND TOTAL	TOTAL U.S. Gov't	Other	Aircraft, Engines, & Parts U.S. Gov't	Other	Missiles, Space, & Rocket Propulsion	Other Aerospace U.S. Gov't	Other	Non-Aerospace
CURRENT DOLLARS									
1977	$ 33,315	$20,704	$12,611	$ 8,848	$ 7,530	$ 5,775	$ 2,839	$2,219	$ 6,104
1978	37,968	21,888	16,080	8,724	10,581	6,380	3,363	2,107	6,813
1979	46,173	23,299	22,944	8,649	16,023	7,197	3,930	2,659	7,715
1980	58,440	26,674	31,766	9,427	20,097	8,393	6,869	2,609	11,045
1981	69,944	33,039	36,905	12,047	21,527	9,722	8,155	3,384	15,109
1982	75,487	42,239	33,248	15,120	16,766	11,980	9,909	4,953	16,759
1983	83,453	49,056	34,397	17,074	18,805	12,745	12,685	2,804	19,340
1984	88,941	55,777	33,164	20,216	17,069	13,624	12,734	2,768	22,530
1985	100,522	63,532	36,990	21,899	22,041	16,741	15,228	2,938	21,675
1986	105,577	65,326	40,251	22,755	25,002	17,535	16,243	3,564	20,478
1987	110,301	68,632	41,669	23,769	25,293	20,715	15,413	3,802	21,309
1988	113,548	68,104	45,444	21,316	29,426	21,514	16,103	3,225	21,964
1989	122,148	72,184	49,964	21,371	32,454	22,643	16,661	3,852	25,167
1990	136,646	73,552	63,094	24,614	41,675	22,040	15,862	4,253	28,202
1991	134,578	66,710	67,868	21,703	46,890	23,793	13,129	4,358	24,705
CONSTANT DOLLARS (1987 = 100)									
1977	$ 61,016	$37,919	$23,097	$16,205	$13,791	$10,577	$ 5,200	$4,064	$11,179
1978	66,031	38,066	27,965	15,172	18,402	11,096	5,849	3,664	11,849
1979	72,713	36,691	36,132	13,620	25,233	11,334	6,189	4,187	12,150
1980	82,776	37,782	44,994	13,353	28,466	11,888	9,729	3,695	15,644
1981	87,980	41,558	46,421	15,153	27,078	12,229	10,258	4,257	19,005
1982	85,878	48,053	37,825	17,201	19,074	13,629	11,273	5,635	19,066
1983	90,513	53,206	37,307	18,518	20,396	13,823	13,758	3,041	20,976
1984	89,119	55,889	33,230	20,257	17,103	13,651	12,760	2,774	22,575
1985	101,846	64,369	37,477	22,187	22,331	16,961	15,429	2,977	21,960
1986	105,789	65,457	40,332	22,801	25,052	17,570	16,276	3,571	20,519
1987	110,301	68,632	41,669	23,769	25,293	20,715	15,413	3,802	21,309
1988	111,431	66,834	44,597	20,919	28,877	21,113	15,803	3,165	21,554
1989	115,125	68,034	47,091	20,142	30,588	21,341	15,703	3,631	23,720
1990	123,662	66,563	57,099	22,275	37,715	19,946	14,355	3,849	25,522
1991	117,433	58,211	59,222	18,938	40,916	20,762	11,456	3,803	21,558

Source: Bureau of the Census, "Aerospace Industry (Orders, Sales, and Backlog),"

ORDERS AND BACKLOG OF MAJOR AEROSPACE COMPANIES
AS REPORTED BY THE BUREAU OF THE CENSUS
Calendar Years 1977 - 1991 (Millions of Dollars)

Year	TOTAL GRAND TOTAL	TOTAL U.S. Gov't	Other	Aircraft, Engines, & Parts U.S. Gov't	Aircraft, Engines, & Parts Other	Missiles, Space, & Rocket Propulsion	Other Aerospace U.S. Gov't	Other Aerospace Other	Non-Aerospace
NET NEW ORDERS									
1977	$ 38,922	$ 22,682	$ 16,240	$ 9,369	$ 11,193	$ 6,232	$ 3,554	$2,170	$ 6,404
1978	49,819	25,992	23,827	11,150	16,961	7,072	4,631	2,450	7,555
1979	67,561	28,107	37,101	8,762	30,695	7,609	5,184	4,487	8,471
1980	69,624	33,496	36,128	16,555	18,123	9,818	8,528	4,081	12,519
1981	74,922	42,431	32,491	16,946	17,911	12,376	9,350	3,250	15,089
1982	89,168	58,849	30,319	20,547	13,591	13,988	13,643	4,762	20,369
1983	91,647	60,290	31,357	22,171	16,428	14,248	15,209	2,641	20,950
1984	104,863	66,968	37,895	25,829	21,273	16,485	14,050	3,461	23,765
1985	110,968	70,240	40,728	23,751	26,191	20,328	14,730	2,800	23,168
1986	110,836	68,001	42,835	21,642	26,315	20,445	16,439	3,907	22,088
1987	121,224	66,264	54,960	17,019	35,328	26,272	13,899	4,658	24,048
1988	147,128	67,850	79,278	19,611	62,537	20,240	18,174	3,293	23,273
1989	173,635	80,633	93,002	25,421	71,170	26,820	17,713	4,046	28,465
1990	145,965	56,264	89,701	15,541	66,845	20,207	13,014	3,487	26,871
1991	132,644	66,668	65,976	22,654	43,517	25,512	10,675	5,110	25,176
BACKLOG AS OF DECEMBER 31									
1977	$ 45,309	$ 26,119	$ 19,190	$12,471	$ 12,592	$ 6,743	$ 2,761	$3,447	$ 7,295
1978	57,160	30,223	26,937	14,897	18,972	7,557	4,029	3,668	8,037
1979	78,548	36,136	42,123	17,316	33,168	7,388	5,613	5,112	9,662
1980	89,732	37,199	52,533	17,435	39,800	8,941	8,421	5,127	10,008
1981	94,710	46,591	48,119	21,292	35,022	11,255	9,052	4,940	13,149
1982	108,391	63,201	45,190	26,644	31,920	13,262	13,268	4,269	16,760
1983	116,585	74,435	42,150	30,688	29,684	14,962	18,489	3,684	19,078
1984	132,507	85,626	46,881	36,312	33,877	17,823	19,684	4,498	20,313
1985	142,953	92,334	50,619	38,150	38,041	21,410	18,937	4,609	21,806
1986	148,212	95,009	53,203	37,041	38,350	24,320	19,133	4,952	23,416
1987	158,650	92,439	66,211	30,323	49,692	30,544	17,888	5,653	24,550
1988	191,518	92,394	99,124	28,412	82,868	29,078	19,822	5,496	25,842
1989	252,401	107,797	144,604	36,320	122,830	33,771	23,558	8,280	27,642
1990	250,079	82,017	168,062	26,911	146,029	31,648	17,865	5,635	21,991
1991	247,597	84,827	162,770	30,859	141,426	32,981	15,199	6,027	21,105

Source: Bureau of the Census, "Aerospace Industry (Orders, Sales, and Backlog)"

SALES OF AIRCRAFT, ENGINES, AND PARTS
Calendar Years 1977 - 1991 (Millions of Dollars)

Year	GRAND TOTAL	TOTAL U.S. Gov't	Other	Complete Aircraft & Parts U.S. Gov't	Other	Aircraft Engines & Parts U.S. Gov't	Other
CURRENT DOLLARS							
1977	$16,378	$ 8,848	$ 7,530	$ 6,855	$ 5,670	$1,993	$ 1,860
1978	19,305	8,724	10,581	6,853	7,873	1,871	2,708
1979	24,672	8,649	16,023	6,378	12,701	2,271	3,322
1980	29,524	9,427	20,097	6,724	15,901	2,703	4,196
1981	33,574	12,047	21,527	8,197	16,877	3,850	4,650
1982	31,886	15,120	16,766	10,903	12,316	4,217	4,450
1983	35,879	17,074	18,805	12,898	14,419	4,176	4,386
1984	37,285	20,216	17,069	15,136	13,121	5,080	3,948
1985	43,940	21,899	22,041	17,783	16,466	4,116	5,575
1986	47,757	22,755	25,002	18,788	19,177	3,967	5,825
1987	49,062	23,769	25,293	18,131	18,899	5,638	6,394
1988	50,742	21,316	29,426	15,278	20,433	6,038	8,993
1989	53,825	21,371	32,454	15,340	23,056	6,031	9,398
1990	66,289	24,614	41,675	18,970	30,925	5,644	10,750
1991	68,593	21,703	46,890	16,029	36,950	5,674	9,940
CONSTANT DOLLARS (1987 = 100)							
1977	$29,996	$16,205	$13,791	$12,555	$10,385	$3,650	$ 3,407
1978	33,574	15,172	18,402	11,918	13,692	3,254	4,710
1979	38,854	13,620	25,233	10,044	20,002	3,576	5,231
1980	41,819	13,353	28,466	9,524	22,523	3,829	5,943
1981	42,231	15,153	27,078	10,311	21,229	4,843	5,849
1982	36,275	17,201	19,074	12,404	14,011	4,797	5,063
1983	38,914	18,518	20,396	13,989	15,639	4,529	4,757
1984	37,360	20,257	17,103	15,166	13,147	5,090	3,956
1985	44,519	22,187	22,331	18,017	16,683	4,170	5,648
1986	47,853	22,801	25,052	18,826	19,215	3,975	5,837
1987	49,062	23,769	25,293	18,131	18,899	5,638	6,394
1988	49,796	20,919	28,877	14,993	20,052	5,925	8,825
1989	50,730	20,142	30,508	14,458	21,730	5,684	8,858
1990	59,990	22,275	37,715	17,167	27,986	5,108	9,729
1991	59,854	18,938	40,916	13,987	32,243	4,951	8,674

Source: Bureau of the Census, "Aerospace Industry (Orders, Sales, and Backlog)"

ORDERS AND BACKLOG OF AIRCRAFT, ENGINES, AND PARTS
Calendar Years 1977 - 1991 (Millions of Current Dollars)

Year	GRAND TOTAL	TOTAL U.S. Gov't	Other	Complete Aircraft & Parts U.S. Gov't	Other	Aircraft Engines & Parts U.S. Gov't	Other
NET NEW ORDERS							
1977	$ 20,562	$ 9,369	$ 11,193	$ 6,507	$ 8,406	$2,862	$ 2,787
1978	28,111	11,150	16,961	9,055	14,229	2,095	2,732
1979	39,457	8,762	30,695	8,762	25,084	2,348	5,611
1980	34,678	16,555	18,123	11,606	14,427	4,949	3,696
1981	34,857	16,946	17,911	11,760	12,621	5,186	5,290
1982	34,138	20,547	13,591	15,978	10,540	4,569	3,051
1983	38,599	22,171	16,428	17,402	11,688	4,769	4,740
1984	47,102	25,829	21,273	19,228	18,148	6,601	3,125
1985	49,942	23,751	26,191	20,062	20,153	3,689	6,038
1986	47,957	21,642	26,315	17,361	20,083	4,281	6,232
1987	52,347	17,019	35,328	12,742	26,411	4,277	8,917
1988	82,148	19,611	62,537	12,862	46,393	6,749	16,144
1989	96,591	25,421	71,170	20,172	56,016	5,249	15,154
1990	82,386	15,541	66,845	10,572	54,565	4,969	12,280
1991	66,171	22,654	43,517	18,119	33,474	4,535	10,043
BACKLOG AS OF DECEMBER 31							
1977	$ 25,063	$12,471	$ 12,592	$ 9,557	$ 10,152	$2,914	$ 2,440
1978	33,869	14,897	18,972	11,759	16,508	3,138	2,464
1979	50,484	17,316	33,168	13,331	27,955	3,985	5,213
1980	57,235	17,435	39,800	12,702	33,258	4,733	6,542
1981	56,314	21,292	35,022	15,626	27,683	5,666	7,339
1982	58,564	26,644	31,920	20,626	25,980	6,018	5,940
1983	60,372	30,688	29,684	24,091	23,377	6,597	6,307
1984	70,189	36,312	33,877	28,183	28,404	8,129	5,473
1985	76,191	38,150	38,041	30,462	32,091	7,688	5,950
1986	76,391	37,041	39,350	29,035	32,997	8,006	6,353
1987	80,015	30,323	49,692	23,645	40,849	6,678	8,843
1988	111,280	28,412	82,868	21,083	66,782	7,329	16,086
1989	159,150	36,320	122,830	29,182	102,814	7,138	20,016
1990	172,940	26,911	146,029	20,382	126,000	6,529	20,029
1991	172,285	30,859	141,426	24,509	123,042	6,350	18,384

Source: Bureau of the Census, "Aerospace Industry (Orders, Sales, and Backlog)"

U.S. AIRCRAFT PRODUCTION - CIVIL
Calendar Years 1969 - 1991

Year	TOTAL	Domestic Shipments			Export Shipments		
		Trans-ports	Heli-copters	General Aviation	Trans-ports	Heli-copters	General Aviation
1969	13,505	332	282	9,996	182	252	2,461
1970	8,076	127	150	5,246	184	332	2,037
1971	8,158	50	171	5,900	173	298	1,566
1972	10,576	79	319	7,702	148	256	2,072
1973	14,709	143	342	10,482	151	428	3,163
1974	15,326	91	433	9,903	241	395	4,263
1975	15,251	127	528	10,804	188	336	3,268
1976	16,429	64	442	12,232	158	315	3,218
1977	17,913	54	527	13,441	101	321	3,469
1978	18,962	130	536	14,346	111	368	3,471
1979	18,460	176	570	13,177	200	459	3,878
1980	13,634	150	841	8,703	237	525	3,178
1981	10,916	132	619	6,840	255	453	2,617
1982	5,085	111	333	3,326	121	254	940
1983	3,356	133	187	2,172	129	216	519
1984	2,999	102	143	2,013	83	233	425
1985	2,691	126	247	1,545	152	137	484
1986	2,156	171	120	1,031	159	210	464
1987	1,800	187	116	598	170	242	487
1988	1,949	206	103	500	217	280	643
1989	2,448	138	221	225	260	294	1,310
1990	2,268	215	254	335	306	349	809
1991	2,181	204	253	487	385	318	534

Source: Aerospace Industries Association, based on company reports; General Aviation Manufacturers Association; and Department of Commerce, International Trade Administration.

CIVIL AIRCRAFT SHIPMENTS
Calendar Years 1977 - 1991

Year	TOTAL	Transport Aircraft	Helicopters	General Aviation
NUMBER OF AIRCRAFT SHIPPED				
1977	17,913	155	848	16,910
1978	18,962	241	904	17,817
1979	18,460	376	1,029	17,055
1980	13,634	387	1,366	11,881
1981	10,916	387	1,072	9,457
1982	5,085	232	587	4,266
1983	3,356	262	403	2,691
1984	2,999	185	376	2,438
1985	2,691	278	384	2,029
1986	2,155	330	330	1,495
1987	1,800	357	358	1,085
1988	1,949	423	383	1,143
1989	2,448	398	515	1,535
1990	2,268	521	603	1,144
1991	2,181	589	571	1,021
VALUE--Millions of Dollars				
1977	$ 4,451	$ 2,649	$251	$ 1,551
1978	6,458	4,308	328	1,822
1979	10,644	8,030	403	2,211
1980	13,058	9,895	656	2,507
1981	13,223	9,706	597	2,920
1982	8,610	6,246	365	1,999
1983	9,773	8,000	303	1,470
1984	7,717	5,689	330	1,698
1985	10,385	8,448	506	1,431
1986	11,858	10,308	288	1,262
1987	12,148	10,507	277	1,364
1988	15,855	13,603	334	1,918
1989	17,129	15,074	251	1,804
1990	24,477	22,215	254	2,008
1991	29,035	26,856	211	1,968

Source: Aerospace Industries Association, based on company reports and General Aviation Manufacturers' Association.

SHIPMENTS OF CIVIL TRANSPORT AIRCRAFT
Calendar Years 1987 - 1991

Company and Model	1987	1988	1989	1990	1991
TOTAL					
Number of Aircraft Shipped	**357**	**423**	**398**	**521**	**589**
Value (Millions of Dollars)	**$10,507**	**$13,690**	**$15,074**	**$22,215**	**$26,856**
Boeing--TOTAL	**257**	**289**	**279**	**379**	**420**
B-737	161	165	146	174	214
B-747	23	24	45	68	64
B-757	40	48	51	77	80
B-767	33	52	37	60	62
Lockheed--TOTAL	**2**	**5**	**-**	**-**	**-**
L-100	2	5	-	-	-
McDonnell Douglas--TOTAL	**98**	**129**	**119**	**142**	**169**
DC-10	3	8	1	-	-
MD-11	-	-	-	3	31
MD-80/90	95	121	118	139	138

Source: Aerospace Industries Association, based on company reports.

PERCENT OF CIVIL TURBOJET ENGINE MARKET
BY MANUFACTURER AND AIRCRAFT MODEL
as of December 1991

Aircraft Manufacturer and Model	Total Installed Engines	Engine Manufacturers					
		P&W	GE	RR	CFM	IAE	Other
TOTAL ENGINES	32,132	15,631	3,262	2,788	2,788	108	7,555
PERCENT SHARE	100.0%	48.7%	10.1%	8.7%	8.7%	0.3%	23.5%
Airbus A300	448	26%	74%	-%	-%	-%	-%
Airbus A300B4-200	254	12	88	-	-	-	-
Airbus A310	382	36	64	-	-	-	-
Airbus A320	460	-	-	-	77	23	-
Antonov AN-72	12	-	-	-	-	-	100
Antonov AN-74	2	-	-	-	-	-	100
Antonov AN-124	124	-	-	-	-	-	100
AS Corvette	16	100	-	-	-	-	-
AS Caravelle	112	71	-	29	-	-	-
AS/BAe CONCORDE	56	-	-	100	-	-	-
BAe 1-11	326	-	-	100	-	-	-
BAe 146	664	-	-	-	-	-	100
BAe HS Trident	96	-	-	100	-	-	-
BAe HS 125	42	-	-	43	-	-	57
Beech 400 Beechjet	2	100	-	-	-	-	-
Boeing B-707	176	93	-	7	-	-	-
Boeing B-707-320C	648	100	-	-	-	-	-
Boeing B-720	48	100	-	-	-	-	-
Boeing B-727 series	444	100	-	-	-	-	-
Boeing B-727	564	100	-	-	-	-	-
Boeing B-727C	351	100	-	-	-	-	-
Boeing B-727-200	840	100	-	-	-	-	-
Boeing B-727-200 ADV	2,694	100	-	-	-	-	-
Boeing B-737	494	48	-	-	52	-	-
Boeing B-737-200	378	100	-	-	-	-	-
Boeing B-737-200 ADV	1,440	100	-	-	-	-	-
Boeing B-737-300	1,394	-	-	-	100	-	-
Boeing B-737-400	380	-	-	-	100	-	-
Boeing B-747	1,560	47	33	20	-	-	-
Boeing B-747-100	660	95	-	5	-	-	-
Boeing B-747-200B	1,140	61	26	13	-	-	-
Boeing B-757	82	63	-	37	-	-	-
Boeing B-757-200	734	45	-	55	-	-	-
Boeing B-767	368	32	60	8	-	-	-
Boeing B-767-200	196	34	66	-	-	-	-
Boeing B-767-200ER	228	52	48	-	-	-	-
Canadair CL 600	2	-	-	-	-	-	100
Canadair CL 601	2	-	100	-	-	-	-
Cessna 500s	96	100	-	-	-	-	-
Cessna 650	8	-	-	-	-	-	100
Convair CV 880	8	-	100	-	-	-	-
Convair CV-990	4	-	100	-	-	-	-
Dassault Falcon	106	-	85	-	-	-	15
Dassault Mercure 100	22	100	-	-	-	-	-
Fokker F-28	178	-%	-%	100%	-%	-%	-%
Fokker F-28-4000	220	-	-	100	-	-	-
Fokker 100	192	-	-	100	-	-	-
Learjet 23	8	-	100	-	-	-	-
Learjet 24	20	-	100	-	-	-	-
Learjet 25	14	-	100	-	-	-	-
Learjet 35	42	-	-	-	-	-	100
Learjet 36	4	-	-	-	-	-	100
Learjet 55	2	-	-	-	-	-	100
Gulfstream II	16	-	-	100	-	-	-
Gulfstream III	10	-	-	100	-	-	-
IAI 1100s	26	-	-	-	-	-	100

(Continued on next page)

PERCENT OF CIVIL TURBOJET ENGINE MARKET
BY MANUFACTURER AND AIRCRAFT MODEL (Continued)
as of December 1991

Aircraft Manufacturer and Model	Total Installed Engines	Engine Manufacturers					
		P&W	GE	RR	CFM	IAE	Other
TOTAL ENGINES	**32,132**	**15,631**	**3,262**	**2,788**	**2,788**	**108**	**7,555**
PERCENT SHARE	**100.0%**	**48.7%**	**10.1%**	**8.7%**	**8.7%**	**0.3%**	**23.5%**
Ilyushin IL-62	336	-	-	-	-	-	100
Ilyushin IL-62M	572	-	-	-	-	-	100
Ilyushin IL-76	1,228	-	-	-	-	-	100
Ilyushin IL-86	316	-	-	-	-	-	100
Ilyushin IL-96-300	8	-	-	-	-	-	100
Lockheed JetStar	32	88	-	-	-	-	12
Lockheed L-1011	393	-	-	100	-	-	-
Lockheed L-1011-1	297	-	-	100	-	-	-
Douglas DC-8	1,200	66	-	-	34	-	-
Douglas DC-9	612	100	-	-	-	-	-
Douglas DC-9-30	1,064	100	-	-	-	-	-
Douglas DC-10	312	39	61	-	-	-	-
Douglas DC-10-10	348	-	100	-	-	-	-
Douglas DC-10-30	444	-	100	-	-	-	-
MDC MD-11	108	47	53	-	-	-	-
MDC MD-82	1,014	100	-	-	-	-	-
MDC MD-83	296	100	-	-	-	-	-
MDC MD-80s	586	100	-	-	-	-	-
Rockwell Sabre	6	100	-	-	-	-	-
Tupolev TU-134	172	-	-	-	-	-	100
Tupolev TU-134A	894	-	-	-	-	-	100
Tupolev TU-154	510	-	-	-	-	-	100
Tupolev TU-154B	339	-	-	-	-	-	100
Tupolev TU-154B2	936	-	-	-	-	-	100
Tupolev TU-154M	441	-	-	-	-	-	100
Yakolev YAK-40 series	27	-	-	-	-	-	100
Yakolev YAK-40	594	-	-	-	-	-	100
Yakolev YAK-42	249	-	-	-	-	-	100

Source: Aerospace Industries Association, based on data from Aviation Data Service.
Data for major (100 or more aircraft) series excluded and reported separately.

KEY: AS = Aerospatiale; BAe = British Aerospace; CFM = CFM International;
GE = General Electric; IAE = International Aero Engines;
IAI = Israel Aircraft Industries; MDC = McDonnell Douglas;
P&W = Pratt & Whitney; RR = Rolls-Royce.

U.S. AIRCRAFT PRODUCTION--MILITARY
Calendar Years 1969 - 1991

Year	TOTAL	U.S. Military Agencies	Exports Total	FMS	Direct
1969	4,290	3,644	646	NA	NA
1970	3,720	3,085	635	NA	NA
1971	2,914	2,232	682	NA	NA
1972	2,530	1,993	537	124	413
1973	1,821	1,243	578	129	449
1974	1,513	799	714	365	349
1975	1,779	844	935	525	410
1976	1,318	625	693	518	175
1977	1,134	454	680	408	272
1978	996	467	529	256	273
1979	837	531	306	203	103
1980	1,047	625	422	194	228
1981	1,062	703	359	215	144
1982	1,159	690	469	68	401
1983	1,053	766	287	70	217
1984	936	561	375	71	304
1985	919	643	276	134	142
1986	1,107	708	399	110	289
1987	1,210	725	485	133	352
1988	1,305	687	618	138	480
1989	1,261	614	647	92	555
1990	1,052	664	388	99	289
1991	877	514	363	93	270

Source: Aerospace Industries Association, based on USAF, USN, and USA survey responses and Department of Commerce, International Trade Administration.

Note: FMS = Foreign Military Service

MILITARY AIRCRAFT ACCEPTED BY U.S. MILITARY AGENCIES
Number and Flyaway Value Calendar Years 1977 - 1991

Year	TOTAL	Bomber/ Patrol/ Command/ Control	Fighter/ Attack	Trans- port/ Tanker	Trainer	Heli- copter	Other
NUMBER							
1977	862	44	488	25	12	273	20
1978	723	30	478	28	-	166	21
1979	734	17	529	16	-	158	14
1980	819	16	551	15	18	189	30
1981	918	19	649	17	60	158	15
1982	758	26	478	14	60	172	8
1983	836	34	421	22	120	233	6
1984	632	34	298	18	30	240	12
1985	777	34	409	25	-	306	3
1986	818	52	424	76	-	266	-
1987	858	74	483	36	-	265	-
1988	842	55	509	31	-	247	-
1989	706	24	408	21	-	253	-
1990	763	24	454	25	-	260	-
1991	607	18	394	23	-	172	-
FLYAWAY VALUE--Millions of Dollars							
1977	$ 4,364	$ 499	$3,190	$ 331	$14	$ 316	$14
1978	4,664	689	3,496	237	-	225	17
1979	5,470	442	4,660	136	-	219	13
1980	6,514	475	5,282	178	32	516	31
1981	8,446	526	6,518	509	32	825	19
1982	8,605	886	6,383	410	42	872	12
1983	9,640	1,259	6,708	575	79	1,009	10
1984	9,308	1,270	5,774	627	18	1,597	22
1985	14,122	3,640	7,923	838	-	1,715	6
1986	20,903	8,177	8,004	2,665	-	2,057	-
1987	21,459	8,569	8,900	2,218	-	1,772	-
1988	16,031	2,911	8,953	2,314	-	1,853	-
1989	11,968	1,423	7,735	743	-	2,067	-
1990	13,036	1,499	8,731	605	-	2,201	-
1991	11,774	1,447	8,579	437	-	1,311	-

Source: Aerospace Industries Association, based on USAF, USN, and USA survey responses.

MILITARY AIRCRAFT ACCEPTANCES BY UNITED STATES AIR FORCE
Calendar Years 1990 - 1991 (Costs in Millions of Dollars)

Type and Model	Number		Flyaway Cost		Weapon System Cost	
	1990	1991	1990	1991	1990	1991
AIR FORCE--TOTAL	**285**	**198**	**$5,150**	**$4,426**	**NA**	**NA**
Fighter/Attack--TOTAL	**239**	**171**	**$4,000**	**$3,597**	**NA**	**$4,907**
F-15	39	30	1,240	1,504	$1,492	2,210
F-16	200	141	2,760	2,093	NA	2,697
Bombers--TOTAL	**1**	**1**	**424**	**424**	**NA**	**NA**
B-2A	1	1	424	424	NA	NA
Transports/Tankers--TOTAL	**23**	**19**	**561**	**349**	**NA**	**NA**
C-27A	-	4	-	62	-	62
C-130H	21	11	433	232	NA	NA
MC-130H	-	4	-	55	-	74
VC-25A	2	-	128	-	152	-
Helicopters--TOTAL	**22**	**7**	**165**	**56**	**NA**	**NA**
MH-60G	22	7	165	56	NA	NA

Source: Department of the Air Force.
NOTE: Costs shown are approximate. Calendar year acceptances may derive from pro-
curement quantities funded in more than one fiscal year.
Air Force acceptances for own use; excludes FMS/MAP shipments.
Flyaway Cost includes airframe, engines, electronics, communications, armament,
other installed equipment, and non-recurring costs associated with the manufacture
of aircraft.
Weapon system cost includes flyaway costs, peculiar ground equipment, training
equipment, and technical data.
NA Not available.

MILITARY AIRCRAFT ACCEPTANCES BY UNITED STATES ARMY
Calendar Years 1990 - 1991 (Costs in Millions of Dollars)

Type and Model	Number		Flyaway Cost		Weapon System Cost	
	1990	1991	1990	1991	1990	1991
ARMY--TOTAL	**168**	**137**	**$1,283**	**$1,000**	**$1,468**	**NA**
Helicopters--TOTAL	**168**	**137**	**$1,283**	**$1,000**	**$1,468**	**NA**
UH-60A	72	72	372	353	427	NA
AH-64	96	65	911	647	1,041	NA

Source: Department of the Army.
Army acceptances for own use; excludes FMS/MAP shipments.
Flyaway cost includes airframes, engines, electronics, communications, armament and other installed equipment.
Weapon System Cost includes flyaway cost, initial spares, ground equipment, training equipment and other support items.
NA Not available.

MILITARY AIRCRAFT ACCEPTANCES BY UNITED STATES NAVY
Calendar Years 1990 - 1991 (Costs in Millions of Dollars)

Type and Model	Number		Flyaway Cost		Weapon System Cost	
	1990	1991	1990	1991	1990	1991
NAVY--TOTAL	**209**	**999**	**$5,249**	**$9,999**	**$6,708**	**$9,999**
Patrol--TOTAL	**23**	**17**	**$1,075**	**$1,023**	**$1,448**	**$1,221**
E-2C	7	6	340	317	427	380
E-6A	4	5	372	465	455	569
EA-6B	12	6	363	241	566	272
Fighter/Attack--TOTAL	**134**	**130**	**3,482**	**3,520**	**4,361**	**4,391**
F-14A	15	16	683	768	1,172	1,116
F/A-18	89	80	2,282	2,092	2,397	2,175
AV-8B	24	22	378	382	520	556
A-6E	6	12	139	278	272	544
Transports/Tankers--TOTAL	**2**	**4**	**44**	**88**	**45**	**98**
C-130T	-	2	-	41	-	48
KC-130	2	2	44	47	45	50
Helicopters--TOTAL	**52**	**999**	**648**	**9,999**	**854**	**9,999**
AH-1W	7	22	40	127	50	158
CH/MH-53E	14	6	209	128	238	152
HH-60H	3	99	32	999	35	999
SH-2F	-	99	-	999	-	999
SH-60B	9	99	121	999	176	999
SH-60F	19	99	241	999	355	999

Source: Department of the Navy.
Navy acceptances for own use; excludes FMS shipments.
Flyaway Cost includes airframe, engines, electronics, communications, armament, other installed equipment, non-recurring costs, and ancillary equipment.
Weapons System Cost (Investment Cost) includes flyaway cost, initial spares, ground equipment, training equipment, and other support items.

MILITARY AIRCRAFT ACCEPTANCES
FOR REIMBURSABLE PROGRAMS
Calendar Years 1990 - 1991 (Millions of Dollars)

Accepting Agency, Type, and Model	Number of Aircraft Accepted		Flyaway Cost	
	1990	1991	1990	1991
TOTAL ACCEPTANCES FOR REIMBURSABLE PROGRAMS	99	93	$1,354	$1,462
AIR FORCE--TOTAL	69	87	$ 973	$1,329
Fighter Attack--TOTAL	69	87	973	1,329
F-15	-	7	-	258
F-16 C/D	69	80	973	1,071
NAVY--TOTAL	17	6	$ 314	$ 133
Fighter/Attack--TOTAL	12	6	276	133
F/A-18	12	6	276	133
Helicopters--TOTAL	5	-	38	-
AH-1	5	-	38	-
ARMY--TOTAL	13	-	$ 67	$ -
Helicopters--TOTAL	13	-	67	-
UH-60	13	-	67	-

Source: Aerospace Industries Association, based on USAF, USN, and USA survey responses. Foreign government aircraft purchases through the Department of Defense Foreign Military Sales program.
Flyaway cost includes airframes, engines, electronics, communications, armament, other installed equipment, and nonrecurring costs associated with the manufacture of the aircraft.

MILITARY AIRCRAFT PROGRAM PROCUREMENT
Fiscal Years 1991, 1992, and 1993 (Millions of Dollars)

Agency and Model	1991		1992		1993	
	No.	Cost	No.	Cost	No.	Cost
AIR FORCE						
B-1B	-	$ 20.8	-	$ 62.5	-	$ 214.9
B-2 Stealth Bomber	2	2,348.4	1	2,798.2	4	2,686.6
C-17	-	260.0	4	1,695.9	8	2,719.5
C-27A	5	79.5	-	-	-	-
C-130H Hercules	-	-	9	289.8	8	300.4
Civil Air Patrol Aircraft	38	1.9	27	2.0	27	2.0
E-8A JSTARS	-	-	-	125.4	1	361.2
EFS	-	-	38	14.0	42	12.3
F-15E Eagle	42	2,158.0	3	773.5	-	11.5
F-16 Falcon	108	2,062.3	48	1,150.6	24	683.2
KC-135 Re-engining/Modern	30	612.8	26	541.3	28	375.2
MC-130H Combat Talon II	-	74.8	-	113.0	-	54.0
MH-60G Pave Hawk	4	36.9	6	23.5	10	30.1
T-1A (TTTS)	28	155.8	34	156.1	36	158.6
ARMY						
AH-64 Attack Helicopter	-	$ 88.4	-	$ 206.9	-	$ 147.8
CH-47 Modernization	-	300.0	-	283.9	-	15.0
OH-58D AHIP Modification	-	28.4	-	228.8	-	96.2
UH-60L Black Hawk	48	152.4	60	507.5	60	406.9
NAVY						
AH-1W Sea Cobra	8	$ 79.0	20	217.3	12	$ 123.9
AV-8B Harrier	21	490.2	6	230.0	-	-
CH/MH-53E Super Stallion	12	323.8	20	499.4	20	513.1
E-2C Hawkeye	6	392.6	6	499.5	-	96.2
EA-6B Prowler	1	349.6	-	115.2	3	530.0
F-14D Tomcat	12	996.4	-	172.5	-	143.1
F/A-18 Hornet	48	1,771.7	48	2,171.6	48	1,808.6
HH-60H	-	6.0	-	-	7	117.4
SH-60B Seahawk LAMPS MK-111	6	157.9	13	260.4	12	262.8
SH-60F CV ASW	18	269.9	12	250.0	12	261.5
T-45 Goshawk	-	157.8	12	325.9	12	303.5
SPECIAL OPERATIONS						
MH-47E	1	28.5	19	202.0	21	-
MH-60K	5	70.9	6	109.4	-	-

Source: Department of Defense Budget, "Program Acquisition Costs by Weapon System" (Annually).

NOTE: Total Obligational Authority for procurement, excluding initial spares.

NET PROFIT AFTER TAXES
AS A PERCENT OF SALES, ASSETS, AND EQUITY
FOR ALL MANUFACTURING CORPORATIONS
AND THE AEROSPACE INDUSTRY
Calendar Years 1977 - 1991

PERCENT OF SALES

Year	All Manufacturing Corporations	Non-Durable Goods	Durable Goods	Aerospace Industry
1977	5.3%	5.3%	5.3%	4.2%
1978	5.4	5.4	5.5	4.4
1979	5.7	6.1	5.2	5.0
1980	4.8	5.6	4.0	4.3
1981	4.7	5.1	4.3	4.4
1982	3.5	4.6	2.4	3.3
1983	4.1	4.9	3.1	3.5
1984	4.6	4.8	4.4	4.1
1985	3.8	4.1	3.4	3.1
1986	3.7	4.6	2.9	2.8
1987	4.9	5.2	4.5	4.1
1988	6.0	6.7	5.2	4.3
1989	5.0	5.8	4.1	3.3
1990	4.0	4.9	3.0	3.4
1991	2.5	4.2	0.6	1.8

PERCENT OF ASSETS AND EQUITY

Year	Percent of Assets		Percent of Equity	
	All Manufacturing	Aerospace Industry	All Manufacturing	Aerospace Industry
1977	7.6%	5.7%	14.2%	14.9%
1978	7.8	5.5	15.0	15.7
1979	8.4	6.3	16.5	18.4
1980	6.9	5.2	13.9	16.0
1981	6.7	5.2	13.6	16.0
1982	4.5	3.7	9.2	12.0
1983	5.1	4.1	10.5	12.1
1984	6.0	4.7	12.5	14.1
1985	4.6	3.6	10.1	11.1
1986	4.2	3.1	9.5	9.4
1987	5.6	4.4	12.8	14.6
1988	6.9	4.4	16.2	14.9
1989	5.6	3.3	13.7	10.7
1990	4.3	3.4	10.7	11.5
1991	2.6	1.9	6.4	6.1

Source: Bureau of the Census, "Quarterly Financial Report for Manufacturing, Mining, and Trade Corporations" (Quarterly).
Based on a sample of corporate entities classified in SIC codes 372 and 376, having as their principal activity the manufacture of aircraft, guided missiles, space vehicles, their propulsion, and parts.

INCOME STATEMENT AND OPERATING RATIOS
FOR AEROSPACE COMPANIES
Calendar Years 1988 - 1991 (Millions of Dollars)

INCOME STATEMENT	1988	1989	1990	1991
Net Sales, Receipts, Operating Revenues	$112,846	$118,297	$133,618	$135,175
Less: Depreciation, Depletion, & Amortization of Property, Plant, and Equipment	3,775	4,014	4,250	4,353
Less: All Other Operating Costs & Expenses, Including Selling Costs & General & Administrative Expenses	103,098	108,824	122,678	123,208
Income (or Loss) from Operations	$ 5,972	$ 5,460	$ 6,692	$ 7,614
Net Non-Operating Income (Expense)	739	(20)	(544)	(3,432)
Income (or Loss) Before Income Taxes (= Total Income)	$ 6,711	$ 5,439	$ 6,147	$ 4,181
Less: Provision for Current & Deferred Domestic Income Taxes	1,828	1,574	1,660	1,698
Income (or Loss) after Income Taxes (= Net Profit)	$ 4,883	$ 3,866	$ 4,487	$ 2,484
Cash Dividends Charged to Retained Earnings	1,465	1,806	1,823	1,678
Net Income Retained in Business	$ 3,417	$ 2,060	$ 2,665	$ 806
Retained Earnings at Beginning of Year	24,139	27,508	28,227	30,694
Adjustments to Retained Earnings	(66)	(931)	(350)	(707)
Retained Earnings at End of Year	$ 27,490	$ 28,637	$ 30,541	$ 30,793

OPERATING RATIOS

	1988	1989	1990	1991
Income before Taxes as Percent of Net Sales	5.9%	4.6%	4.6%	3.1%
Provision for Current & Deferred Domestic Income Taxes as Percent of Income before Taxes (Total Income)	27.2	28.9	27.0	40.6
Income after Taxes (Net Profit) as Percent of Net Sales	4.3	3.3	3.4	1.8
Income after Taxes (Net Profit) as Percent of Stockholders' Equity	14.9	10.7	11.5	6.1
Income after Taxes (Net Profit) as Percent of Total Assets	4.4	3.3	3.4	1.9

Source: Bureau of the Census, "Quarterly Financial Report for Manufacturing, Mining, and Trade Corporations" (Quarterly).

NOTE: Detail may not add to totals because of rounding.

Based on sample of corporate entities classified in SIC codes 372 and 376, having as their principal activity the manufacture of aircraft, guided missiles, space vehicles, and their propulsion, and parts.

Beginning-of-year retained earnings for any particular year do not equal end-of-year retained earnings for the previous year because of rotation of small companies in survey sample.

Other direct credits (or charges) to retained earnings (net), including stock and other non-cash dividends, etc.

Retained Earnings at End of Year CALCULATED AS Retained Earnings at Beginning of Year PLUS Income (Loss) after Income Taxes MINUS Cash Dividends Charged to Retained Earnings PLUS Adjustments to Retained Earnings.

Average of four quarters.

BALANCE SHEET FOR AEROSPACE COMPANIES
December 31, 1988 - 1991 (Millions of Dollars)

	1988	1989	1990	1991
Assets:				
Current Assets				
Cash	$ 2,156	$ 1,480	$ 2,172	$ 2,950
Securities, Com'l Paper, & Other				
Short-term Financial Investments	3,328	1,785	2,920	3,468
Total Cash and U.S. Gov't				
and Other Securities	$ 5,484	$ 3,264	$ 5,092	$ 6,418
Receivables (Total)	16,102	18,732	19,620	17,812
Inventories (Gross)	45,558	49,944	50,423	49,973
Other Current Assets	1,576	2,391	2,327	2,166
Total Current Assets	$ 68,720	$ 74,332	$ 77,463	$ 76,370
Net Plant, Property, & Equipment	22,211	24,506	26,161	26,557
Other Non-Current Assets	18,614	23,053	28,199	28,012
Total Assets	$109,545	$121,892	$131,823	$130,939
Liabilities:				
Current Liabilities				
Short Term Loans	$ 1,369	$ 3,799	$ 2,677	$ 1,943
Trade Accts. & Notes Payable	10,424	10,898	12,445	12,188
Income Taxes Accrued	3,519	1,925	2,002	1,151
Installments Due on				
Long Term Debts	751	1,269	1,392	1,767
Other Current Liabilities	40,825	43,813	44,690	44,823
Total Current Liabilities	$ 56,888	$ 61,704	$ 63,205	$ 61,871
Long Term Debt	12,447	16,191	20,979	20,682
Other Non-Current Liabilities	6,342	7,081	7,741	8,123
Total Liabilities	$ 75,676	$ 84,976	$ 91,926	$ 90,676
Stockholders' Equity:				
Capital Stock	$ 6,379	$ 8,661	$ 9,510	$ 9,681
Retained Earnings	27,490	28,255	30,386	30,581
Total Stockholders' Equity	$ 33,869	$ 36,916	$ 39,896	$ 40,262
Total Liabilities				
& Stockholders' Equity	$109,545	$121,892	$131,823	$130,939
Net Working Capital	$ 11,832	$ 12,628	$ 14,257	$ 14,499

Source: Bureau of the Census, "Quarterly Financial Report for Manufacturing, Mining, and Trade Corporations" (Quarterly).

NOTE: Detail may not add to totals because of rounding.
Based on sample of corporate entities classified in SIC codes 372 and 376, having as their principal activity the manufacture of aircraft, guided missiles, space vehicles, their propulsion, and parts.

NEW PLANT AND EQUIPMENT EXPENDITURES
Calendar Years 1964 - 1992 (Billions of Dollars)

			Aerospace		
Year	All Industries	All Manufacturing Industries	Durable Goods	Current Dollars	Constant Dollars (1982=100)
1964	$ 51.26	$ 21.23	$10.98	$0.41	$1.23
1965	59.52	25.41	13.49	0.53	1.57
1966	70.40	31.37	17.23	1.17	3.38
1967	72.75	32.25	17.83	1.25	3.49
1968	76.42	32.34	17.93	1.23	3.32
1969	85.74	36.27	19.97	1.29	3.37
1970	91.91	36.99	19.80	0.88	2.19
1971	92.91	33.60	16.78	0.63	1.51
1972	103.40	35.42	18.22	0.68	1.59
1973	120.03	42.35	22.63	0.79	1.79
1974	139.67	52.48	26.77	1.21	2.40
1975	142.42	53.66	25.37	1.19	2.04
1976	158.44	58.53	27.50	1.02	1.64
1977	184.82	67.48	32.77	1.14	1.72
1978	216.81	78.13	39.02	1.77	2.48
1979	255.26	95.13	47.72	2.71	3.50
1980	286.40	112.60	54.82	3.60	4.20
1981	324.73	126.68	58.93	3.40	3.59
1982	326.19	123.97	54.58	3.45	3.45
1983	321.16	117.35	51.61	2.95	2.87
1984	373.83	139.61	64.57	3.63	3.45
1985	410.12	152.88	70.87	3.51	3.27
1986	399.36	137.95	65.68	3.86	3.52
1987	410.52	141.06	68.03	3.60	3.22
1988	455.49	163.45	77.04	3.49	3.05
1989	507.40	183.80	82.56	4.17	3.51
1990	532.61	192.61	82.58	4.02	3.27
1991	529.20	183.61	77.95	4.04	3.19
1992 E	553.68	182.81	78.18	3.89	NA

Source: Bureau of the Census, "Plant and Equipment Expenditures and Plans" (Quarterly). Data are company-based (not establishment nor product-based) and represent corporate entities whose principal activity falls in SIC codes 372 and 376.
E Estimate.
NA Not Available.

AEROSPACE SALES AND THE NATIONAL ECONOMY
Calendar Years 1977 - 1991 (Billions of Dollars)

	Industry Sales				Aerospace Sales As Percent of		
Year	Gross Domestic Product	Manufac- turing	Durable Goods	Aerospace	GDP	Manufac- turing	Durable Goods

CURRENT DOLLARS

Year	Gross Domestic Product	Manufac- turing	Durable Goods	Aerospace	GDP	Manufac- turing	Durable Goods
1977	$1,974.1	$1,358.4	$710.0	$32.2	1.6	2.4	4.5
1978	2,232.7	1,522.9	812.8	37.7	1.7	2.5	4.6
1979	2,488.6	1,727.2	911.1	45.4	1.8	2.6	5.0
1980	2,708.0	1,852.7	929.0	54.7	2.0	3.0	5.9
1981	3,030.6	2,017.5	1,004.7	64.0	2.1	3.2	6.4
1982	3,149.6	1,960.2	950.5	67.8	2.2	3.5	7.1
1983	3,405.7	2,070.6	1,025.8	80.0	2.3	3.9	7.8
1984	3,777.2	2,288.2	1,175.3	83.5	2.2	3.6	7.1
1985	4,038.7	2,334.5	1,215.4	96.6	2.4	4.1	7.9
1986	4,268.6	2,335.9	1,238.9	106.2	2.5	4.5	8.6
1987	4,539.9	2,475.9	1,297.5	110.0	2.4	4.4	8.5
1988	4,900.4	2,682.5	1,415.9	114.6	2.3	4.3	8.1
1989	5,244.0	2,792.7	1,460.4	120.5	2.3	4.3	8.3
1990	5,513.7	2,873.5	1,468.6	134.4	2.4	4.7	9.1
1991	5,676.4	2,821.7	1,422.6	138.9	2.4	4.9	9.8

CONSTANT DOLLARS (1987 = 100)

Year	Gross Domestic Product	Manufac- turing	Durable Goods	Aerospace	GDP	Mfg.	Durs.	Aero.
1977	$3,533.4	$2,431.4	$1,270.8	$59.0	4.5%	7.2%	9.3%	0.8%
1978	3,703.3	2,525.9	1,348.1	65.6	4.8	3.9	6.1	11.2
1979	3,796.5	2,635.0	1,390.0	71.5	2.5	4.3	3.1	9.1
1980	3,776.3	2,583.6	1,295.5	77.5	(0.5)	(2.0)	(6.8)	8.3
1981	3,843.0	2,558.4	1,274.1	80.5	1.8	(1.0)	(1.7)	3.9
1982	3,760.3	2,340.3	1,134.8	77.1	(2.2)	(8.5)	(10.9)	(4.2)
1983	3,907.4	2,375.6	1,176.9	86.7	3.9	1.5	3.7	12.5
1984	4,148.5	2,513.1	1,290.8	83.7	6.2	5.8	9.7	(3.6)
1985	4,279.6	2,473.7	1,287.9	97.8	3.2	(1.6)	(0.2)	17.0
1986	4,404.3	2,410.1	1,278.2	106.4	2.9	(2.6)	(0.7)	8.7
1987	4,539.9	2,475.9	1,297.5	110.0	3.1	2.7	1.5	3.4
1988	4,718.7	2,583.0	1,363.4	112.4	3.9	4.3	5.1	2.2
1989	4,836.7	2,575.8	1,347.0	113.6	2.5	(0.3)	(1.2)	1.0
1990	4,885.0	2,545.9	1,301.2	121.6	1.0	(1.2)	(3.4)	7.0
1991	4,847.5	2,409.6	1,214.8	121.2	(0.8)	(5.3)	(6.6)	(0.3)

Source: Bureau of Economic Analysis, "Business Statistics" and "Survey of Current Business" (Monthly); and Aerospace Industries Association.

GROSS DOMESTIC PRODUCT, FEDERAL BUDGET, AND DEFENSE BUDGET
Fiscal Years 1962 - 1993 (Billions of Dollars)

Year	Fiscal Year GDP	Federal Budget Outlays Net Total	Federal Budget Outlays National Defense	Defense Outlays as percent of GDP	Defense Outlays as percent of Federal Budget
1962	$ 554.3	$ 106.8	$ 52.3	9.4%	49.0%
1963	585.0	111.3	53.4	9.1	48.0
1964	626.5	118.5	54.8	8.7	46.2
1965	671.4	118.2	50.6	7.5	42.8
1966	738.6	134.5	58.1	7.9	43.2
1967	791.3	157.5	71.4	9.0	45.4
1968	849.8	178.1	81.9	9.6	46.0
1969	926.6	183.6	82.5	8.9	44.9
1970	985.6	195.6	81.7	8.3	41.8
1971	1,051.6	210.2	78.9	7.5	37.5
1972	1,145.8	230.7	79.2	6.9	34.3
1973	1,278.0	245.7	76.7	6.0	31.2
1974	1,403.3	269.4	79.3	5.7	29.5
1975	1,511.0	332.3	86.5	5.7	26.0
1976	1,685.1	371.8	89.6	5.3	24.1
1977	1,919.7	409.2	97.2	5.1	23.8
1978	2,156.4	458.7	104.5	4.8	22.8
1979	2,431.9	503.5	116.3	4.8	23.1
1980	2,644.5	590.9	134.0	5.1	22.7
1981	2,964.7	678.2	157.5	5.3	23.2
1982	3,124.9	745.8	185.3	5.9	24.8
1983	3,317.0	808.4	209.9	6.3	26.0
1984	3,696.7	851.8	227.4	6.2	26.7
1985	3,970.9	946.4	252.7	6.4	26.7
1986	4,219.6	990.3	273.4	6.5	27.6
1987	4,453.3	1,003.9	282.0	6.3	28.1
1988	4,810.0	1,064.1	290.4	6.0	27.3
1989	5,170.1	1,144.2	303.6	5.9	26.5
1990	5,459.5	1,251.8	299.3	5.5	23.9
1991	5,626.6	1,323.0	273.3	4.9	20.7
1992	5,865.0	1,475.4	307.3	5.2	20.8
1993	6,231.6	1,515.3	291.4	4.7	19.2

Source: Office of Management and Budget, "The Budget of the United States Government" (Annually).

FUNDS FOR INDUSTRIAL RESEARCH AND DEVELOPMENT
ALL INDUSTRIES AND THE AEROSPACE INDUSTRY
By Funding Source
Calendar Years 1976 - 1990 (Millions of Dollars)

Year	All Industries			Aerospace Industry		
	Total	Federal Funds	Company Funds	Total	Federal Funds	Company Funds
CURRENT DOLLARS						
1976	$ 26,997	$ 9,561	$17,436	$ 6,339	$ 4,921	$1,418
1977	29,825	10,485	19,340	7,033	5,486	1,547
1978	33,304	11,189	22,115	7,536	5,713	1,823
1979	38,226	12,518	25,708	8,041	5,840	2,201
1980	44,505	14,029	30,476	9,198	6,628	2,570
1981	51,810	16,382	35,428	11,968	8,528	3,440
1982	58,650	18,545	40,105	14,451	10,265	4,186
1983	65,268	20,680	44,588	15,406	11,396	4,010
1984	74,800	23,396	51,404	18,858	14,094	4,764
1985	84,239	27,196	57,043	22,231	16,582	5,649
1986	87,823	27,891	59,932	21,050	14,984	6,066
1987	92,155	30,752	61,403	24,458	18,519	5,939
1988	97,889	32,117	65,772	25,900	19,877	6,023
1989	101,854	31,292	70,562	25,638	19,633	6,005
1990	104,344	30,580	73,764	25,357	19,217	6,140
CONSTANT DOLLARS (1987 = 100)						
1976	$ 51,649	$18,292	$33,358	$12,127	$ 9,415	$2,713
1977	53,383	18,767	34,616	12,588	9,819	2,769
1978	55,240	18,559	36,681	12,500	9,476	3,024
1979	58,316	19,097	39,219	12,267	8,909	3,358
1980	62,062	19,564	42,499	12,827	9,243	3,584
1981	65,699	20,774	44,925	15,176	10,814	4,362
1982	70,021	22,141	47,881	17,253	12,255	4,998
1983	74,883	23,726	51,156	17,676	13,075	4,601
1984	82,153	25,696	56,457	20,712	15,479	5,232
1985	89,265	28,818	60,446	23,557	17,571	5,986
1986	90,614	28,777	61,837	21,719	15,460	6,259
1987	92,155	30,752	61,403	24,458	18,519	5,939
1988	94,260	30,926	63,334	24,940	19,140	5,800
1989	93,944	28,862	65,082	23,647	18,108	5,539
1990	92,446	27,093	65,353	22,466	17,026	5,440

Source: National Science Foundation, "Annual Survey of Industrial Research and Development" (Annually).
NOTE: Detail may not add to totals because of rounding.
Includes all manufacturing industries, plus those non-manufacturing industries known to conduct or finance research and development.
Companies classified in SIC codes 372 and 376, having as their principal activity the manufacture of aircraft, guided missiles, space vehicles, and parts.
Company funds include all funds for industrial R&D work performed within company facilities except funds provided by the Federal Government. Excluded are company-financed research and development contracted to outside organizations such as research institutions, universities and colleges, or other non-profit organizations.
Based on GDP implicit price deflator.

RESEARCH AND DEVELOPMENT FUNDS AS PERCENT OF NET SALES
ALL MANUFACTURING INDUSTRIES AND THE AEROSPACE INDUSTRY
Calendar Years 1978 - 1990

	All Manufacturing Industries		Aerospace Industry	
Year	Total R&D Funds as Percent of Net Sales	Company R&D Funds as Percent of Net Sales	Total R&D Funds as Percent of Net Sales	Company R&D Funds as Percent of Net Sales
1978	2.9%	2.0%	13.3%	3.2%
1979	2.6	1.9	12.9	3.5
1980	3.0	2.1	13.7	3.8
1981	3.1	2.2	16.0	4.6
1982	3.8	2.6	17.1	5.1
1983	3.9	2.6	15.2	4.1
1984	3.9	2.6	15.4	4.0
1985	4.4	3.0	14.9	3.9
1986	4.7	3.2	13.4	4.0
1987	4.6	3.1	14.7	3.6
1988	4.7	3.1	15.6	3.6
1989	4.6	3.2	15.3	3.6
1990	4.6	3.2	14.3	3.5

Source: National Science Foundation, "Annual Survey of Industrial Research and Development" (Annually).
Includes all manufacturing industries known to conduct or finance research and development.
Companies classified in SIC codes 372 and 376, having as their principal activity the manufacture of aircraft, guided missiles, space vehicles, and parts.

FUNDS FOR INDUSTRIAL RESEARCH AND DEVELOPMENT IN THE AEROSPACE INDUSTRY
By Type of Research and Funding Source Calendar Years 1963 - 1990 (Millions of Dollars)

Year	TOTAL AERO-SPACE	Total	Federal Funds	Company Funds	Total	Federal Funds	Company Funds	Total	Federal Funds	Company Funds
1963	$ 4,712	$ 59	$ 31	$ 28	$ 735	$ 585	$ 150	$ 3,917	$ 3,634	$ 283
1964	5,078	67	34	34	766	607	159	4,244	3,948	296
1965	5,148	71	41	30	735	563	172	4,342	3,921	421
1966	5,526	69	36	33	773	563	210	4,685	4,162	523
1967	5,669	71	33	38	726	490	236	4,871	4,071	800
1968	5,765	68	26	42	677	426	251	5,021	4,145	876
1969	5,882	65	24	41	597	347	250	5,220	4,216	1,004
1970	5,219	63	20	43	565	352	213	4,591	3,718	873
1971	4,881	54	37	17	461	279	182	4,365	3,583	782
1972	4,950	60	44	16	451	267	184	4,438	3,722	716
1973	5,052	50	21	29	512	308	204	4,491	3,633	858
1974	5,278	51	19	32	609	360	249	4,617	3,735	882
1975	5,713	54	17	37	614	381	233	5,044	4,119	925
1976	6,339	54	21	33	666	365	301	5,619	4,521	1,098
1977	7,033	56	25	31	753	419	334	6,223	5,017	1,206
1979	8,041	86	44	42	880	499	381	7,076	5,314	1,762
1981	11,968	131	60	71	1,484	897	587	10,353	7,738	2,615
1983	13,853	146	NA	NA	3,466	NA	NA	10,241	7,668	2,573
1984	16,033	247	NA	NA	3,067	NA	NA	12,718	9,870	2,848
1985	17,619	304	162	142	3,785	2,776	1,009	13,530	10,483	3,047
1986	21,050	311	208	103	3,198	1,571	1,627	17,541	13,205	4,336
1987	24,488	425	335	90	2,949	1,709	1,239	21,115	16,475	4,640
1988	25,900	366	263	104	2,997	1,915	1,082	22,537	17,700	4,838
1989	25,638	653	537	116	3,065	2,107	957	21,921	16,989	4,932
1990	25,357	645	506	139	3,322	1,928	1,394	21,390	16,783	4,607

Source: National Science Foundation, "Annual Survey of Industrial Research and Development" (Annually).
NOTE: Detail may not add to totals because of rounding.
Break-outs by Research Type and Funding Source available only for odd-numbered years between 1977 and 1983.
NA Not available.

FEDERAL AERONAUTICS RESEARCH AND DEVELOPMENT
Fiscal Years 1973 - 1991 (Millions of Dollars)

Year	TOTAL	NASA	DOD	DOT
BUDGET AUTHORITY				
1973	$ 2,187	$ 313	$1,799	$ 75
1974	2,030	278	1,678	74
1975	2,015	314	1,627	74
1976	2,351	325	1,941	85
1977	2,727	378	2,256	93
1978	3,338	437	2,807	94
1979	2,850	519	2,240	91
1980	2,991	560	2,336	95
1981	3,286	526	2,653	106
1982	3,581	516	2,984	81
1983	3,871	547	3,221	103
1984	4,087	600	3,224	263
1985	4,355	648	3,422	265
1986	6,660	601	4,927	1,132
1987	5,824	698	4,179	946
1988	6,974	723	4,989	1,262
1989	10,656	872	8,240	1,544
1990	10,690	932	7,867	1,891
1991	9,416	968	6,148	2,300
OUTLAYS				
1982	$ 3,309	$ 563	$2,657	$ 89
1983	3,554	563	2,920	71
1984	3,727	586	2,995	146
1985	4,010	643	3,101	266
1986	6,071	648	4,373	1,050
1987	5,866	622	4,182	1,062
1988	6,340	679	4,448	1,213
1989	8,491	855	6,420	1,216
1990	10,009	889	7,649	1,471
1991	9,679	1,017	6,792	1,870

Source: NASA, "Aeronautics and Space Report of the President" (Annually).

TOTAL U.S. FUNDS FOR RESEARCH AND DEVELOPMENT
BY SOURCE AND PERFORMER
Calendar Years 1989 - 1992 (Millions of Current Dollars)

PERFORMER

Source of Funds	TOTAL All Per- formers	Federal Govern- ment	Industry	Colleges & Univers- ities	Federally- Funded Research & Develop- ment Centers	Non- Profit Insti- tutions
All Sources -- TOTAL	**$140,763**	**$15,121**	**$101,854**	**$15,009**	**$4,729**	**$4,050**
Federal Government	62,626	15,121	31,292	8,984	4,729	2,500
Industry	72,110	-	70,562	998	-	550
Colleges & Universities	3,946	-	-	3,946	-	-
Nonprofit Institutions	2,081	-	-	1,081	-	1,000
1990						
All Sources -- TOTAL	**$146,153**	**$16,003**	**$104,344**	**$16,325**	**$4,831**	**$4,650**
Federal Government	63,925	16,003	30,580	9,611	4,831	2,900
Industry	75,499	-	73,764	1,135	-	600
Colleges & Universities	4,356	-	-	4,356	-	-
Nonprofit Institutions	2,373	-	-	1,223	-	1,150
1991						
All Sources -- TOTAL	**$150,800**	**$16,500**	**$106,750**	**$17,450**	**$5,000**	**$5,100**
Federal Government	65,200	16,500	30,400	10,100	5,000	3,200
Industry	78,250	-	76,350	1,250	-	650
Colleges & Universities	4,750	-	-	4,750	-	-
Nonprofit Institutions	2,600	-	-	1,350	-	1,250
1992						
All Sources -- TOTAL	**$157,400**	**$17,600**	**$110,300**	**$19,000**	**$5,100**	**$5,400**
Federal Government	68,200	17,600	31,300	10,900	5,100	3,300
Industry	81,050	-	79,000	1,350	-	700
Colleges & Universities	5,250	-	-	5,250	-	-
Nonprofit Institutions	2,900	-	-	1,500	-	1,400

Source: National Science Foundation, "Annual Survey of Industrial Research and Development" (Annually).

FEDERAL OUTLAYS
DEFENSE, NASA, AND AEROSPACE PRODUCTS & SERVICES
Fiscal Years 1965 - 1993 (Millions of Dollars)

| Year | TOTAL National Defense | NASA | Federal Outlays for Aerospace Products & Services | | | Aerospace as Percent of Total National Defense and NASA |
			TOTAL	DOD	NASA	
1965	$ 50,620	$ 5,093	$11,858	$ 7,296	$ 4,562	21.3%
1966	58,111	5,933	14,065	8,704	5,361	22.0
1967	71,417	5,426	15,478	10,341	5,137	20.1
1968	81,926	4,724	16,279	11,681	4,598	18.8
1969	82,497	4,252	15,872	11,686	4,186	18.3
1970	81,692	3,753	14,559	10,860	3,699	17.0
1971	78,872	3,382	12,918	9,580	3,338	15.7
1972	79,174	3,423	12,309	8,936	3,373	14.9
1973	76,681	3,315	11,360	8,089	3,271	14.2
1974	79,347	3,256	11,168	7,987	3,181	13.5
1975	86,509	3,267	11,544	8,373	3,181	12.9
1976	89,619	3,669	12,364	8,816	3,548	13.3
1977	97,241	3,945	13,229	9,389	3,840	13.1
1978	104,495	3,983	13,926	10,067	3,859	12.8
1979	116,342	4,197	16,686	12,622	4,064	13.8
1980	133,995	4,852	20,269	15,558	4,711	14.6
1981	157,513	5,421	24,276	19,002	5,274	14.9
1982	185,309	6,035	29,501	23,575	5,926	15.4
1983	209,903	6,664	35,364	28,808	6,556	16.3
1984	227,413	7,048	39,663	32,723	6,940	16.9
1985	252,748	7,318	44,483	37,335	7,148	17.1
1986	273,375	7,404	49,773	42,558	7,215	17.7
1987	281,999	7,591	51,871	44,429	7,442	17.9
1988	290,361	9,092	48,848	39,922	8,926	16.3
1989	303,559	11,052	52,933	42,072	10,861	16.8
1990	299,331	12,429	53,202	40,992	12,210	17.1
1991	273,292	13,878	53,640	40,098	13,551	18.7
1992	307,304	13,819	50,214	36,843	13,371	15.6
1993	291,353	14,088	46,382	32,786	13,596	15.2

Source: Office of Management and Budget, "The Budget of the United States Government" (Annually)
Department of Defense, "Status of Funds" (Annual Summaries); and NASA, "Pocket Statistics" (Annually).

FEDERAL OUTLAYS FOR AEROSPACE PRODUCTS AND SERVICES
Fiscal Years 1965 - 1993 (Millions of Dollars)

Year	TOTAL	Department of Defense			NASA
		TOTAL	Aircraft	Missiles	
1965	$11,858	$ 7,296	$ 5,200	$ 2,096	$ 4,562
1966	14,065	8,704	6,635	2,069	5,361
1967	15,478	10,341	8,411	1,930	5,137
1968	16,279	11,681	9,462	2,219	4,598
1969	15,872	11,686	9,177	2,509	4,186
1970	14,559	10,860	7,948	2,912	3,699
1971	12,918	9,580	6,549	3,031	3,338
1972	12,309	8,936	5,927	3,009	3,373
1973	11,360	8,089	5,066	3,023	3,271
1974	11,168	7,987	5,006	2,981	3,181
1975	11,554	8,373	5,484	2,889	3,181
1976	12,364	8,816	6,520	2,296	3,548
1977	13,229	9,389	6,608	2,781	3,840
1978	13,926	10,067	6,971	3,096	3,859
1979	16,686	12,622	8,836	3,786	4,064
1980	20,269	15,558	11,124	4,434	4,711
1981	24,276	19,002	13,193	5,809	5,274
1982	29,501	23,575	16,793	6,782	5,926
1983	35,364	58,808	21,013	7,795	6,556
1984	39,663	32,723	23,196	9,527	6,940
1985	44,483	37,335	26,586	10,749	7,148
1986	49,773	42,558	30,828	11,730	7,215
1987	51,871	44,429	32,956	11,473	7,442
1988	48,848	39,922	28,246	11,676	8,926
1989	52,933	42,072	27,569	14,503	10,861
1990	53,202	40,992	26,142	14,851	12,210
1991	53,640	40,089	25,689	14,400	13,551
1992	50,214	36,843	23,950	12,893	13,371
1993	46,382	32,786	20,899	11,887	13,596

Source: Department of Defense, "Status of Funds" (Annual Summaries); Office of Management and Budget, "The Budget of the United States Goverment" (Annually); and NASA, "Pocket Statistics" (Annually).

FEDERAL OUTLAYS FOR CONDUCT OF
RESEARCH AND DEVELOPMENT
Fiscal Years 1979 - 1993 (Millions of Dollars)

Year		TOTAL	DOD	NASA	Energy	Other
CURRENT DOLLARS						
1979		$26,325	$11,045	$4,064	$ 4,692	$ 6,524
1980		30,235	13,469	4,711	4,808	7,247
1981		34,168	15,739	5,279	4,381	8,769
1982		34,660	18,363	3,220	5,178	7,899
1983		35,900	20,566	2,538	4,924	7,872
1984		40,986	23,850	3,539	5,182	8,415
1985		47,216	28,165	2,970	6,954	9,127
1986		52,141	33,396	3,432	5,392	9,921
1987		53,256	34,732	3,250	5,262	10,012
1988		56,100	35,605	3,832	5,332	11,331
1989		60,760	37,819	4,975	5,681	12,285
1990		63,810	38,247	6,325	5,957	13,281
1991		65,965	35,330	7,072	9,674	13,889
1992	E	70,855	36,934	7,272	11,121	15,528
1993	E	74,457	38,912	7,710	11,179	16,656
CONSTANT DOLLARS (1987 = 100)						
1979		$40,160	$16,850	$6,200	$ 7,158	$ 9,953
1980		42,163	18,783	6,570	6,705	10,106
1981		43,327	19,958	6,694	5,555	11,120
1982		41,380	21,923	3,844	6,182	9,431
1983		41,189	23,596	2,912	4,577	9,032
1984		45,015	26,194	3,887	4,346	9,242
1985		50,033	29,845	3,147	4,393	9,672
1986		53,798	34,457	3,541	4,109	10,236
1987		53,256	34,732	3,250	3,967	10,012
1988		54,020	34,285	3,690	5,134	10,911
1989		56,041	34,882	4,589	5,240	11,331
1990		56,534	33,886	5,604	5,278	11,767
1991		56,332	30,171	6,039	8,261	11,861
1992	E	58,679	30,587	6,022	9,210	12,860
1993	E	59,656	31,177	6,177	8,957	13,345

Source: Office of Management and Budget, "The Budget of the United States Government" (Annually).

NOTE: Detail may not add to totals because of rounding.
E Estimate. Latest year reflects Administration's budget proposal.

Section VI.

Department
of
Defense

100 COMPANIES LISTED ACCORDING
TO NET VALUE OF PRIME CONTRACT AWARDS
FISCAL YEAR 1992

RANK	COMPANIES	Thousands of Dollars	Percent of Total	Cumu-lative Percent of Total
	TOTAL	121,437,966	100	100
	TOTAL, 100 COMPANIES	75,412,007	62.09	62.09
1	MCDONNELL DOUGLAS CORPORATION	5,311,151	4.37	4.37
2	NORTHROP CORPORATION	4,851,015	3.99	8.36
3	LOCKHEED CORPORATION	4,650,404	3.82	12.19
4	GENERAL DYNAMICS CORPORATION	4,463,771	3.67	15.87
5	GENERAL ELECTRIC COMPANY	4,007,764	3.3	19.17
6	GENERAL MOTORS CORPORATION	3,694,122	3.04	22.21
7	RAYTHEON COMPANY	2,840,686	2.33	24.55
8	UNITED TECHNOLOGIES CORP	2,802,878	2.3	26.86
9	BOEING COMPANY THE	2,495,191	2.05	28.91
10	MARTIN MARIETTA CORPORATION	2,356,077	1.94	30.85
11	LITTON INDUSTRIES INC	2,334,436	1.92	32.78
12	GRUMMAN CORPORATION	2,182,739	1.79	34.57
13	LORAL CORP	1,815,445	1.49	36.07
14	AMERICAN TELEPHONE & TELG CO	1,337,587	1.1	37.17
15	ROCKWELL INTERNATIONAL CORP	1,233,435	1.01	38.18
16	TEXTRON INC	1,161,072	0.95	39.14
17	BATH HOLDING COMPANY	1,148,372	0.94	40.09
18	WESTINGHOUSE ELECTRIC CORP	1,147,210	0.94	41.03
19	TRW INC	1,012,521	0.83	41.86
20	INTERNATIONAL BUS MCHS CORP	931,776	0.76	42.63
21	UNISYS CORPORATION	834,011	0.68	43.32
22	ITT CORPORATION	797,273	0.65	43.98
23	FOUNDATION HEALTH CORPORATION	761,262	0.62	44.6
24	TEXAS INSTRUMENTS INCORPORATED	730,601	0.6	45.2
25	GTE CORPORATION	724,489	0.59	45.8
26	SCIENCE APPLICATIONS INTL CORP	685,606	0.56	46.37
27	ALLIANT TECHSYSTEMS INC	609,676	0.5	46.87
28	TENNECO INC	584,822	0.48	47.35
29	OLIN CORPORATION	573,331	0.47	47.82
30	E-SYSTEMS INC	500,880	0.41	48.23
31	COMPUTER SCIENCES CORPORATION	495,419	0.4	48.64
32	RENCO GROUP INC	462,143	0.38	49.02
33	ALLIED SIGNAL INC	458,869	0.37	49.4
34	BOEING CO & SIKORSKY ACFT JV	458,619	0.37	49.78
(1) 35	CFM INTERNATIONAL INC	447,648	0.36	50.15
36	FMC CORPORATION	447,616	0.36	50.51
37	DYNCORP	446,517	0.36	50.88
38	MITRE CORPORATION	434,253	0.35	51.24
39	JOHNS HOPKINS UNIVERSITY	406,267	0.33	51.57
40	TELEDYNE INC	401,602	0.33	51.91
41	MASSACHUSETTS INST OF TCHNOLOG	389,144	0.32	52.23
42	JOHNSON CONTROLS INC	373,738	0.3	52.53
43	OSHKOSH TRUCK CORP	372,228	0.3	52.84
44	HERCULES INCORPORATED	364,165	0.29	53.14
45	MOTOROLA INC	353,384	0.29	53.43
46	ROYAL DUTCH SHELL GROUP OF COS	344,473	0.28	53.71
47	GENCORP INC	339,078	0.27	53.99
48	HARRIS CORPORATION	307,604	0.25	54.25
49	COASTAL CORPORATION	306,543	0.25	54.5
50	EXXON CORPORATION	306,116	0.25	54.75
51	PENN CENTRAL CORPORATION	297,826	0.24	55
52	HONEYWELL INC	296,372	0.24	55.24
53	STEWART & STEVENSON SVCS INC	270,414	0.22	55.46
54	HARSCO CORPORATION	264,731	0.21	55.68
55	BLACK & DECKER CORPORATION	259,951	0.21	55.9
56	ATLANTIC RICHFIELD COMPANY	258,479	0.21	56.11
(2) 57	FED EX PAN AM NORTHWEST ETAL	256,349	0.21	56.32
58	AEROSPACE CORPORATION	253,455	0.2	56.53

(Continued on next page)

100 COMPANIES LISTED ACCORDING
TO NET VALUE OF PRIME CONTRACT AWARDS (Continued)
FISCAL YEAR 1992

RANK	COMPANIES	Thousands of Dollars	Percent of Total	Cumulative Percent of Total
59	THIOKOL CORPORATION	245,998	0.2	56.73
60	MIP INSTANDSETZUNGSBETRIC	231,506	0.19	56.92
61	EG&G INC	229,430	0.18	57.11
62	GENERAL ELECTRIC CO PLC THE	223,504	0.18	57.29
63	BDM HOLDINGS INC	209,348	0.17	57.47
64	NV PHILIPS GLOEILAMPENFABREIKN	199,224	0.16	57.63
65	NATIONAL STEEL & SHIPBLDG CO	197,300	0.16	57.79
66	MONTECATINI EDISON SPA IN ABBR	197,270	0.16	57.96
67	LOGICON INC	197,157	0.16	58.12
68	C A E INDUSTRIES LTD	196,060	0.16	58.28
69	CERIDIAN CORPORATION	195,749	0.16	58.44
70	ASTRONAUTICS CORP OF AMERICA	192,035	0.15	58.6
71	ESCO ELECTRONICS CORPORATION	186,818	0.15	58.75
72	HALLIBURTON COMPANY	185,876	0.15	58.91
73	AVONDALE INDUSTRIES INC	173,080	0.14	59.05
(3) 74	WORLD ROSEN KEY AMER ETAL J V	169,054	0.13	59.19
75	TRACOR INC	167,004	0.13	59.32
76	MARTIN MARIETTA WESTNGHSE JV	163,985	0.13	59.46
77	EASTMAN KODAK COMPANY	161,972	0.13	59.59
78	SOUTHWEST MARINE INC	161,607	0.13	59.73
79	CHRYSLER CORPORATION	156,721	0.12	59.86
80	AMOCO CORPORATION	154,737	0.12	59.98
81	CSX CORPORATION	143,854	0.11	60.1
(4) 82	NATIONAL PROGRAM OFFICE	142,804	0.11	60.22
83	BOOZ ALLEN & HAMILTON INC	141,446	0.11	60.34
84	SEQUA CORPORATION	140,330	0.11	60.45
85	INTERNATIONAL SHIPHOLDING CORP	136,113	0.11	60.56
86	DIGITAL EQUIPMENT CORPORATION	134,175	0.11	60.67
87	MOBIL CORPORATION	131,554	0.1	60.78
88	FORSTMANN LITTLE & CO	130,836	0.1	60.89
89	CHEVRON CORPORATION	126,700	0.1	60.99
90	ISRAEL AIRCRAFT INDUSTRIES LTD	126,543	0.1	61.1
91	EATON CORPORATION	125,777	0.1	61.2
92	BLACKSTONE GROUP LP	124,677	0.1	61.3
93	XEROX CORP	123,714	0.1	61.41
94	DUCHOSSOIS ENTERPRISES INC	123,444	0.1	61.51
95	ENSERCH CORPORATION	121,509	0.1	61.61
96	TRINITY INDUSTRIES INC	121,360	0.09	61.71
97	UNITED INDUSTRIAL CORPORATION	118,687	0.09	61.81
98	ARVIN INDUSTRIES INC	117,605	0.09	61.9
99	VINNELL CORP BRONW & ROOT JV	116,446	0.09	62
100	INFOTEC DEVELOPMENT INC	116,392	0.09	62.09

FOOTNOTES:
(1)   A joint venture of SNECMA (France) and General Electric Co., USA.
(2)   A joint venture of Federal Express, Northwest Airlines, Pan Am World Airways, Tower Air, and United Parcel Services.
(3)   A joint venture of World Airways, Rosenbalm Aviation, Evergreen International Airlines, and Emery Worldwide.
(4)   A joint venture of McDonnell Douglas Corp., Rockwell International Corp., General Dynamics Corp., Rocketdyne, and Pratt and Whitney.

INDEX OF 100 PARENT COMPANIES WHICH RECEIVED
THE LARGEST DOLLAR VOLUME OF PRIME CONTRACT AWARDS
FISCAL YEAR 1992

RANK	PARENT COMPANY	RANK	PARENT COMPANY
58	AEROSPACE CORPORATION	20	INTERNATIONAL BUS MCHS CORP
27	ALLIANT TECHSYSTEMS INC	85	INTERNATIONAL SHIPHOLDING CORP
33	ALLIED SIGNAL INC	90	ISRAEL AIRCRAFT INDUSTRIES LTD
14	AMERICAN TELEPHONE & TELG CO	22	ITT CORPORATION
80	AMOCO CORPORATION	39	JOHNS HOPKINS UNIVERSITY
98	ARVIN INDUSTRIES INC	42	JOHNSON CONTROLS INC
70	ASTRONAUTICS CORP OF AMERICA	11	LITTON INDUSTRIES INC
56	ATLANTIC RICHFIELD COMPANY	3	LOCKHEED CORPORATION
73	AVONDALE INDUSTRIES INC	67	LOGICON INC
17	BATH HOLDING CORPORATION	13	LORAL CORP
63	BDM HOLDINGS INC	10	MARTIN MARIETTA CORPORATION
55	BLACK & DECKER CORPORATION	76	MARTIN MARIETTA WESTNGHSE JV
92	BLACKSTONE GROUP LP	41	MASSACHUSETTS INST OF TCHNOLOG
34	BOEING CO & SIKORSKY ACFT JV	1	MCDONNELL DOUGLAS CORPORATION
83	BOOZ ALLEN & HAMILTON INC	60	MIP INSTANDSETZUNGSBETRIC
68	C A E INDUSTRIES LTD	38	MITRE CORPORATION
69	CERIDIAN CORPORATION	87	MOBIL CORPORATION
35	CFM INTERNATIONAL INC	66	MONTECATINI EDISON SPA IN ABBR
89	CHEVRON CORPORATION	45	MOTOROLA INC
79	CHRYSLER CORPORATION	82	NATIONAL PROGRAM OFFICE
49	COASTAL CORPORATION	65	NATIONAL STEEL & SHIPBLDG CO
31	COMPUTER SCIENCES CORPORATION	2	NORTHROP CORPORATION
81	CSX CORPORATION	64	NV PHILIPS GLOEILAMPENFABREIKN
86	DIGITAL EQUIPMENT CORPORATION	29	OLIN CORPORATION
94	DUCHOSSOIS ENTERPRISES INC	43	OSHKOSH TRUCK CORP
37	DYNCORP	51	PENN CENTRAL CORPORATION
30	E-SYSTEMS INC	7	RAYTHEON COMPANY
77	EASTMAN KODAK COMPANY	32	RENCO GROUP INC
91	EATON CORPORATION	15	ROCKWELL INTERNATIONAL CORP
61	EG&G INC	46	ROYAL DUTCH SHELL GROUP OF COS
95	ENSERCH CORPORATION	26	SCIENCE APPLICATIONS INTL CORP
71	ESCO ELECTRONICS CORPORATION	84	SEQUA CORPORATION
50	EXXON CORPORATION	78	SOUTHWEST MARINE INC
57	FED EX PAN AM NORTHWEST ETAL	53	STEWART & STEVENSON SVCS INC
36	FMC CORPORATION	40	TELEDYNE INC
88	FORSTMANN LITTLE & CO	28	TENNECO INC
23	FOUNDATION HEALTH CORPORATION	24	TEXAS INSTRUMENTS INCORPORATED
47	GENCORP INC	16	TEXTRON INC
4	GENERAL DYNAMICS CORPORATION	59	THIOKOL CORPORATION
62	GENERAL ELECTRIC CO PLC THE	75	TRACOR INC
5	GENERAL ELECTRIC COMPANY	96	TRINITY INDUSTRIES INC
6	GENERAL MOTORS CORPORATION	19	TRW INC
12	GRUMMAN CORPORATION	21	UNISYS CORPORATION
25	GTE CORPORATION	97	UNITED INDUSTRIAL CORPORATION
72	HALLIBURTON COMPANY	8	UNITED TECHNOLOGIES CORP
48	HARRIS CORPORATION	99	VINNELL CORP BROWN & ROOT JV
54	HARSCO CORPORATION	18	WESTINGHOUSE ELECTRIC CORP
44	HERCULES INCORPORATED	74	WORLD ROSEN KEY AMER ETAL J V
52	HONEYWELL INC	93	XEROX CORP
100	INFOTEC DEVELOPMENT INC		

THE DEFENSE BUDGET
($ in billions)

	FY 1991 ACTUAL	FY 1992 ESTIMATE	FY 1993 ESTIMATE
Budget Authority	290.9	270.9	267.6
Outlays	262.4	282.6	272.8

DoD's Slice of the Dollar
DEFENSE OUTLAYS AS A PERCENT OF
FISCAL YEAR

	GROSS NATIONAL PRODUCT	FEDERAL OUTLAYS	NET PUBLIC SPENDING*
1955	9.1	51.5	34.5
1960	8.2	45	28.8
1965	6.8	38.8	23.8
1970	7.8	39.4	23.6
1975	5.6	25.5	15.1
1980	5	22.5	13.8
1981	5.2	23	14.4
1982	5.8	24.5	15.5
1983	6.2	25.4	16.1
1984	6	25.9	16.3
1985	6.2	25.9	16.4
1986	6.3	26.8	16.6
1987	6.2	27.3	16.5
1988	5.9	26.5	16.4
1989	5.7	25.6	16
1990	5.3	23.2	14.4
1991	5.0	20.4	13
1992**	4.8	19.8	12.4
1993**	4.5	18	11.3

* Federal, state and local net spending excluding government enterprises (such as the U.S. Postal Service and public utilities) except for any support these activities receive from tax funds.
**Estimated.

BREAKOUT OF THE BUDGET
(Current $ in billions)
BUDGET AUTHORITY
APPROPRIATION TITLE

	FY 1991* ACTUAL	FY 1992* ESTIMATE	FY 1993 ESTIMATE
Military Personnel (includes retired pay)	84.2	79.2	77
Operation and Maintenance	131.9	92.5	86.5
Procurement	71.7	60.5	54.4
Research, Development, Test and Evaluation	36.1	37	38.8
Military Construction	5.2	4.9	6.2
Family Housing	3.3	3.7	4
Defensewide Contingency			-0.1
Revolving and Management Funds	2.7	4.3	1.6
Trust and Receipts	-44.3	-5.7	-0.8
Deduct, Intragov't. Receipt	-0.03	-0.18	-0.03
Total	**290.9**	**276.3**	**267.6**

* In fiscal 1991-92, increases in budget authority, especially Operations and Maintenance, were due to incremental costs of Operation Desert Shield/Storm. The fiscal 1991-92 rise in receipts reflects offsetting allied contributions. Fiscal 1991 Operations and Maintenance also includes $15 billion appropriated for the Persian Gulf Regional Defense Fund. From this fund, only $300 million was spent, and that was for refugee assistance, not for Operation Desert Shield/Storm.

(Continued on next page)

THE DEFENSE BUDGET (Continued)

**DoD's Budget by Component
(Current $ in millions)
BUDGET AUTHORITY**

	FY 1991** ACTUAL	FY 1992** ESTIMATE	FY 1993** ESTIMATE
Department of the Army*	91,825	71,163	63,325
Department of the Navy*	103,470	87,090	84,586
Department of the Air Force*	91,257	82,545	83,859
Defense Agencies/OSD/JCS	21,134	29,389	21,548
Defensewide	-16,781	6,110	14,311
Total	**290,904**	**276,297**	**267,628**

* Figures include retired pay accrual.

**Data includes Gulf War incremental costs. Fiscal 1991-92 defensewide entries include appropriations that made available allied cash contributions to offset these incremental costs; entries also include $15 billion appropriated for the Persian Gulf Regional Defense Fund. In fiscal 1992, $9.1 billion was shifted from the military services to defense agencies/OSD for the new Defense Medical Program. In fiscal 1993, that program is in the defensewide line and totals $9.5 billion.

DEPARTMENT OF DEFENSE
APPROPRIATIONS FOR RESEARCH,
DEVELOPMENT, TEST, AND EVALUATION
Fiscal Years 1991 - 1993 (Millions of Dollars)

	1991	1992 E	1993 E
TOTAL--APPROPRIATIONS FOR RDT&E	**$34,870**	**$38,340**	**$38,813**
BY APPROPRIATION			
Army	$ 5,573	$ 6,453	$ 5,414
Navy	7,989	8,550	8,518
Air Force	11,975	13,591	14,532
Defense Agencies	9,082	9,522	10,053
Director of Test & Evaluation, Defense	237	210	282
Director of Operational Test & Evaluation, Defense	14	13	13
BY RESEARCH CATEGORIES			
Research	$ 1,157	$ 1,024	$ 1,124
Exploratory Development	2,730	2,892	2,986
Advanced Development	10,769	10,642	11,373
Engineering Development	8,702	10,302	8,994
Management and Support	2,866	2,886	2,899
Operational Systems Development	8,646	10,593	11,436
RECAP OF BUDGET ACTIVITIES			
Technology Base	$ 3,886	$ 3,914	$ 4,084
Advanced Technology Development	5,298	6,471	7,682
Strategic Programs	4,375	4,312	4,647
Tactical Programs	12,611	14,530	13,241
Intelligence and Communications	4,471	5,201	5,011
Defensewide Mission Support	4,230	3,912	4,147
RECAP OF FYDP PROGRAMS			
Strategic Forces	$ 831	$ 535	$ 394
General Purpose Forces	2,050	2,635	2,925
Intelligence and Communications	5,241	7,029	7,653
Airlift/Sealift	13	11	20
Research and Development (FYDP Program 6)	26,224	27,747	27,376
Central Supply and Maintenance	306	109	172
Administration and Associated Activities	9	5	5
Support of Other Nations	6	3	4
Special Operations Forces	191	265	263

Source: Department of Defense Budget, "R,D,T&E Programs (R-1)" (Annually).
NOTE: Detail may not add to totals because of rounding.
E Estimate. Latest year reflects Administration's budget proposal.

DEPARTMENT OF DEFENSE
TOTAL MILITARY OUTLAYS BY FUNCTIONAL TITLE
Fiscal Years 1984 - 1993 (Millions of Dollars)

	1984	1985	1986	1987	1988	1989	1990	1991	1992	1993
TOTAL	$220,928	$245,154	$265,480	$273,966	$281,935	$294,880	$289,755	$306,806	$300,621	$279,077
Procurement--TOTAL	61,879	70,381	76,517	80,744	77,166	81,620	80,972	82,028	73,952	67,263
Aircraft	23,196	26,586	30,828	32,956	28,246	27,569	26,142	25,689	23,950	20,899
Missiles	9,527	10,749	11,730	11,473	11,676	14,503	14,851	14,400	12,893	11,887
Ships	8,487	9,145	9,501	9,316	8,878	10,587	11,016	11,512	10,275	9,355
Weapons	3,691	3,801	4,343	4,962	4,727	4,384	3,873	3,716	3,689	2,632
Ammunition	1,826	2,080	1,933	2,111	2,250	1,993	2,003	2,103	1,545	1,465
Other	15,152	18,020	18,182	19,926	21,389	22,585	23,088	24,609	21,600	21,025
Military Personnel--TOTAL	64,158	67,842	71,511	72,020	76,337	80,676	75,622	83,439	79,289	76,952
Active Forces	42,732	60,344	63,139	63,810	67,642	71,571	66,541	74,571	69,735	67,831
Reserve Forces	4,923	7,498	8,373	8,210	8,694	9,104	9,081	8,868	9,554	9,122
Retired Pay	16,503									
Research, Development, Test, & Evaluation	23,117	27,103	32,283	33,596	34,792	37,002	37,458	34,589	36,145	37,914
Operations & Maintenance	67,388	72,371	75,288	76,205	84,475	87,001	88,340	101,769	97,887	87,639
Military Construction	3,706	4,260	5,067	5,853	5,874	5,275	5,080	3,497	4,541	6,264
Family Housing	2,413	2,642	2,819	2,908	3,082	3,257	3,501	3,296	3,404	3,652
Other	(1,732)	553	1,995	2,640	210	50	(1,218)	(1,812)	5,403	(607)

Source: Department of Defense, "Status of Funds" (Annual Summaries) and Office of Management and Budget, "The Budget of the United States Government" (Annually).

DEPARTMENT OF DEFENSE
OUTLAYS FOR AIRCRAFT PROCUREMENT
By Agency
Fiscal Years 1962 - 1993 (Millions of Dollars)

Year	TOTAL AIRCRAFT PROCUREMENT	Air Force	Navy	Army
1962	$ 6,659	$ 4,387	$ 2,102	$ 170
1963	6,309	3,747	2,328	234
1964	6,053	3,894	1,859	300
1965	5,200	3,115	1,739	346
1966	6,635	4,074	2,021	540
1967	8,411	4,842	2,607	962
1968	9,462	5,079	3,244	1,139
1969	9,177	5,230	2,821	1,126
1970	7,948	4,623	2,488	837
1971	6,631	3,960	2,125	546
1972	5,927	3,191	2,347	389
1973	5,066	2,396	2,557	113
1974	5,006	2,078	2,806	122
1975	5,484	2,211	3,137	136
1976	6,520	3,323	3,061	136
1977	6,608	3,586	2,721	301
1978	6,971	3,989	2,602	380
1979	8,836	5,138	3,140	558
1980	11,124	6,647	3,689	787
1981	13,193	7,941	4,397	855
1982	16,793	9,624	5,872	1,297
1983	21,013	11,799	7,490	1,724
1984	23,196	12,992	8,040	2,165
1985	26,586	15,619	8,263	2,705
1986	30,828	18,919	8,922	2,987
1987	32,956	20,036	9,614	3,306
1988	28,246	15,961	9,407	2,878
1989	27,569	14,662	10,073	2,834
1990	26,142	14,303	9,031	2,808
1991	25,689	13,794	9,055	2,840
1992 E	23,950	13,675	8,151	2,124
1993 E	20,899	11,451	7,576	1,872

Source: Department of Defense Budget (Annually).
NOTE: Detail may not add to totals because of rounding.
E Estimate. Latest year reflects Administration's budget proposal.

DEPARTMENT OF DEFENSE
OUTLAYS FOR RESEARCH, DEVELOPMENT, TEST, AND EVALUATION
Fiscal Years 1972 - 1993 (Millions of Dollars)

Year	TOTAL, All RDT&E Functions	Air Force	Navy	Army	Other
1972	$ 7,881	$ 3,205	$2,427	$1,779	$ 470
1973	8,157	3,362	2,404	1,912	479
1974	8,582	3,240	2,623	2,190	529
1975	8,866	3,308	3,021	1,964	573
1976	8,923	3,338	3,215	1,842	528
1977	9,795	3,618	3,481	2,069	627
1978	10,508	3,626	3,825	2,342	715
1979	11,152	4,080	3,826	2,409	837
1980	13,127	5,017	4,382	2,707	1,021
1981	15,278	6,341	4,783	2,958	1,196
1982	17,729	7,794	5,240	3,230	1,465
1983	20,554	9,182	5,854	3,658	1,861
1984	23,117	10,353	6,662	3,812	2,289
1985	27,103	11,573	8,054	3,950	3,527
1986	32,283	13,417	9,667	3,984	5,215
1987	33,596	13,347	9,176	4,721	6,352
1988	34,792	14,302	8,828	4,624	7,038
1989	37,002	14,912	9,291	4,966	7,833
1990	37,458	14,443	9,160	5,513	8,342
1991	34,589	13,050	7,586	5,559	8,371
1992 E	36,145	12,773	7,959	5,844	9,554
1993 E	37,914	14,063	8,165	5,675	9,998

Source: Office of Management and Budget, "The Budget of the United States Government" (Annually).
E Estimate. Latest year reflects Administration's budget proposal.

DEPARTMENT OF DEFENSE
OUTLAYS FOR MISSILE PROCUREMENT
By Agency
Fiscal Years 1962 - 1993 (Millions of Dollars)

Year	TOTAL MISSILE PROCUREMENT	Air Force	Navy	Army
1962	$ 3,442	$2,385	$ 593	$ 464
1963	3,817	2,676	718	423
1964	3,577	2,100	981	496
1965	2,096	1,320	522	254
1966	2,069	1,313	512	244
1967	1,930	1,278	432	220
1968	2,219	1,388	436	395
1969	2,509	1,382	534	593
1970	2,912	1,467	702	743
1971	3,140	1,497	791	852
1972	3,009	1,334	831	844
1973	3,023	1,454	628	941
1974	2,981	1,537	541	903
1975	2,889	1,602	615	672
1976	2,296	1,549	584	163
1977	2,781	1,501	905	374
1978	3,096	1,376	1,302	418
1979	3,786	1,537	1,702	547
1980	4,434	1,810	1,973	651
1981	5,809	2,366	2,297	1,146
1982	6,782	3,069	2,444	1,269
1983	7,795	3,383	2,812	1,600
1984	9,527	4,640	2,809	2,079
1985	10,749	5,409	2,941	2,399
1986	11,731	6,473	2,780	2,478
1987	11,473	6,002	3,157	2,314
1988	11,676	6,046	3,392	2,239
1989	14,503	7,349	4,445	2,709
1990	14,851	7,951	4,446	2,453
1991	14,400	6,906	4,954	2,540
1992	12,893	6,294	4,331	2,268
1993	11,887	5,742	4,086	2,060

Source: Office of Management and Budget, "The Budget of the United States Government" (Annually).
NOTE: Detail may not add to totals because of rounding.

DEPARTMENT OF DEFENSE MAJOR CONTRACTORS
Fiscal Years 1987 - 1991
Listed by rank according to net value of
prime contracts awarded during last fiscal year (Millions of Dollars)

Company	1987	1988	1989	1990	1991
TOTAL CONTRACTS	**$142,483**	**$137,049**	**$128,958**	**$130,758**	**$136,640**
McDonnell Douglas Corp.	$ 7,715	$ 8,003	$ 8,617	$ 8,211	$ 8,057
General Dynamics Corp.	7,041	6,522	6,899	6,306	7,848
General Electric Co.	5,802	5,701	5,771	5,589	4,866
General Motors Corp.	4,082	3,550	3,692	4,107	4,427
Raytheon Co.	3,820	4,055	3,761	4,071	4,090
Northrop Corp.	1,068	533	631	746	3,319
United Technologies Corp.	3,587	3,508	3,556	2,856	2,825
Martin Marietta Corp.	3,726	3,715	3,337	3,492	2,689
Lockheed Corp.	5,574	3,538	3,652	3,553	2,667
Grumman Corp.	3,393	2,848	2,373	2,697	2,363
Westinghouse Electric Corp.	1,684	2,185	1,650	2,243	1,812
Rockwell International Corp.	2,238	2,184	2,133	2,217	1,708
Litton Industries Inc.	2,035	2,561	1,437	1,576	1,601
FMC Corp.	744	862	796	634	1,467
Unisys Corp.	2,268	1,380	1,245	1,376	1,379
Loral Corp.	692	494	451	618	1,283
LTV Corp.	1,308	942	757	1,183	1,255
The Boeing Co.	3,547	3,018	2,868	2,267	1,166
TRW Inc.	1,135	1,250	1,294	1,087	1,092
Textron Inc.	1,546	1,276	908	1,190	997
Texas Instruments Inc.	1,109	1,232	946	704	982
ITT Corp.	995	769	1,163	870	948
Bath Holding Corp.			218	734	872
Federal Express, et al. JV				253	829
Alliant Techsystems Inc.					827
GTE Corp.	1,475	423	2,342	1,294	801
IBM Corp.	1,822	1,065	1,309	1,286	773
AT&T Co.	509	791	754	769	699
Allied Signal Inc.	943	711	906	725	689
Harsco Corp.	299	496	433	178	621

Source: Department of Defense, "100 Companies Receiving the Largest Dollar Volume of
Prime Contract Awards" (Annually).

DEPARTMENT OF DEFENSE
PRIME CONTRACT AWARDS OVER $25,000
FOR SELECTED MAJOR MILITARY HARD GOODS
By Geographic Region
Fiscal Years 1989, 1990, and 1991

	Millions of Dollars			Percent of Program Total		
Program and Region	**1989**	**1990**	**1991**	**1989**	**1990**	**1991**
AIRCRAFT--TOTAL	**$27,565**	**$27,107**	**$26,227**	**100.0%**	**100.0%**	**100.0%**
New England	3,872	3,098	3,206	14.0	11.4	12.2
Middle Atlantic	2,738	3,226	2,442	9.9	11.9	9.3
East North Central	2,797	2,648	1,877	10.1	9.8	7.2
West North Central	5,082	5,227	4,513	18.4	19.3	17.2
South Atlantic	2,142	2,344	2,504	7.8	8.6	9.5
East South Central	222	324	379	0.8	1.2	1.4
West South Central	4,458	3,909	4,515	16.2	14.4	17.2
Mountain	1,175	1,909	730	4.3	7.0	2.8
Pacific	5,079	4,423	6,062	18.4	16.3	23.1
MISSILE & SPACE SYSTEMS--TOTAL	**$20,655**	**$18,630**	**$17,990**	**100.0%**	**100.0%**	**100.0%**
New England	3,075	2,220	2,516	14.9	11.9	14.0
Middle Atlantic	1,263	1,252	1,489	6.1	6.7	8.3
East North Central	102	57	140	0.5	0.3	0.8
West North Central	1,034	521	1,169	5.0	2.8	6.5
South Atlantic	1,525	1,707	1,243	7.4	9.2	6.9
East South Central	921	658	748	4.5	3.5	4.2
West South Central	1,255	1,470	1,632	6.1	7.9	9.1
Mountain	3,584	3,459	3,077	17.4	18.6	17.1
Pacific	7,896	7,285	5,977	38.2	39.1	33.2
ELECTRONICS & COMMUNICATIONS EQUIPMENT--TOTAL	**$19,369**	**$19,876**	**$17,470**	**100.0%**	**100.0%**	**100.0%**
New England	3,464	3,053	1,680	17.9	15.4	9.6
Middle Atlantic	3,222	3,270	3,444	16.6	16.5	19.7
East North Central	1,345	1,002	1,292	6.9	5.0	7.4
West North Central	938	901	800	4.8	4.5	4.6
South Atlantic	4,430	5,110	4,595	22.9	25.7	26.3
East South Central	94	221	210	0.5	1.1	1.2
West South Central	1,014	989	1,013	5.2	5.0	5.8
Mountain	900	866	485	4.6	4.4	2.8
Pacific	3,962	4,464	3,951	20.5	22.5	22.6

Source: Department of Defense, "Prime Contract Awards by Region and State" (Annually).
NOTE: Detail may not add to totals because of rounding.

DEPARTMENT OF DEFENSE
PRIME CONTRACT AWARDS
FOR RESEARCH, DEVELOPMENT, TEST, AND EVALUATION
Fiscal Years 1987 - 1991 (Millions of Dollars)

Program Categories	1987	1988	1989	1990	1991
TOTAL--RDT&E	**$21,809**	**$22,543**	**$23,206**	**$22,319**	**$20,898**
Research	1,730	1,444	1,429	994	1,063
Exploratory Development	1,524	1,623	1,581	1,813	2,288
Other Development	17,964	18,937	18,966	18,697	16,424
Management & Support	592	538	1,230	815	1,124
Aircraft--TOTAL	**$ 3,561**	**$ 5,055**	**$ 4,689**	**$ 4,364**	**$ 3,143**
Research	437	139	11	(191)	13
Exploratory Development	103	125	85	82	83
Other Development	3,007	4,777	4,563	4,431	3,002
Management & Support	14	14	30	42	45
Missile and Space Systems--TOTAL	**7,943**	**7,800**	**6,962**	**6,865**	**6,649**
Research	64	106	260	175	95
Exploratory Development	356	340	331	308	710
Other Development	7,401	7,218	6,277	6,291	5,759
Management & Support	122	135	95	91	86
Electronics & Communications Equipment--TOTAL	**4,637**	**3,854**	**3,744**	**3,925**	**3,814**
Research	162	137	182	188	127
Exploratory Development	280	251	289	327	299
Other Development	4,117	3,417	3,190	3,337	3,323
Management & Support	79	49	83	73	64
All Other--TOTAL	**5,668**	**5,834**	**7,811**	**7,165**	**7,292**
Research	1,067	1,062	976	822	827
Exploratory Development	785	907	876	1,097	1,196
Other Development	3,439	3,525	4,936	4,637	4,341
Management & Support	377	340	1,022	609	928

Source: Department of Defense, "Prime Contract Awards by Service Category and Federal Supply Classification" (Annually).

NOTE: Detail may not add to totals because of rounding.
"All Other" includes ships, tank-automotive, weapons, ammunition, services, and other.
( ) Reflects net cancellations.

DEPARTMENT OF DEFENSE
NET VALUE OF PRIME CONTRACT AWARDS OVER $25,000
FOR RESEARCH, DEVELOPMENT, TEST, AND EVALUATION
By Region and Type of Contractor Fiscal Year 1991

		Type of Contractor		
REGION	**TOTAL**	**Educational Institutions**	**Other Non-Profit Institutions**	**Business Firms**
TOTAL--Millions of Dollars	**$20,268**	**$504**	**$2,043**	**$17,721**
New England	$1,804	$ 45	$ 676	$ 1,082
Middle Atlantic	3,233	88	104	3,041
East North Central	1,328	56	70	1,202
West North Central	659	10	5	644
South Atlantic	3,298	82	612	2,604
East South Central	596	18	3	576
West South Central	1,271	21	48	1,202
Mountain	2,509	87	2	2,420
Pacific	5,570	98	522	4,950
PERCENT OF TOTAL	**100.0%**	**100.0%**	**100.0%**	**100.0%**
New England	8.9%	9.0%	33.1%	6.1%
Middle Atlantic	16.0	17.4	5.1	17.2
East North Central	6.6	11.0	3.4	6.8
West North Central	3.3	2.0	0.2	3.6
South Atlantic	16.3	16.2	30.0	14.7
East South Central	2.9	3.5	0.1	3.2
West South Central	6.3	4.1	2.4	6.8
Mountain	12.4	17.3	0.1	13.7
Pacific	27.5	19.4	25.6	27.9

Source: Department of Defense, "Prime Contract Awards by Region and State" (Annually).
NOTE: Detail may not add to totals because of rounding.
Includes contracts with other government agencies.
Includes Alaska and Hawaii.

MISSILE PROGRAMS
RESEARCH, DEVELOPMENT, TEST, AND EVALUATION
By Agency and Model
Fiscal Years 1991, 1992, and 1993 (Millions of Dollars)

Agency and Model	1991	1992 E	1993 E
AIR FORCE			
ACM	$ 51.8	$ 28.6	$ 82.3
AGM-130	14.4	24.7	8.2
AMRAAM	21.6	32.9	38.3
Peacekeeper (M-X)	371.7	2.9	1.0
Small ICBM	87.8	-	-
SRAM II	144.5	-	-
SRAM-T	26.7	-	-
NAVY			
AAAM	$101.0	$ 88.5	$ -
HARM	-	4.0	-
Harpoon	2.1	-	-
Standard	48.0	36.5	34.9
Tomahawk	12.2	33.1	3.7
Trident II	68.7	53.3	65.9
ARMY			
AATWS-M	$ 75.9	$119.8	$ 91.4
Avenger	-	2.5	4.8
BAT	26.8	115.7	121.5
Laser Hellfire	34.4	20.7	5.0
LOSAT	53.4	139.8	122.8
LOS-F-H	94.8	107.3	-
NLOS/FOG-M	68.3	-	-
Patriot	45.9	37.9	38.4
Stinger	-	3.0	5.2
TOW 2	18.2	33.5	-

Source: Department of Defense Budget, "Program Acquisition Costs by Weapon System" (Annually).

NOTE: See Missile Programs Chapter for missile program procurement authorization data.
E Estimate. Latest year reflects Administation's budget proposal.
NA Not Available

Missile Program Acronyms:

AAAM – Advanced Air-to-Air Missile
ACM – Advanced Cruise Missile
BAT – Brilliant Anti-Armor Submunition
HARM – High-Speed Anti-Radiation Missile
LOSAT – Line-of-Sight Anti-Tank
NLOS – Non-Line of Sight
SRAM-T – Short Range Attack Missile Tactical
TOW – Tube-launched Optically-tracked Wire command link guided missile

AATWS-M – Advanced Anti-Tank Weapon System Medium
AMRAAM – Advanced Medium Range Air-to-Air Missile
FOG-M – Fiber Optic Guided Missile
ICBM – InterContinential Ballistic Missile
LOS-F-H – Line Of Sight-Forward-Heavy
SRAM – Short Range Attack Missile

MILITARY AIRCRAFT PROGRAMS
RESEARCH, DEVELOPMENT, TEST, AND EVALUATION
By Agency, Type, and Model
Fiscal Years 1991, 1992, and 1993 (Millions of Dollars)

Agency, Type, and Model	1991	1992 E	1993 E
AIR FORCE			
B-1B	$ -	$ 1.4	$ 90.7
B-2 Advanced Technology Bomber	1,715.7	1,546.0	1,261.4
C-17	732.2	372.5	210.0
E-3 AWACS	125.4	205.2	130.9
E-8A JSTARS	216.1	311.3	355.9
F-15E Eagle	66.4	111.7	54.0
F-16 Falcon	26.4	158.3	183.8
F-22 Lightning (ATF)	943.5	1,621.1	2,224.3
KC-135 Re-engining/modernization	3.5	12.8	16.7
National Aerospace Plane	161.5	200.0	175.0
T-1A (TTTS)	2.4	4.2	4.7
NAVY			
AH-1W Sea Cobra	$ 14.3	$ 11.3	$ 5.4
AV-8B Harrier	30.2	9.2	11.1
AX Advanced Strike	137.5	-	165.6
CH/MH-53E Super Stallion	17.7	9.2	12.5
E-2C Hawkeye	35.7	6.3	6.7
EA-6B Prowler	14.3	10.8	23.9
F-14D Tomcat	119.8	115.3	101.2
F/A-18 Hornet	84.3	416.9	1,133.6
Medium Lift Replacement	-	-	9.7
SH-60B Seahawk (LAMPS MK-III)	16.6	30.1	31.8
SH-60F Carrier ASW	12.0	19.7	40.7
T-45 Goshawk	14.7	23.1	32.0
V-22 Osprey	234.6	790.0	-
ARMY			
LONGBOW	$ 197.0	$ 232.2	$ 281.8
OH-58D AHIP	21.6	9.3	-
RAH-66 Comanche	333.7	538.8	443.0
UAVs	91.6	66.9	129.1
SPECIAL OPERATIONS			
MC-130H Combat Talon II	$ 3.6	$ 3.3	$ -
MH-47E	7.4	14.4	0.8
MH-60K	10.4	13.2	0.7

Source: Department of Defense Budget, "Program Acquisition Costs by Weapon System"
(Annually).
E Estimate. Latest year reflects Administration's budget proposal.

MISSILE PROGRAM PROCUREMENT
Fiscal Years 1991, 1992, and 1993 (Millions of Dollars)

Agency and Model	1991		1992 E		1993 E	
	No.	Cost	No.	Cost	No.	Cost
AIR FORCE						
ACM	85	$ 454.2	120	$ 500.1	-	$ -
AGM-130	48	38.4	120	70.0	149	76.1
AMRAAM	810	821.7	821	737.8	1,190	868.9
HARM	3,481	776.4	1,214	323.1	846	250.1
HAVE NAP	26	25.8	32	34.5	-	-
Peacekeeper	-	398.2	-	194.5	-	-
SRAM II	-	10.1	-	-	-	-
NAVY						
Harpoon	167	$ 249.0	-	$ 37.2	-	$ -
Penguin	40	45.0	42	44.2	-	-
Standard	405	287.6	330	331.1	330	256.8
Tomahawk	678	1,045.9	176	411.2	200	404.2
Trident II	52	1,511.1	28	1,195.4	21	986.8
ARMY						
AAWS-M	-	$ -	5	$ 12.9	26	$ 46.5
ATACMS	373	236.9	300	170.9	340	188.2
Avenger	88	117.6	144	183.6	144	148.2
Laser Hellfire$Sd	6,709	234.6	112	19.7	3,158	154.0
MLRS	56,286	438.4	3,714	61.7	-	2.2
Patriot	1,100	1,002.8	97	156.1	-	25.2
Stinger	6,922	252.2	-	38.2	-	9.5
TOW 2	18,309	320.3	12,168	240.4	9,440	183.1

Source: Department of Defense, "Program Acquisition Costs by Weapon System" (Annually). Total Obligational Authority excluding initial spares and RDT&E.
E   Estimate. Latest year reflects Administration's budget proposal.

MAJOR WEAPON SYSTEMS AND COMBAT FORCES
(Highlights)

Strategic	FY 1989	FY 1990	FY 1991	FY 1992 ESTIMATE	FY 1993 ESTIMATE
STRATEGIC OFFENSE					
Land-Based ICBMs*					
Minutemen	950	950	950	880	802
Peacekeeper	50	50	50	50	50
Strategic Bombers**					
B-52G/H***	173	154	138	125	84
B-1B	90	90	90	84	84
Fleet Ballistic Launchers (SLBMs)*					
Poseidon (C-3 and C-4)	384	368	352	176	160
Trident (C-4 and D-5)	192	216	264	288	288
STRATEGIC DEFENSE INTERCEPTORS					
(PAA**/Squadrons)					
Active	36/2	18/1	18/1	- -	
Air National Guard	216/12	216/12	216/12	216/12	216/12

* Number on-line
** PAA = Primary Aircraft Authorized
*** Does not include conventional B-52 force.

General Purpose	FY 1988	FY 1990	FY 1991	FY 1992 ESTIMATE	FY 1993 ESTIMATE
LAND FORCES					
Army Divisions:					
Active	18	18	16	14	14
Reserve	10	10	10	10	8
Marine Corps Divisions:					
Active	3	3	3	3	3
Reserve	1	1	1	1	1
Army Separate Brigades:*					
Active	8	8	8	7	7
Reserve	20	19	18	16	11
Army Special Forces Groups:					
Active	4	5	5	5	5
Reserve	4	4	4	4	4
Army Ranger Regiment	1	1	1	1	1

(Continued on next page)

MAJOR WEAPON SYSTEMS AND COMBAT FORCES (Continued)
(Highlights)

TACTICAL AIR FORCES (Primary Aircraft Authorized/Squadrons)	FY 1989	FY 1990	FY 1991	FY 1992 ESTIMATE	FY 1993 ESTIMATE
Air Force Attack/Fighter:					
Active	1,868/79	1,722/76	1,560/71	1,230/57	1,158/56
Reserve	909/43	873/43	861/43	873/43	831/41
Conventional Bombers					
B-52G	0	33	33	33	33
Navy Attack/Fighter:					
Active	706/67	622/57	654/59	678/61	620/56
Reserve	110/10	97/9	116/10	116/10	116/10
Marine Corps Attack/Fighter:					
Active	341/25	334/24	361/26	336/23	326/22
Reserve	96/8	84/8	60/7	84/8	72/7
NAVAL FORCES					
Strategic Forces Ships	43	39	40	33	27
Battle Forces Ships	437	412	392	362	361
Support Forces Ships	60	65	62	60	55
Reserve Forces Ships	25	31	32	19	19
Total Ship Battle Forces	565	547	526	474	462
Surface Combatants/Mine Warfare Ships	21	16	13	16	14
Support Ships	5	3	3	2	2
Total Other Forces	**26**	**19**	**16**	**18**	**16**

Airlift and Sealift	FY 1989	FY 1990	FY 1991	FY 1992 ESTIMATE	FY 1993 ESTIMATE
INTERTHEATER AIRLIFT*					
C-5	110	109\	109	109	109
C-141	234	234	234	234	214
KC-10	57	57	57	57	57
C-17	0	0	0	0	6
INTRATHEATRE AIRLIFT*					
Air Force					
C-130	492	460	462	445	428
SEALIFT SHIPS**					
Active					
Tankers	29	28	20	20	20
Cargo	40	40	39	39	39
Reserve					

(Continued on next page)

MAJOR WEAPON SYSTEMS AND COMBAT FORCES (Continued)
(Highlights)

RRF***	93	96	96	99	112
NDRF****	128	121	121	122	122

* Primary Aircraft Authorized
** Includes fast sealift ships, afloat pre-positioned force ships and common user (charter) ships
*** Ready Reserve Force (assigned to 5-, 10- or 20-day reactivation groups)
**** National Defense Reserve Fleet (beginning in fiscal 1988, specific NDRF ships were designated militarily useful ships).
Differences from previous year's defense report are due to operational changes (damaged aircraft and actual long-term ship charters) and congressional direction/funding (retention of C-130s in the reserve components and underfunding of the Ready Reserve Force acquisitions).

SPECIFICATIONS OF U.S. MILITARY AIRCRAFT
On Order or in Production as of 1991

Primary Mission, DOD Designation, & Popular Name	Manufacturer	U.S. Military Service	Crew (000's lbs)	Empty Weight	Engines	Performance Typical for Primary Mission	Remarks
ATTACK							
A-6E Intruder	Grumman	USN/USMC	2	30	2xP&W J52	Mach 0.8 at sea level	Also EA-6A/B & KA-6D
AV-8B Harrier 2	MDC/BAe	USMC	1	13	1xRR F402	Mach 0.91	Graphite/epoxy super-critical wing
BOMBERS							
B-2 Stealth Bomber	Northrop	USAF	2	-	4xGE F118	-	Radar eluding tactical bomber
ELECTRONIC WARFARE							
EA-6B Prowler	Grumman	USN/USMC	4	33	2xP&W J52	493 n.m. standoff radius	Tactical jamming system
FIGHTERS							
F-14A Tomcat	Grumman	USN	2	40	2xP&W TF30	Mach 2.3 class	Missile, gun fleet defense
F14A+ Super Tomcat	Grumman	USN	2	42	2xGE F110	Mach 2.3 class	F-14A with upgraded engines and radar
F14D	Grumman	USN	2	42	2xGE F110	Mach 2.3 class	F14A+ with improved avionics and infrared track and search system
F-15C/D Eagle	MDC	USAF	1-2	31	2xP&W F100	Mach 2.5 class	Air superiority, defense, guns, missiles; 15D=2 seat trainer
F-15E Eagle	MDC	USAF	2	37	2xP&W F100	Mach 2.5 class	Dual role fighter/long range interdiction
F-16 A/B Fighting Falcon	GD	USAF	1-2	16	1xP&W F100	Mach 2+ class	Multirole fighter; fully fly-by-wire; missiles, guns.
F-16 C/D Fighting Falcon	GD	USAF	1-2	18	1xP&W F100/ 1xGE F110	Mach 2+ class	Provisions for AMRAAM, LANTIRN, and new EW Nav. Comm. Systems
F/A-18 Hornet	MDC/Northrop	USN/USMC	1-2	23	2xGE F404	Mach 1.7 class	Missiles, guns; also export

(Continued on next page)

SPECIFICATIONS OF U.S. MILITARY AIRCRAFT (Continued)
On Order or in Production as of 1991

Primary Mission, DOD Designation, & Popular Name	Manufacturer	U.S. Military Service	Crew	Empty Weight (000's lbs)	Engines	Performance Typical for Primary Mission	Remarks
COMMAND/CONTROL AND PATROL							
E-2C Hawkeye	Grumman	USN	5	38	2xAll T56	6 hr. mission duration	AEW command & control; passive detection
E-6A Tacamo	Boeing	USN	18	167	4xCFM56	Long endurance	AEW command & control
CARGO-TRANSPORT							
C/HC-130 Hercules	Lockheed	USAF/USN	4	74-78	4xAll T56	363 mph; 2,038 n.m.	92-128 troops or 39-43,000 lbs
C-17A	MDC	USAF	3	267	4xP&W F117	Mach 0.77; 3,000 n.m.	102 troops or 172,000 lbs.
TRAINING							
T-45A Goshawk	MDC/BAe	USN	2	9	1xRR F405	Mach 1.04 at 25,000 ft.	Next generation trainer
T-1A Jayhawk	Beech	USAF	3	10	2xP&W JT-15D	Max 538 mph	Tanker/Transport Trainer
HELICOPTERS							
AH-1W Super Cobra	Bell-Textron	USN	2	10	2xGE T700	Max 218 mph; 395 mi.	TOW, hellfire, sidewinder
AH-64 Apache	MDC	Army	2	11	2xGE T700	Max 197 mph; 445 mi.	Attack helicopter
CH/MH-53E	Sikorsky	USN	3-8	33-36	3xGE T64	Max 196 mph; 710 mi.	55 passengers, aux. tanks/ minesweeping
HH-60H	Sikorsky	USN	4-12	14	2xGE T700	Max 135 mph; 500 mi.	Strike and rescue
SH-2F Seasprite	Kaman	USN	3	7	2xGE T58	Max 160 mph; 430 mi.	LAMPS Mk.1 helicopter
SH-60B Seahawk	Sikorsky	USN	3	15	2xGE T700	Max 171 mph; 640 mi.	ASW
SH-60F	Sikorsky	USN	4	14	2xGE T700	Max 177 mph; 789 mi.	ASW
UH-60A Black Hawk	Sikorsky	Army/USAF	3	11	2xGE T700	Max 184 mph; 373 mi.	UTTAS

Source: Aviation Week & Space Technology, "Aerospace Forecast & Inventory" (Annually).
KEY: All = Allison Gas Turbine; BAe = British Aerospace; CFM = CFM International; GA = Garrett Engine; GD = General Dynamics; GE = General Electric; Lyc = Textron Lycoming; MDC = McDonnell Douglas; P&W = Pratt & Whitney; PWC = P&W of Canada; RR = Rolls Royce.

ACTIVE U.S. MILITARY AIRCRAFT IN CONTINENTAL U.S.
Fiscal Years 1979 - 1993

Year	Total	Fixed Wing Aircraft				Helicopter
		Total	Jet	Turboprop	Piston	
1979	18,526	11,365	8,656	1,859	850	7,161
1980	18,969	11,362	8,794	1,869	699	7,607
1981	19,363	11,645	9,111	1,943	591	7,718
1982	21,728	12,063	9,647	1,900	516	9,665
1983	18,652	11,603	9,495	1,745	363	7,049
1984	18,833	11,661	9,551	1,777	333	7,172
1985	19,333	11,929	9,640	1,881	408	7,404
1986	20,157	11,919	9,730	1,803	386	8,238
1987	20,514	12,054	9,819	1,865	370	8,460
1988	21,210	12,481	9,954	2,222	305	8,529
1989	19,223	11,938	9,501	2,131	261	7,330
1990	20,037	12,817	10,360	2,199	258	7,220
1991	19,966	12,587	10,221	2,119	247	7,379
1992	19,222	11,945	9,678	2,033	234	7,277
1993	19,310	11,503	9,325	2,006	172	7,807

Source: Office of the Secretary of Defense, as reported in "FAA Aviation Forecasts" (Annually).
Includes Army, Air Force, Navy, and Marine regular service aircraft, as well as Reserve and National Guard Aircraft.
E Estimate.

MAJOR MISSILE PROGRAMS
RESEARCH, DEVELOPMENT, PRODUCTION, OPERATION

Program	Agency	Status	Systems Contractor	Propulsion Manufacturer	Guidance Manufacturer
AIR-TO-AIR					
AMRAAM-120A	USAF/USN	D,P	Hughes/Ray	Hercules	Hughes/Ray
Phoenix-54A	USN	O	Hughes/Ray	Hercules	Hughes
Phoenix-54C	USN	P,O	Hughes/Ray	Hercules	Hughes/Ray
Sidewinder-9J	USAF	O	Loral	Hercules/ Aerojet	Loral
Sidewinder-9L	USN/USAF	O	NASC	Bermite/ Hercules	Raytheon/ Loral
Sidewinder-9M	USN/USAF	P	NASC	MTI/Hercules	Ray/Loral
Sidewinder-9N	USAF	O	Loral/Ray	-	Loral
Sidewinder-9P	USAF	P,O	Loral/Ray	Hercules/ Aerojet	Loral
Sidewinder-9R	USN	P	Loral/Ray	MTI/Hercules	Ray/Loral
Sparrow-7F	USN/USAF	O	NASC	Hercules	Raytheon/GD
Sparrow-7M	USN/USAF	P	Raytheon/GD	Hercules	Raytheon/GD
Sparrow-7P	USN	D	NASC	-	Raytheon
Sparrow-7R	USN	D	NASC	-	Raytheon/GD
AIR-TO-SURFACE					
ALCM-86B	USAF	P	Boeing	WI	Honeywell/ Litton
HARM-88A/B	USN/USAF	P	TI	MTI/Hercules	TI
Harpoon-84A/C/D	USN	P,O	MDC	Teledyne CAE	TI/IBM/LSI/ Northrop
GBU-15	USAF	P	RI	Hughes	-
Maverick-65A/B	USAF	P,O	Hughes	MTI/Aerojet	Hughes
Maverick-65D	USAF	P,O	Hughes/Ray	MTI/Aerojet	Hughes/Ray
Maverick-65E	USMC	P	Hughes	MTI/Aerojet	Hughes
Maverick-65F	USN	P	Hughes/Ray	MTI/Aerojet	Hughes/Ray
Maverick-65G	USAF	D	Hughes/Ray	MTI/Aerojet	Hughes/Ray
Shrike-45A/B	USN/USAF	O	NWC/PMTC	Aerojet/ Hercules	Texas Instruments
Sidearm 1-122A	USMC	P	Motorola	MTI/Hercules	Motorola
SLAM-84E MDC/Hughes/RI	USN	P	MDC	Teledyne CAE	
SRAM-69A	USAF	O	Boeing	Lockheed	Kearfott
Standard ARM-78D	USN/USAF	O	GD	NOSIH	GD
Walleye 1-62	USN	O	MM	-	MM/Hughes
Walleye 1ER-62	USN	R,D	NAC	-	NAC
Walleye 2-62	USN	O	NAC	-	NAC
Walleye 2 (ER/DL)-62	USN	O	NAC	-	NAC
ACM-129	USAF	P	GD/MDC	WI	Kearfott
AGM-130A	USAF	D	RI	Hercules	RI
AGM-130B	USAF	D	RI	Hercules	RI
ANTI-SUBMARINE					
VLA-44A	USN	O	Loral	MTI	Kearfott
SURFACE-TO-AIR					
ADATS LOS-F-H	Army	P	MM	-	MM
Chaparral-72A	Army	O	Loral	Hercules/ Bermite	GE/Raytheon
Chaparral-72E/H	Army	P,O	Loral	AR	Loral
Hawk-23B	Army	P,O	Raytheon	Aerojet	Raytheon
Patriot-104	Army	P	Raytheon	MTI	Raytheon
RAM-116A	USN	D	General Dynamics	Bermite/MTI/ Hercules	General Dynamics

(Continued on next page)

MAJOR MISSILE PROGRAMS (Continued)
RESEARCH, DEVELOPMENT, PRODUCTION, OPERATION

Program	Agency	Status	Systems Contractor	Propulsion Manufacturer	Guidance Manufacturer
SURFACE-TO-AIR (Continued)					
Redeye-43A	Army/USMC	O	GD	AR	GD
Roland-115	Army	O	Hughes/ Boeing	Hercules	Hughes/ Boeing
Sea Sparrow-7M	USN	P,O	Raytheon/GD	Aerojet/ Hercules	Raytheon/GD
Standard 1 MR	USN	P,O	GD	Aerojet/NOSIH	GD
Standard 2 MR	USN	P,O	GD	AR/Aerojet/MTI	GD
Standard 1 ER	USN	O	GD	AR/NOSIH	GD
Standard 2 ER	USN	P,O	GD/Raytheon	AR/NOSIH/MTI	GD/Raytheon
Stinger-92A	Army/USMC	P,O	GD/Raytheon	AR	GD/Raytheon
SURFACE-TO-SURFACE					
Harpoon-84A/C/D	USN	P,O	MDC	Teledyne CAE/ MTI	TI/IBM/LSI/ Northrop
Minuteman 2-30F	USAF	O	AFLC	MTI/Aerojet/ Hercules	Rockwell Autonetics
Minuteman 3-30G	USAF	O	AFLC	MTI/Aerojet	Rockwell Autonetics
Peacekeeper (MX)-118A	USAF	P,O	BMO	MTI/Avco/RI Aerojet/GE/ Hercules	RI/Northrop/ Honeywell/ Litton
Poseidon C3-73A	USN	O	Lockheed	MTI/Hercules	GE/MIT/Ray/ Hughes
Tomahawk (SLCM)	USN	P	GD/MDC	WI/ARC/CSD	MDC/GD
Trident 1 (C-4)	USN	P,O	Lockheed	Hercules/MTI	GE/Draper/ Ray/Hughes/ Kearfott
Trident 2 (D-5)	USN	D,P	Lockheed	Hercules/MTI/ UTC	GE/Draper/ Ray/Hughes/ Kearfott/RI
BATTLEFIELD SUPPORT AND ANTIARMOR					
ATACMS	Army	P	LTV	ARC	-
Dragon-47	Army	P,O	MDC	MDC	MDC
Hellfire-114A	Army/USMC	P	RI	Hercules/MTI	MM
HOMS-114K	Army/USMC	D	MM	Hercules/MTI	-
Javelin (AAWS-M)	Army/USMC	D	TI/MM	ARC	-
Lance-52C	Army	O	LTV	RI/Rocketdyne	E-Systems/ Sys-Donner/ Arma
MLRS-26,-270	Army	P,O	LTV	AR	-
Shillelagh-51C	Army	O	Loral	Hercules	Loral
SMAW	USMC	P,O	MDC	MDC	-
TOW-71A	Army	O	Hughes	Hercules	Emerson El.
ITOW-71C	Army	P,O	Hughes	Hercules	Emerson El.
TOW2-71D	Army	P,O	Hughes	Hercules/MTI	Emerson El./TI
TOW2A-71E	Army	P,O	Hughes	Hercules/MTI	Emerson El./TI
TOW2B-71F	Army	P	Hughes	Aerojet/Thorn	Emerson El./TI

Source: Aerospace Industries Association, based on information from "Aviation Week & Space Technology" Magazine.

Status: R-Research; D-Development; P-Production; O-Operational.

Abb:

AFLC - Air Force Logistics Cmd.	MIT - Massachusetts Institute of Technology	PMTC - Pacific Missile Test Center
AR - Atlantic Research		Ray - Raytheon
BMO - Ballistic Missile Office	MTI - Thiokol	RI - Rockwell International
GD - General Dynamics	NAC - Naval Avionics Center	TI - Texas Instruments
GE - General Electric	NASC - Naval Air Systems Command	USAF - United States Air Force
LSI - Lear Siegler	NOSIH - Naval Ordnance Station, Indian Head	USMC - United States Marine Corps
MM - Martin Marietta		USN - United States Navy
MDC - McDonnell Douglas	NWC - Naval Weapons Center	WI - Williams International

ORDERS, SALES, AND BACKLOG
MISSILE SYSTEMS AND PARTS
Calendar Years 1975 - 1991 (Millions of Dollars)

Year	SALES--Current Dollars	SALES--Constant Dollars
1975	$ 3,548	$ 6,694
1976	3,237	6,347
1977	3,118	5,711
1978	3,264	5,677
1979	3,706	5,836
1980	3,971	5,625
1981	4,662	5,864
1982	5,676	6,457
1983	5,991	6,498
1984	6,094	6,106
1985	7,975	8,080
1986	8,236	8,253
1987	9,671	9,671
1988	9,485	9,308
1989	9,283	8,749
1990	9,102	8,237
1991	8,989	7,844

Year	NET NEW ORDERS	BACKLOG AS OF DECEMBER 31
1975	$ 3,655	$ 4,580
1976	3,036	4,379
1977	3,280	4,541
1978	2,948	4,581
1979	3,724	4,916
1980	4,961	5,558
1981	6,030	6,749
1982	6,034	7,107
1983	7,231	8,406
1984	7,731	10,043
1985	8,122	10,190
1986	11,023	12,754
1987	11,482	14,302
1988	9,437	14,255
1989	8,998	14,005
1990	7,917	12,956
1991	8,072	12,040

Source: Bureau of the Census, "Aerospace Industry (Orders, Sales, and Backlog)," Series MA37D (Annually).

Defense Agencies RDT&E requests for Fiscal Year 1994
(Research, Development, Test and Evaluation)
Dollar figures are in thousands.

Program Element Number	Item	FY 1992	FY 1993	FY 1994
0601101A	In-House Laboratory Indep. Research	8,808	11,729	10,954
0601101D	In-House Lab Indp Res.			3,368
0601101E	Defense Res Sciences	115,790	109,629	79,657
0601101W	In-House Lab Indp Res.	3,193	4,123	
0601102D	Defense Res Sciences	2,021		
0601103D	University Res Initiatives	219,591	317,144	242,611
0601109D	US-JAPAN Mgmt. Training	9,597	9,377	
0601110D	Focused Res Initiatives	29,472		
0602109H	Superconductive Magnetic Energy Storage	40,000	20,000	
0602222D	Counterterror Tech Supp	6,512	9,368	6,169
0602227D	Medical Free Electron Laser	23,000	18,755	19,248
0602228D	Historically Black Colleges & Univ. (HBCU) Science and Engineer	15,000	14,066	
0602301E	Computing Sys & Comm Tech	288,292	349,463	368,589
0602702E	Tactical Technology	128,052	98,019	143,891
0602707E	Particle Beam Technology	2,495		
0602708E	Integrated Cmnd & Cntrl Tech	109,008	152,180	57,214
0602712E	Materials & Elec. Tech	198,409	255,112	198,502
0602714E	Treaty Verification	19,466		
0602715H	Defense Nuclear Agency	357,646	384,722	288,388
0602756D	DoD Software Tech Initiative			43,304
0602787D	Medical Technology	6,737		
0602790C	Small Bus Innovative Res/Small Bus Tech Trnsfr. Pilot Prog		41,513	42,552
0602790D	Small Bus Innovative Res/Small Bus Tech Trnsfr. Pilot Prog		15,249	24,703
0602790E	Small Bus Innovative Res/Small Bus Tech Trnsfr. Pilot Prog		16,182	
0602790H	Small Bus Innovative Res/Small Bus Tech Trnsfr. Pilot Prog		4,868	3,851
0305108K	Command & Control Research			1,847
0901600H	Contract Admin/Audit			6,834
1160279BB	Small Bus Innovative Res/Small Bus Tech Trnsfr. Pilot Prog		1,749	2,281
Technology Base		**1,536,051**	**1,821,519**	**1,571,239**
0602227D	Not Used			
0603002D	Medical Advd Tech			4,701
0603214C	Space Based Interceptors	434,020		
0603215C	Limited Defense System	1,486,722		
0603216C	Theater Missile Defenses	764,913		
0603217C	Other Follow-On Systems	559,891	308,570	354,187
0603218C	Res and Support Acts.	661,351		
0603225D	Joint DoD-DoE Munitions Technology Development	17,422	17,795	16,446
0603226E	Experimental Eval of Major Innovative Technologies	249,495	286,568	512,198
0603227E	Relocatable Target Detection Tech Program	28,240		
0603569E	Advanced Submarine Tech	71,458	52,049	32,556
0603570E	Dual-Use Partnerships	60,000	561,633	324,000
0603704D	Special Tech Support	9,677	8,743	8,841
0603705D	Manufacturing Technology			147,733
0603716D	Strategic Environ Res Prog	77,612	169,940	97,958
0603718D	Medical Research	10,000		

(Continued on next page)

Defense Agencies RDT&E requests for Fiscal Year 1994 (Continued)

Program Element Number	Item	FY 1992	FY 1993	FY 1994
0603719D	Focus Hope	20,000		
0603720D	Environ Special Project	19,315		
0603721D	DoD Environ Studies Devel	5,000		
0603736D	Computer Aided Logistics Supp	15,116	15,206	10,424
0603737D	Balanced Tech Initiative	122,110	141,721	
0603738D	Cooperative DoD/VA Med Res	19,501	28,561	
0603739E	Manufacturing Tech	206,061	219,330	299,597
0603744E	Advd Simulation		28,522	9,207
0603745E	Semiconductor Manu Tech		94,845	100,000
0603755D	High Perf Comp Modern. Prog			122,819
0603756D	Consolidated DoD Sftwr Initia.	25,894	31,940	9,151
0603756E	Consolidated DoD Sftwr Initia.	52,317		
0603832D	Joint Wargaming Simulation Management Office		57,122	67,152
0604704D	Rocket Motor Demilitarization Program	26,377	15,847	12,267
0305108K	Command & Control Research	1,956	1,740	
1160401BB	Special Ops Tech Develop.	3,385	3,730	17,794
1160402BB	Special Ops Advd Tech Dev.	16,314	12,909	9,655
1160407BB	SOF Medical Tech Dev.		541	1,310
Advanced Technology Development		**4,964,146**	**2,057,312**	**2,157,996**
0603214C	Space Based Interceptors		270,000	
0603215C	Limited Defense System		1,714,144	1,195,459
0603217C	Not Used			
0603218C	Research and Support Acts		363,073	358,223
0603711H	Verificaion Tech Demon.	74,979	56,841	46,350
0603734J	Island Sun Support	58,776	20,550	15,822
0603741D	Air Defense Initiative	199,458	162,197	
0302016K	Nat'l Military Comd Sys-Wide Support	10,438	8,756	3,500
0302019K	WWMCCS Systems Engineer	11,798	8,765	9,253
0303131K	Min. Essential Emer Comm Network (MEECN)	3,659	3,195	3,285
0303154J	WWMCCS ADP Modernization		3,900	7,000
0303154K	WWMCCS ADP Modernization	39,881	24,209	
0901600J	Contract Admin/Audit			436
Strategic Programs		**398,989**	**2,635,630**	**1,639,328**
0603216C	Theater Missile Defenses		1,018,110	1,636,304
0603228D	Physical Security Equip	56,473	25,233	20,676
0603709D	Joint Robotics Program	20,106	18,399	22,125
0603710D	Classified Program - C3I	10,115	44,185	9,912
0603714D	Advd Sensor Apps Program	42,333	43,471	25,920
0603715D	AIM-9 Consolidated Program	37,659	14,281	9,593
0603724D	Biological Def-Advd Dev.			26,355
0604225C	Theater Missile Defenses		9,390	50,410
0604705D	Mobile Offshore Base Analysis	2,850	6,563	
0604771D	Joint Tact Information Distribution Sys (JTIDS)	106,180	69,716	67,053
0201135J	CINC C2 Initiatives		1,786	1,193
0201135K	CINC C2 Initiatives	1,803		
0208045K	C3 Interoperability (Joint Tactical C3 Agency)	48,099	24,173	28,088
0208298K	Mgmt Hdqs. (Jnt Tact C3 Agency) 6,893			
0305141D	Joint Remotely Piloted Vehicles Program	99,728	131,046	180,112
0305815D	General Support for SO/LIC	1,897		
0901600BB	Contract Admin/Audit			4,656
0901600K	Contract Admin/Audit			1,283
1160404BB	Special Ops Tact Sys Develop	208,009	118,621	221,305

(Continued on next page)

Defense Agencies RDT&E requests for Fiscal Year 1994 (Continued)

Program Element Number	Item	FY 1992	FY 1993	FY 1994
1160405BB	Special Ops Intelligence Systems Development	10,057	27,254	6,686
1160408BB	SOF Operational Enhancements	55,990	69,754	72,167
Tactical Programs		**708,192**	**1,621,982**	**2,383,838**
0301011G	Cryptologic Activities			
0301301L	Gen Def Intelligence Prog			
0301308L	Missile Intelligence Agency			
0303126K	Long-Haul Comm (DCS)	14,931	12,817	20,720
0303127K	Support of the Nat'l Comm Sys	3,454	3,155	3,839
0303132G	Global Grid Communications			
0303140G	Info Systems Security Prog			
0303401G	Comm Security (COMSEC)			
0304311D	Selected Activities		14,066	
0305098L	Defense Support Act - IPSG			
0305106LC	Consolidated Imagery Acts			
0305107LC	Tact Imagery Activities			
0305139B	DMA Mapping, Charting, & Geodesy (MC&G) Prod Sys	226,521	53,080	66,334
0305154I	Airborne Reconn Supp Prog	122,755	194,135	356,303
0305157I	Land Remote Sensing Satt Sys	5,750	34,506	
0305159B	Defense Reconn Supp Acts	6,248	6,127	11,320
0305159G	Defense Reconn Supp Acts			
0305159I	Defense Reconn Supp Acts	83,539	54,426	81,872
0305167G	Computer Security			
0305190D	C3I Intelligence Programs	8,001	12,084	6,754
0305830K	Center for Info Mgmt.		3,284	
0305884L	Intelligence Plng & Revw Acts			
0305885G	Tact Cryptologic Acts			
0305889D	Intelligence Support to OSD Counternarcotics	17,727		
0305889G	Intelligence Support to OSD Counternarcotics			
0305889L	Intelligence Support to OSD Counternarcotics			
0305898L	Mgmt Hdqs (Auxiliary Forces)			
0901600B	Contract Admin/Audit			1,357
0901600G	Contract Admin/Audit			23,451
0901600I	Contract Admin/Audit			4,825
0901600L	Contract Admin/Audit			260
1160409BB	Other Force Programs	6,080	1,170	
Intelligence and Communications		**1,780,320**	**1,507,562**	**1,892,282**
0603705D	Manufacturing Tech	18,694		
0603708D	Integrated Diagnostics	6,861	10,624	10,441
0603790D	NATO Research & Develop.	10,881	60,180	57,641
0603832D	Joint Wargaming Simulation Management Office	18,831		
0605104D	Tech Studies, Supp & Anal.	36,509	30,799	37,434
0605114E	BLACK LIGHT	4,000	4,770	4,875
0605116D	General Support to C3I	14,457		
0605117D	Foreign Material Acquisition & Exploitation	9,800	9,498	336,176
0605120S	Technical Info Service	7,653	5,657	
0605136D	FCIMS Programs	26,700		
0605137D	Manufacturing Eng Education	22,110		
0605502D	Small Bus Innovative Res	14,316		
0605502E	Small Bus Innovative Res	12,115		
0605798S	Defense Support Acts.	14,200	12,955	12,561
0605872D	Productivity Investments	585		
0605898E	Mgmt Hdqs (R&D)	19,644	22,150	24,005

(Continued on next page)

Defense Agencies RDT&E requests for Fiscal Year 1994 (Continued)

Program Element Number	Item	FY 1992	FY 1993	FY 1994
0305889E	Intelligence Supp to OSD Counternarcotics	32,450		
0708011S	Industrial Preparedness	16,900		
0901600D	Contract Admin/Audit			18,625
0901600E	Contract Admin/Audit			27,873
0901600S	Contract Admin/Audit			235
Defensewide Mission Support		**286,706**	**156,633**	**529,866**
Total Res Develop Test & Eval Defwide		**9,674,405**	**9,800,638**	**10,174,549**

DEFENSE AGENCY PROCUREMENT REQUESTS

No.	Item Nomenclature	FY 1994 Unit Cost	FY 1992 Quan.	Cost	FY 1993 Quan.	Cost	FY 1994 Quan.	Cost
				(Dollars) Millions of Dollars				

BUDGET ACTIVITY 01: Major Equipment

MAJOR EQUIPMENT, Office of the Secretary of Defense

No.	Item Nomenclature	FY 1994 Unit Cost	FY 1992 Quan.	FY 1992 Cost	FY 1993 Quan.	FY 1993 Cost	FY 1994 Quan.	FY 1994 Cost
1	C-20F Aircraft		3	93.0				
2	Motor Vehicles					.2		
3	Major Equip, OSD/WHS			163.1		208.4		62.4
4	Remotely Piloted Vehicles			129.7		137.8		69.3
5	Corp Info Mgmt			87.9		63.5		20.2
6	Cntrct Admin/Audit							6.2

MAJOR EQUIPMENT, National Security Agency

No.	Item Nomenclature	FY 1994 Unit Cost	FY 1992 Quan.	FY 1992 Cost	FY 1993 Quan.	FY 1993 Cost	FY 1994 Quan.	FY 1994 Cost
7	Classified Equip							
8	Cntrct Admin/Audit							7.4

MAJOR EQUIPMENT, Defense Nuclear Agency

No.	Item Nomenclature	FY 1994 Unit Cost	FY 1992 Quan.	FY 1992 Cost	FY 1993 Quan.	FY 1993 Cost	FY 1994 Quan.	FY 1994 Cost
9	Vehicles	19,240	30	.9	26	.3	25	.5
10	Other Cap. Eqp	365,000	231	11.3	37	17.2	10	3.7
11	Cntrct Admin/Audit							1

(Continued on next page)

DEFENSE AGENCY PROCUREMENT REQUESTS (CONTINUED)

		FY 1994	FY 1992 (Dollars)	Millions of Dollars	FY 1993		FY 1994	
No.	Item Nomenclature	Unit Cost	Quan.	Cost	Quan.	Cost	Quan.	Cost
MAJOR EQUIPMENT, Defense Information Systems Agency								
12	World-Wide Military Command and Control System ADP Sys			10.8		8.4		8.7
13	Info Srvcs Trnfr					22.0		
14	Cntrct Admin/Audit							1.4
15	Items Less than $2M			40.4		66.9		44.2
16	Drug Interdiction Supp			3.3				
17	Industr/Depot Maint Equip			13.4				
MAJOR EQUIPMENT, Defense Intelligence Agency								
18	Intelligence & Comm							
19	Cntrct Admin/Audit							.8
MAJOR EQUIPMENT, Defense Logistics Agency								
20	Defense Supp Acts			23.5		1.9		3.4
MAJOR EQUIPMENT, Defense Mapping Agency								
21	Comm Equipment							5.9
22	ADP Equipment							2.5
23	VECTOR Prod Equip							2.8
24	Develp Test Facility							17.5
25	MC&G Mainframe Upgrade							2.2

(Continued on next page)

DEFENSE AGENCY PROCUREMENT REQUESTS (CONTINUED)

No.	Item Nomenclature	FY 1994 Unit Cost	FY 1992 Quan.	FY 1992 Cost	FY 1993 Quan.	FY 1993 Cost	FY 1994 Quan.	FY 1994 Cost
				(Dollars)		Millions of Dollars		
26	Vehicles			2		.3		.4
27	Cntrct Admin/Audit					.		.7
28	Other Capital Equip			19.6		34.3		18.8
29	Geodesy & Geophysical Equip			1.2		2.1		
30	Defense Hydrographic Equip							3.7
MAJOR EQUIPMENT, DIS								
31	Vehicles			2.1		.1		3.2
32	Other Capital Equip			3.2		1.8		1.9
33	Contract Administration/Audit							1
MAJOR EQUIPMENT, USUHS								
34	Items Less Than $2M			.8				
MAJOR EQUIPMENT, Defense Contract Audit Agency								
35	Items Less than $2M			3.6		5.6		
MAJOR EQUIPMENT, Defense Support Project Office								
36	Major Equipment, DSPO					78.5		170.4
37	Major Equipment			178.3		105.8		186.2
38	Contract Administration/Audit							3.3

(Continued on next page)

DEFENSE AGENCY PROCUREMENT REQUESTS (CONTINUED)

No.	Item Nomenclature	FY 1994 Unit Cost	FY 1992 Quan.	Cost	FY 1993 Quan.	Cost	FY 1994 Quan.	Cost
						(Dollars) Millions of Dollars			

ON-SITE INSPECTION AGENCY

MAJOR EQUIPMENT, Office of the Joint Chiefs of Staff

No.	Item Nomenclature	FY 1994 Unit Cost	FY 1992 Cost	FY 1993 Cost	FY 1994 Cost
39	Major Eqpt, OJCS		21.6	22.7	50.3
40	Cntrct Admin/Audit				4
41	Vehicles	26,500	.1	4	.1
42	Other Capital Equip		23.8	6.6	.9
43	Cntrct Admin/Audit				.1

STRATEGIC DEFENSE INITIATIVE ORGANIZATION

No.	Item Nomenclature	FY 1994 Unit Cost	FY 1992 Cost	FY 1993 Cost	FY 1994 Cost
44	Patriot		24.9	75.2	120.7

CENTRAL IMAGERY OFFICE

45 Major Equipment, CJO

			FY 1992	FY 1993	FY 1994
TOTAL MAJOR EQUIPMENT			**1,447.0**	**1,313.9**	**1,279.5**

BUDGET ACTIVITY 02: Special Operations Command

AVIATION PROGRAMS

No.	Item Nomenclature	FY 1994 Unit Cost	FY 1992 Cost	FY 1993 Cost	FY 1994 Cost
46	MC-130H Combat Talon II		113.0	53.5	23.7
47	AC-130U Gunship Acq		77.9		27.5
48	C-130 Mods		157.2	132.5	63.8

(Continued on next page)

DEFENSE AGENCY PROCUREMENT REQUESTS (CONTINUED)

(Dollars) Millions of Dollars

No.	Item Nomenclature	FY 1994 Unit Cost	FY 1992 Quan.	FY 1992 Cost	FY 1993 Quan.	FY 1993 Cost	FY 1994 Quan.	FY 1994 Cost
49	HH-53 Mods			19.6		7.6		13.7
50	MH-47/MH-60 Mods			320.7		10.0		7.6
51	MH-60 Mods					.6		
52	Other Aircraft Mods					3.0		
53	Aircraft Support			15.1		118.3		30.2
SHIPBUILDING								
54	PC,Cyclone Class			4.2		20.0		13.4
55	Sub Conversion							.4
56	MK V Patrol Boat							9.0
AMMUNITION PROGRAMS								
57	Special Operations Forces Pyro/Demo			21.0		13.2		12.6
58	Special Operations Forces Platform Gun Ammo			39.9		38.1		19.0
59	Special Operations Forces Indiv Weapons Ammo			13.1		17.4		12.6
OTHER PROCUREMENT PROGRAMS								
60	Cntrct Admin & Audit Acts							13.7
61	Comm Equip & Elec.			72.2		78.4		40.1
62	Special Operations Forces Intelligence Sys					44.9		26.7
63	Special Operations Forces Small Arms & Wpns.			4.6		4.6		2.2

(Continued on next page)

DEFENSE AGENCY PROCUREMENT REQUESTS (CONTINUED)

(Dollars) Millions of Dollars

No.	Item Nomenclature	FY 1994 Unit Cost	FY 1992 Quan.	FY 1992 Cost	FY 1993 Quan.	FY 1993 Cost	FY 1994 Quan.	FY 1994 Cost
64	Special Warfare Equip			23.1		18.6		17.7
65	Misc Equip			43.0		10.3		4.2
66	Special Operations Forces Plng & Rehearsal System (SOFP)							10.5
67	Classified Programs			132.0		109.3		95.7
68	Psyop Equipment			.1				
	TOTAL SPECIAL OPERATIONS COMMAND			**1,056.7**		**680.2**		**450.7**
	TOTAL PROCUREMENT, DEFENSEWIDE			**2,503.7**		**1,994.1**		**1,730.2**

DEPARTMENT OF DEFENSE SPACE PROGRAMS
PROCUREMENT (INCLUDING INITIAL SPARES) AND RDT&E
Fiscal Years 1991, 1992, and 1993 (Millions of Dollars)

Agency and Program	1991		1992 E		1993 E	
	Procurement	RDT&E	Procurement	RDT&E	Procurement	RDT&E
AIR FORCE						
DMSP	$ 48.7	$ 48.5	$106.8	$ 28.2	$ 32.1	$ 23.8
DSCS	63.9	16.1	55.5	13.8	25.5	15.7
Defense Support Program	337.7	270.1	71.0	51.5	297.1	74.4
LANDSAT	-	-	-	30.0	-	6.0
Medium Launch Vehicle	269.7	221.0	221.3	42.8	226.6	42.7
Milstar	193 5	760.0	320.3	1,043.0	272.0	1,261.9
National Launch System	-	25.0	-	54.3	-	125.0
Navstar GPS	222.7	59.6	330.4	66.4	350.7	69.0
Space Boosters	207.0	128.3	290.5	140.9	382.2	145.9
NAVY						
FSC	244.4	-	283.1	-	326.0	-
JOINT PROGRAMS						
SDI	-	2,692	-	3,282	-	4,315

Source: Department of Defense, "Program Acquisition Costs by Weapon System" (Annually).
E Estimate. Latest year reflects Administration's budget proposal.
KEY: DMSP   = Defense Meteorological Satellite Program
DSCS   = Defense Satellite Communications System
FSC   = Fleet Satellite Communications
GPS   = Global Positioning System
LANDSAT = Land Remote Sensing Satellite System
SDI   = Strategic Defense Initiative

STRATEGIC DEFENSE INITIATIVE ORGANIZATION
FUNDING BY PROJECT NUMBER
Fiscal Years 1989 - 1993 (Millions of Dollars)

Project Number and Title		1989	1990	1991	1992 E	1993 E
1101	Passive Sensors	$ 70	$ 57	$ 35	$ 34	$ 56
1102	Microwave Radar	14	-	5	12	18
1103	Laser Radar Technology	80	59	30	13	13
1104	Signal Processing	80	67	45	30	45
1105	Discrimination	179	134	122	89	126
1106	Sensor Studies & Experiments	166	182	159	184	208
1109	Theater Defense Discrimination	-	-	-	10	11
1110	Sensors/Integration	-	-	-	21	54
1201	Interceptor Component Technology	93	86	100	31	63
1202	Interceptor Integration Technology	67	95	129	126	79
1203	Hypervelocity Technology	24	20	15	6	11
1204	Interceptor Studies & Analysis	48	13	54	15	18
1205	Foreign Technology Support	15	6	12	-	-
1206	Advanced TMD Weapons	61	85	31	18	14
1208	Discriminating Interceptor	-	-	-	7	50
1209	Endoatmospheric Interceptor Technology	-	-	-	57	63
1210	Navy LEAP Technology Demonstration	-	-	-	8	35
1212	D-2 Program	-	-	-	6	19
1301	Free Electron Laser	203	130	29	23	24
1302	Chemical Laser Technology	99	117	91	104	175
1303	Neutral Particle Beam Technology	108	116	105	80	76
1304	Nuclear Directed Energy Technology	21	13	10	5	-
1305	Acquisition, Tracking, Pointing & Fire Control Technology	237	274	80	67	47
1307	Directed Energy Demonstration	-	-	-	-	24
1405	Communications Engineering	-	6	6	11	24
1501	Survivability Technology	103	107	57	68	135
1502	Lethality and Target Hardening	62	39	27	51	50
1503	Power & Power Conditioning	109	84	49	6	47
1504	Materials & Structures	31	36	27	24	58
1505	Launch Planning, Development and Demonstration	57	32	16	-	-
1601	Innovative Science & Technology	114	113	66	70	83
1602,3	New Concepts Development	-	-	25	40	41
1701	Launch Services	-	-	24	71	68
1702	Special Test Activities	-	-	23	17	36
2101	Boost Surveillance and Tracking System	233	300	-	-	-
2102	Brilliant Eyes	93	78	48	116	278
2103	Ground-Based Surveillance & Tracking System	10	40	47	118	112
2104	Ground-Based Radar	71	89	39	82	212
2106	Advanced Contingency Theater Sensor	-	-	-	28	90
2201	Space-Based Interceptor	116	73	35	9	-
2202	Ground-Based Exoatmospheric Interceptor Development	163	128	85	173	160
2203	HEDI (E2I)	113	66	103	66	-
2204	DEW Concept Definition	23	8	4	2	5
2205	Brilliant Pebbles	46	129	392	390	450
2207	PATRIOT Multi-mode Missile	-	-	-	160	171
2208	Extended Range Interceptor (ERINT)	-	-	-	160	129
2209	Arrow Continuation Experiments (ACES)	$ -	$ -	$ -	$ 60	$ 58
2210	THAAD	-	-	-	100	243
2212	CORPS SAM	-	-	-	25	25
2213	Sea-Based TMD Interceptor	-	-	-	30	26
2300	Command Center	116	88	39	74	1,204
2304	System Software Engineering	-	-	4	8	8
3100	Systems Engineering	48	69	-	-	-
3102	System Engineering	-	-	65	74	199
3104	Integrated Logistics Support	8	7	4	4	7
3105	Producibility & Manufacturing	7	10	9	9	20
3107	Environment, Siting & Facilities	6	4	14	11	16
3109	System Security Engineering	-	-	7	11	12
3111	Surveillance Engineering	-	-	7	10	11

(Continued on next page)

STRATEGIC DEFENSE INITIATIVE ORGANIZATION (Continued)
FUNDING BY PROJECT NUMBER
Fiscal Years 1989 - 1993 (Millions of Dollars)

Project Number and Title	1989	1990	1991	1992 E	1993 E
3112 System Engineering Support	-	-	-	27	29
3113 Ground Communications	-	-	-	15	13
3201 Architecture and Analysis	10	13	7	3	5
3202 Operations Interface	3	7	7	7	6
3203 Intelligence Threat Development	5	12	10	10	10
3204 Countermeasures Integration	17	17	19	17	22
3205 Theater Missile Defense Special Studies	20	14	30	68	32
3206 System Threat	-	-	7	8	7
3207 System Architecture	-	-	20	24	-
3211 C4I and Operational Analysis	-	-	-	16	19
3301 SDIO Test Data Centers	-	-	-	11	22
3302 System Test Environment	113	125	104	83	116
3303 Independent Test & Evaluation	5	4	4	6	6
3304 Targets	15	47	65	147	217
3305 Theater Test Bed	8	27	38	55	37
3306 Computer Resources and Engineering	13	14	12	29	29
3307 Airborne Surveillance Test Bed	88	56	44	38	45
3308 System Simulating (Level 1 and Level 2)	-	-	5	9	7
3309 System Test Planning and Execution	-	-	-	24	133
3310 Test and Evaluation Facilities and Launch Support	-	-	-	49	57
3311 Mobile Test Assets	-	-	-	12	14
3312 System Test Environment Support	-	-	-	15	15
4000 Operational Support Costs	209	247	228	407	351
4305 Miniaturized Accelerators for PET	17	20	0	1	1
Other programs	0	0	0	0	0
TOTAL DETAILED PROJECTS	**$3,628**	**$3,572**	**$2,878**	**$4,168**	**$6,435**

Source: Strategic Defense Initiative Organization, "1992 Report to the Congress on the Strategic Defense Initiative" (Annually).
E Estimate. Represents Administration's budget request.

CHAIRMEN OF THE JOINT CHIEFS OF STAFF (1949-PRESENT)

	From	To
General of the Army Omar N. Bradley, USA	16-Aug-49	August 15, 1953
Adm. Arthur W. Radford, USN	15-Aug-53	15-Aug-57
Gen. Nathan F. Twining, USAF	15-Aug-57	30-Sep-60
Gen. Lyman L. Lemnitzer, USA	1-Oct-60	30-Sep-62
Gen. Maxwell D. Taylor, USA	1-Oct-62	1-Jul-64
Gen. Earle G. Wheeler, USA	3-Jul-64	2-Jul-70
Adm. Thomas H. Moorer, USN	2-Jul-70	1-Jul-74
Gen. George S. Brown, USAF	1-Jul-74	20-Jun-78
Gen. David C. Jones, USAF	21-Jun-78	18-Jun-82
Gen. John W. Vessey Jr., USA	18-Jun-82	30-Sep-85
Adm. William J. Crowe Jr., USN	1-Oct-85	30-Sep-89
Gen. Colin L. Powell, USA	1-Oct-89	Present

Section VII.

U.S. Air Force

AIR FORCE RDT&E REQUESTS FOR FISCAL YEAR 1994
(Research, Development, Test and Evaluation)
Dollar figures are in thousands.

Program Element Number	Item	FY 1992	FY 1993	FY 1994
0601101F	In-House Lab Independent Research	6,029	4,730	5,155
0601102F	Defense Research Sciences	201,347	235,751	241,317
0602101F	Geophysics	38,137	36,981	30,252
0602102F	Materials	65,284	71,768	70,805
0602201F	Aerospace Flight Dynamics	62,937	67,898	64,238
0602202F	Human Systems Tech	61,811	61,087	51,392
0602203F	Aerospace Propulsion	66,159	68,660	78,100
0602204F	Aerospace Avionics	74,838	70,935	74,835
0602205F	Personnel, Trng & Simulation	28,251	31,208	28,942
0602206F	Civil Eng. & Environ. Quality	7,068	11,276	7,187
0602302F	Rocket Propulsion & Astronautics Tech	36,601	36,012	40,031
0602601F	Advanced Weapons	38,468	46,327	32,961
0602602F	Conventional Munitions	40,981	36,079	46,653
0602702F	Command Control & Comms.	79,280	82,453	95,957
0602790F	Small Bus. Innovative Res./ Small Bus. Tech Transfer Pilot Program		137,525	140,976
Technology Base		**807,191**	**998,690**	**1,008,801**
0603106F	Logistics Systems Tech	12,917	14,014	14,318
0603112F	Advd Materials for Weapon Sys	12,573	16,679	15,825
0603202F	Aerospace Propulsion Subsys. Integration	25,279	27,279	28,004
0603203F	Advd Avionics for Aero Veh.	32,818	40,146	49,226
0603205F	Aerospace Veh. Technology	13,222	15,273	13,114
0603211F	Aerospace Structures	14,436	15,907	12,641
0603216F	Aero Propulsion & Power Tech	35,893	38,148	36,614
0603227F	Personnel, Trng & Simul Tech	7,877	8,968	8,818
0603231F	Crew Systems & Personnel Protection Technology	15,078	17,909	10,460
0603238F	Global Surveillance/Air Def/Prec. Strike Tech Dem.		4,676	14,999
0603245F	Advd Fighter Tech Integration	11,729	17,650	15,613
0603250F	Lincoln Laboratory	25,233	24,958	22,908
0603253F	Advd Avionics Integration	15,240	23,077	30,384
0603269F	Nat'l Aero Space Plane Tech	198,114	141,244	43,259
0603270F	Electronic Warfare Technology	27,922	30,658	25,689
0603302F	Space & Missile Rocket Prop	6,758	12,640	10,027
0603311F	Ballistic Missile Tech	59,179	59,211	58,980
0603319F	Airborne Laser Tech			3,845
0603363F	Armament Tech Integration	1,782		
0603401F	Advd Spacecraft Tech	23,461	27,121	24,275
0603410F	Space Sys Environ Interactions Technology	4,217	3,912	3,500
0603428F	Space Subsystems Technology	4,533	4,117	
0603601F	Conventional Weapons Tech	27,311	25,257	25,964
0603605F	Advd Radiation Tech	62,330	81,251	55,415
0603707F	Weather Systems Tech	4,725	5,002	4,452
0603723F	Civil & Environ Eng. Tech	10,990	10,552	8,435
0603726F	C3I Subsystem Integration	8,385	8,838	15,882
0603728F	Advd Computing Tech	8,793	16,029	19,619
0603789F	C3 Advanced Development	7,715	9,173	17,066
Advanced Technology Development		**678,510**	**699,689**	**589,332**
0603105F	Olympic			
0603110F	Special Evaluation Program			
0603111F	Meridian			
0603425F	Advanced Warning System	76,460		
0604226F	B-1B (H)	5,322	80,597	93,543
0604240F	B-2 Advanced Tech Bomber	1,522,346	1,189,290	790,497
0604312F	ICBM Modernization	119,060	51,304	

(Continued on next page)

AIR FORCE RDT&E REQUESTS FOR FISCAL YEAR 1994 (Continued)

Program Element Number	Item	FY 1992	FY 1993	FY 1994
0604711F	Systems Surv. (Nuclear Eff.)	6,566	6,038	3,643
0101113F	B-52 Squadrons	3,921	20,472	
0101120F	Advanced Cruise Missile	39,300	19,543	25,393
0101142F	KC-135 Squadrons	5,730	3,499	
0101213F	Minuteman Squadrons	57,697	26,750	184,335
0101312F	PACCS and WWABNCP System EC-135 Class V Mods	1,298	1,978	
0101313F	Strat War Plng Sys-USSTRATCOM	3,792	6,862	
0101815F	Advd Strategic Programs			
0102310F	NCMC - TW/AA Systems	129,080		
0102325F	Joint Surveillance System	7,945	4,355	3,246
0102411F	Surveillance Radar Sta/Sites	11,473	7,695	8,306
0102412F	Distant Early Warning (DEW) Radar Stations	2,195	2,444	2,578
0102423F	Ballistic Missile Early Warning System (BMEWS)	14,399		
0102424F	SPACETRACK	34,231		
0102431F	Defense Support Program	51,459		
0102432F	Sub-Launched Ballistic Missile (SLBM) Radar Warning System	966		
0102433F	NUDET Detection System	6,812		
0303131F	Min. Essential Emer Comm. Network (MEECN)	19,938	12,975	35,634
0303152F	World-Wide Military Command & Cntrl Systems, Info Sys	602		
0303601F	Milstar Satellite Comm Sys (AF Terminals)	156,285	1,138,567	973,162
0303603F	Milstar Satellite Comm Sys	886,146		
0303606F	UHF Satellite Comm.			11,457
0305124F	Special Apps Program			
0305145F	Arms Control Implementation	14,639	4,266	7,107
0305172F	Combined Advd Applications			
0305181F	Wstrn Space Launch Fac. (WSLF)			9,546
0305182F	Estrn Space Launch Fac. (ESLF)			41,242
0305892F	Special Analysis Activities			
0305905F	Improved Space Based TW/AA		236,557	214,794
0305906F	NCMC - TW/AA System		151,101	141,841
0305909F	Ballistic Missile Early Wrng System (BMEWS)		3,856	599
0305910F	SPACETRACK		71,502	45,246
0305911F	Defense Support Program		49,081	66,777
0305912F	Sub-Launched Ballistic Missile (SLBM) Radar Warning Sys		944	
0305913F	NUDET Detection System		4,897	9,359
0401218F	KC-135s			20,811
Strategic Programs		**3,553,016**	**3,443,775**	**2,986,396**
0603107F	Technical Eval System			
0603260F	Intelligence Advd Dev.	4,374	7,170	6,134
0603307F	Air Base Operability Avd Dev	3,315	3,520	3,739
0603617F	Command, Cntrl & Comm Apps	3,159	3,861	9,395
0603714F	DoD Physical Sec. Eqpt-Ext.	730	525	2,971
0603742F	Combat ID Technology	23,674	18,169	28,759
0603801F	Special Programs			
0604201F	Aircraft Avionics Eqpt Dev	14,039	15,094	6,637
0604212F	Aircraft Eqpt Dev.	3,919	3,804	1,532
0604218F	Engine Model Derivative Program (EMDP)	3,950	945	863
0604222F	Nuclear Weapons Support	5,735	5,296	5,475
0604231F	C-17 Program	256,855	168,717	179,799
0604233F	Specialized Undergraduate Pilot Training	4,196	4,392	36,835
0604237F	Variable Stability In-Flight Simulator Test Aircraft	2,062	2,066	5,838
0604239F	F-22 EMD	1,606,804	1,925,199	2,250,997
0604242F	Advd Interdiction Aft (AX)	1,999	1,897	3,835
0604249F	Night/Precision Attack	3,125	25,094	82,210

(Continued on next page)

AIR FORCE RDT&E REQUESTS FOR FISCAL YEAR 1994 (Continued)

Program Element Number	Item	FY 1992	FY 1993	FY 1994
0604268F	Aircraft Engine Component Improvement Program	110,854	104,096	102,704
0604270F	Electronic Warfare Development	194,705	154,496	143,433
0604321F	Joint Tactical Fusion Prog	4,922	2,883	4,221
0604327F	Hardened Target Munitions	12,789	5,377	
0604601F	Chemical/Biological Def Eqpt	11,509	15,356	9,874
0604602F	Armament/Ordnance Dev.	5,425	5,839	11,407
0604604F	Submunitions	4,990	7,063	3,835
0604607F	Wide-Area, Anti-Armor Muni.	9,697		
0604617F	Air Base Operability	9,886	13,586	11,023
0604618F	Joint Direct Attack Muni.		24,648	87,822
0604703F	Aeromedical/Chemical Def Sys	6,368	6,559	10,260
0604704F	Common Support Eqpt Dev	16,435	8,871	4,793
0604706F	Life Support Systems	12,384	12,044	11,024
0604708F	Civil, Fire, Environ, Shelter Engineering	2,583	2,513	4,524
0604727F	Joint Standoff Weapons Sys		5,463	24,614
0604733F	Surface Defense Suppression	20,577	7,722	1,917
0604740F	Comp. Res Tech Trans (CRTT)	17,690	20,192	7,137
0604750F	Intelligence Equipment	2,919	2,853	2,875
0604754F	Joint Tact Info Dist Sys(JTIDS) 15,415	15,140	16,113	
0604756F	Side Looking Airborne Radar		3,899	
0604770F	Joint Surveillance/Target Attack Radar Sys (JSTARS)	307,363	313,463	295,228
0604779F	Joint Interoperability of Tact Comm & Cntrl Sys (JINTACCS)	4,818	6,670	4,793
0207129F	F-111 Squadrons	27,823	27,639	25,679
0207130F	F-15A/B/C/D Squadrons		10,453	
0207131F	A-10 Squadrons	12,719		
0207133F	F-16 Squadrons	147,661	109,409	116,947
0207134F	F-15E Squadrons	92,988	49,496	91,497
0207136F	Manned Destructive Suppress.	4,958	8,362	20,496
0207137F	CONSTANT HELP			
0207141F	F-117A Squadrons	64,182	1,224	6,778
0207160F	Tri-Srvc Standoff Attck Miss.			
0207161F	Tactical AIM Missiles			33,887
0207163F	Advd Med Range Air-to-Air Missile (AMRAAM)	30,309	33,273	69,785
0207217F	Follow-On Tact Reconn. Sys	87,391	58,362	65,338
0207247F	AF TENCAP	858	4,099	14,722
0207248F	Special Eval Program	1,421	855	120,711
0207411F	Overseas Air Weapon Cont Sys	1,905		19,570
0207412F	Tact Air Control Systems	19,281	24,543	28,913
0207417F	Air Wrng & Cont Sys (AWACS)	201,974	62,927	87,066
0207419F	Tact Air Comm & Cont Sys	6,389	5,958	
0207423F	Advd Comm Systems	3,441	471	478
0207424F	Eval & Analysis Program	14,159	15,389	75,384
0207431F	Tact Air Intell. Sys Acts.			
0207433F	Advd Program Technology	30,552	22,454	148,114
0207438F	Theater Battle Mgmt (TBM) C4I			12,518
0207579F	Advd Systems Improvements	54,089	51,975	129,164
0207590F	Seek Eagle	18,248	28,976	15,171
0207591F	Advd Program Evaluation	167,328	92,046	89,604
0208006F	Mission Planning Systems	12,383	14,480	24,249
0208010F	Joint Tact Comms Prog (TRI-TAC)	5,971	11,755	
0208021F	Electronic Combat Support			
0208042F	HAVE FLAG			
0303605F	Satellite Comms Terminals	1,544	4,145	1,399
0305137F	Nat'l Airspace Sys (NAS) Plan	4,176	6,672	18,773
0305142F	Applied Tech & Integration			
0305158F	CONSTANT SOURCE	11,893	7,062	3,245
0305887F	Elec Combat Intell. Support	1,841	1,790	2,004
0401840F	MAC Command & Control System	10,975	12,082	11,361
Tactical Programs		**4,420,435**	**4,180,317**	**4,910,864**

(Continued on next page)

AIR FORCE RDT&E REQUESTS FOR FISCAL YEAR 1994 (Continued)

Program Element Number	Item	FY 1992	FY 1993	FY 1994
0102830F	Classified Program			
0301305F	Intelligence Prod. Acts.			
0301310F	Foreign Tech Division			
0301313F	Defense Dissemination Prog.			
0301314F	Infrared/Electro-Optical/Dir. Energy Weapons Processing & Exploitation			
0301315F	Missile & Space Tech Coll.			
0301317F	SENIOR YEAR Operations			
0301324F	FOREST GREEN			
0301339F	Intell Telecom & Defense Special Sec. Sys			
0301357F	NUDET Detection System		2,564	
0303110F	Defense Satellite Comm Sys	13,765	12,856	25,522
0303126F	Long-Haul Comm (DCS)	3,253	3,196	
0303140F	Info Sys Security Prog			15,418
0303144F	Electromagnetic Compatibility Analysis Ctr (ECAC)	10,029	9,733	9,978
0303401F	Comms Security (COMSEC)			
0304111F	Special Activities			
0305114F	Air Traffic Control, Approach, & Landing Sys (ATCALS)	9,524	12,030	9,304
0305159F	Def Reconnaissance Supp Acts.			
0305164F	NAVSTAR Global Positioning Sys (User Eqpt.)	14,687	20,434	16,164
0305165F	NAVSTAR Global Positioning Sys (Space & Cntrl Segments)	51,293	56,112	38,990
Intelligence and Communications		**2,183,018**	**2,409,816**	**2,291,341**
0603402F	Space Test Program	43,706	50,745	50,465
0603438F	Satellite Sys Survivability	9,782	5,126	10,732
0604211F	Advd Aerial Target Dev.	19,349	20,559	
0604227F	Training Sys Dev.	41,142	35,118	30,015
0604243F	Manpower, Personnel & Trng Dev	2,134	3,115	4,838
0604256F	Threat Simulator Dev			34,362
0604258F	Target Systems Dev			10,154
0604408F	Nat'l Launch System	48,673	9,435	53,906
0604609F	R&M Maturation/Tech Insert.	18,253	21,692	20,593
0604707F	Weather Systems - Eng Dev	5,497	6,119	9,379
0604735F	Range Improvement	68,619	52,908	15,714
0604747F	Electromagnetic Radiation Test Facilities	3,508	3,699	
0604755F	Improved Capability for Dev. Test and Evaluation	46,959	51,383	
0604759F	Major T&E Investment			55,798
0605101F	RAND Project Air Force	22,488	22,380	26,748
0605306F	Ranch Hand II Epidemiology Study	9,580	8,925	3,707
0605502F	Small Bus Innovative Res (H)	66,775		
0605708F	Nav/Radar/Sled Trck Test Supp	25,682	26,074	28,313
0605712F	Initial Operational Test and Evaluation	21,807	26,016	32,811
0605807F	Test & Evaluation Support	369,903	386,983	399,930
0605808F	Development Planning	8,697	9,611	9,796
0605856F	Environmental Compliance		12,014	39,575
0605863F	RDT&E Aircraft Support	43,503	41,647	42,157
0605876F	Minor Constrc. (RPM)-RDT&E		2,899	7,739
0605878F	Maint & Repair (RPM)-RDT&E		50,918	46,020
0605894F	Real Property Maint-RDT&E (H)	102,493		
0605896F	Base Operations - RDT&E	75,808	103,934	121,974
0207601F	USAF Wargaming & Simulation			11,573
0305110F	Satellite Control Network	108,344	97,053	110,164
0305119F	Medium Launch Vehicles	40,404	49,659	58,502
0305130F	AFSCN Operations	14,916		
0305138F	Upper Stage Space Vehicles	5,731	43,362	4,141
0305144F	Tital Space Launch Vehicles	140,717	120,811	330,740
0305160F	Def Meteorological Satellite Program (DMSP)	28,179	21,938	31,953
0701112F	Inventory Control Point Ops	2,941	1,131	

(Continued on next page)

AIR FORCE RDT&E REQUESTS FOR FISCAL YEAR 1994 (Continued)

Program Element Number	Item	FY 1992	FY 1993	FY 1994
0702207F	Depot Maint (Non-IF)	2,928	2,729	1,830
0708011F	Industrial Preparedness	59,994	98,553	
0708012F	Logistics Supp Acts.	6,543	5,811	6,336
0708026F	Prod., Reliability, Avail., Maint. Prog Ofc (PRAMPO)	23,677	22,307	18,068
0708054F	Pollution Prevention			25,518
0804734F	Cryptologic/SIGINT-Related Skill Training			1,926
0901218F	Civilian Compensation Prog	5,199	5,135	5,775
0901600F	Contract Admin/Audit			243,178
1001004F	Int'l Activities	3,023	3,522	3,820
Defensewide Mission Support		**1,496,954**	**1,423,311**	**1,908,250**
Research Development Test & Eval AF		**13,139,124**	**13,155,598**	**13,694,984**

AIR FORCE AIRCRAFT AND MISSILE PROCUREMENT REQUESTS

No.	Item Nomenclature	FY 1994 Unit Cost	FY 1992 Quan.	FY 1992 Cost	FY 1993 Quan.	FY 1993 Cost	FY 1994 Quan.	FY 1994 Cost
				Millions of Dollars				

BUDGET ACTIVITY 01: Combat Aircraft

STRATEGIC OFFENSIVE

No.	Item Nomenclature	FY 1994 Unit Cost	FY 1992 Quan.	FY 1992 Cost	FY 1993 Quan.	FY 1993 Cost	FY 1994 Quan.	FY 1994 Cost
1	B-1B (MYP)			154.5		165.8		162.5
2	B-2A (MYP)1			(1746.1)	4	(3407.1)		(604.3)
	Less:Adv Pro (PY)			(-164.9)		(-747.0)		
				1581.2		**2660.1**		**604.3**
3	B-2A (MYP) Adv Pro (CY)			717.0				

TACTICAL FORCES

No.	Item Nomenclature	FY 1994 Unit Cost	FY 1992 Quan.	FY 1992 Cost	FY 1993 Quan.	FY 1993 Cost	FY 1994 Quan.	FY 1994 Cost
4	F-15 E		3	694.6		11.4		28.7
5	F-16 C/D (MYP)	33,045,833	48	(1274.4)	24	(733.8)	24	(793.1)
	Less:Adv Pro (PY)			(-201.8)		(-125.0)		(-68.4)
				1072.7		**608.8**		**724.7**
6	F-16 C/D (MYP) Adv Pro (CY)			78.1		67.7		70.8

OTHER COMBAT AIRCRAFT

No.	Item Nomenclature	FY 1994 Unit Cost	FY 1992 Quan.	FY 1992 Cost	FY 1993 Quan.	FY 1993 Cost	FY 1994 Quan.	FY 1994 Cost
7	E-3A		4					

TOTAL COMBAT AIRCRAFT — 4,298.5 — 3,513.8 — 1,591.1

GLOSSARY

MYP = multiyear procurement **PY = prior year** **Adv Pro (CY) = current year**
TRNR = Trainer

(Continued on next page)

AIR FORCE AIRCRAFT AND MISSILE PROCUREMENT REQUESTS (Continued)

Millions of Dollars

No.	Item Nomenclature	FY 1994 Unit Cost	FY 1992 Quan.	FY 1992 Cost	FY 1993 Quan.	FY 1993 Cost	FY 1994 Quan.	FY 1994 Cost
BUDGET ACTIVITY 02: Airlift Aircraft								
STRATEGIC AIRLIFT								
8	DOD Financial Sys							144.3
TACTICAL AIRLIFT								
9	C-17 ( )	387,284,833	4	(1683.1)	6	(1961.2)	6	(2323.7)
	Less:Adv Pro(PY)			(-159.3)		(-172.4)		(-250.9)
				1523.8		**1788.8**		**2072.8**
10	C-17 (MYP) Adv Proc (CY)			172.4		252.5		245.5
11	HC-130			70.9		99.0		
OTHER AIRLIFT								
12	C-130H		9	289.4	8	297.4		53.8
13	LC130		2	92.0				
TOTAL AIRLIFT AIRCRAFT				**2,148.6**		**2,437.7**		**2,516.4**
BUDGET ACTIVITY 03: Trainer Aircraft								
14	Enhanced Flght Screener	301,575	38		42	12.1	33	10.0
15	Tanker, Trans.,Trnr System	∠,210,171	34	156.1	36	157.0	35	147.4
TOTAL TRAINER AIRCRAFT				**156.1**		**169.2**		**157.3**

(Continued on next page)

AIR FORCE AIRCRAFT AND MISSILE PROCUREMENT REQUESTS (Continued)

No. Item Nomenclature	FY 1994 Unit Cost	FY 1992 Quan.	Cost	FY 1993 Quan.	Cost	FY 1994 Quan.	Cost
BUDGET ACTIVITY 04: Other Aircraft							
HELICOPTERS							
16 MH-60G	6			10	29.8		
MISSION SUPPORT AIRCRAFT							
17 CIVIL AIR PATROL A/C	94,185	27	2.0	27	2.7	27	2.5
OTHER AIRCRAFT							
18 E-8B	360,160,000			2	(632.1)	1	(360.2)
Less:Adv Pro(PY)					(-125.4)		(-78.3)
					506.7		**281.8**
19 E-8B							
Adv Pro (CY)					78.3		123.7
20 SOF A/C CSE							18.8
TOTAL OTHER AIRCRAFT			**2.0**		**617.6**		**426.8**
BUDGET ACTIVITY 05: **Modification of Inservice Aircraft**							
STRATEGIC AIRCRAFT							
21 B-2A			1.7				21.9
22 B-1B			97.4		45.4		50.8
23 B-52			45.2		64.1		47.4
24 F-117					24.5		16.3

(Continued on next page)

AIR FORCE AIRCRAFT AND MISSILE PROCUREMENT REQUESTS (Continued)

Millions of Dollars

No. Item Nomenclature	FY 1994 Unit Cost	FY 1992 Quan.	FY 1992 Cost	FY 1993 Quan.	FY 1993 Cost	FY 1994 Quan.	FY 1994 Cost
TACTICAL AIRCRAFT							
25 A-10			20.8		4.0		28.4
26 F/RF-4							2.0
27 F-15			295.9		301.3		282.7
28 F-16			228.1		168.0		120.5
29 EF-111					8.9		
30 F-111			80.7		38.3		19.1
31 T/AT-37			7.3		1.9		3.4
AIRLIFT AIRCRAFT							
32 C-5			40.7		8.2		31.1
33 C-9			1.4		2.0		8.5
34 C-17A							16.5
35 C-21			.3		.1		.3
36 C-STOL			1.6		.1		.1
37 C-137			.8		10.6		3.5
38 C-141			35.9		40.3		29.2
TRAINER AIRCRAFT							
39 T-38			39.4		30.3		12.9

(Continued on next page)

AIR FORCE AIRCRAFT AND MISSILE PROCUREMENT REQUESTS (Continued)

No. Item Nomenclature	FY 1994 Unit Cost	FY 1992 Quan.	FY 1992 Cost	FY 1993 Quan.	FY 1993 Cost	FY 1994 Quan.	FY 1994 Cost
40 T-41 AIRCRAFT			.1		.2		.2
41 T-43			11.6		.3		.3
OTHER AIRCRAFT							
42 KC-10A (ATCA)			11.9		17.6		36.7
43 C-12			.2		.2		.3
44 C-18			.2		.2		.2
45 C-20 MODS			.3		.1		.1
46 VC-25A MOD			.2		.2		.6
47 C-130			84.7		66.1		141.1
48 C-135			570.3		522.2		46.6
49 E-3			57.4		74.6		4.6
50 E-4			6.4		17.8		31.5
51 H-1							.1
52 H-60			.6				29.6
53 Other Aircraft			65.9		78.8		84.0
OTHER MODIFICATIONS							
54 Classified Proj			109.0		44.9		37.6
TOTAL MODIFICATION OF INSERVICE AIRCRAFT			**1,815.9**		**1,570.9**		**1,108.0**

(Continued on next page)

AIR FORCE AIRCRAFT AND MISSILE PROCUREMENT REQUESTS (Continued)

Millions of Dollars

No.	Item Nomenclature	FY 1994 Unit Cost	FY 1992 Quan.	FY 1992 Cost	FY 1993 Quan.	FY 1993 Cost	FY 1994 Quan.	FY 1994 Cost
BUDGET ACTIVITY 06: **Aircraft Spares and Repair Parts**								
AIRCRAFT SPARES & REPAIR PARTS								
55	Spares & Rpr Parts			589.0		495.7		556.1
	TOTAL AIRCRAFT SPARES AND REPAIR PARTS			**589.0**		**495.7**		**556.1**
BUDGET ACTIVITY 07: **Aircraft Support Equipment and Facilities**								
COMMON AGE								
56	Common Age			330.2		439.7		193.5
INDUSTRIAL RESPONSIVENESS								
57	Indust. Resp			15.3		35.1		25.1
WAR CONSUMABLES								
58	War Consum.			25.4		27.8		31.9
OTHER PRODUCTION CHARGES								
59				518.7		595.0		670.2

(Continued on next page)

AIR FORCE AIRCRAFT AND MISSILE PROCUREMENT REQUESTS (Continued)

Millions of Dollars

No.	Item Nomenclature	FY 1994 Unit Cost	FY 1992 Quan.	FY 1992 Cost	FY 1993 Quan.	FY 1993 Cost	FY 1994 Quan.	FY 1994 Cost
COMMON ECM EQUIPMENT								
60	Common Ecm Equip			187.3		98.0		24.5
	TOTAL AIRCRAFT SUPPORT EQUIPMENT AND FACILITIES			**1,076.9**		**1,195.5**		**945.3**
	TOTAL AIRCRAFT PROCUREMENT, AIR FORCE			**10,086.9**		**10,000.3**		**7,301.0**

APPROPRIATION: 3020F MISSILE PROCUREMENT, AIR FORCE

No.	Item Nomenclature	FY 1994 Unit Cost	FY 1992 Quan.	FY 1992 Cost	FY 1993 Quan.	FY 1993 Cost	FY 1994 Quan.	FY 1994 Cost
				Millions of Dollars				
BUDGET ACTIVITY 01: Ballistic Missiles								
STRATEGIC								
1	Peacekeeper (M-X0			120.4		27.1		
MISSILE REPLACEMENT EQUIP - BALLIST								
2	Missile Repl EQ-Ballist			45.9		39.8		27.1
TOTAL BALLISTIC MISSILES				**166.3**		**66.9**		**27.1**
BUDGET ACTIVITY 02: Other Missiles								
STRATEGIC								
3	DOD Fin. Sys							91.5
4	Have NAP		32	34.5		23.6		
5	Tri-Srvc Attck Miss.							195.9
6	Advd Cruise Miss. Less:Adv Pro(PY)		57	(263.8) (-88.3) **175.5**		(99.0) **99.0**		(59.4) **59.4**
7	Advd Cruise Miss. Adv Pro (CY)			16.5				
8	Have FLAG							

(Continued on next page)

APPROPRIATION: 3020F MISSILE PROCUREMENT, AIR FORCE (Continued)

(Dollars) Millions of Dollars

No.	Item Nomenclature	FY 1994 Unit Cost	FY 1992 Quan.	FY 1992 Cost	FY 1993 Quan.	FY 1993 Cost	FY 1994 Quan.	FY 1994 Cost
TACTICAL								
9	AMRAAM	669,731	630	532.3	900	623.1	749	501.6
10	AGM-130 Powered GBU-15	724,323	120	71.2	102	74.9	102	73.9
11	AGM-65D Maverick			3.9				
12	AGM-88A HARM	465		110.3	846	215.1		
TARGET DRONES								
13	MQM107 Subscale Drone	438,566			84	29.4	60	26.3
14	Target Drones	42		24.7	40	36.4		
15	QF-4 Full Scale Aerial Drone							4.7
INDUSTRIAL FACILITIES								
16	None					9.8	6.3	
MISSILE REPLACEMENT EQUIPMENT - OTHER								
17	Miss. Replacement Equipment			28.8		28.0		21.4
18	Classified Prog							
TOTAL OTHER MISSILES				**1,000.6**		**1,139.3**		**980.9**

(Continued on next page)

APPROPRIATION: 3020F MISSILE PROCUREMENT, AIR FORCE (Continued)

(Dollars) Millions of Dollars

No.	Item Nomenclature	FY 1994 Unit Cost	FY 1992		FY 1993		FY 1994	
			Quan.	Cost	Quan.	Cost	Quan.	Cost
BUDGET ACTIVITY 03: Modification of Inservice Missiles								
CLASS IV								
19	Have NAP			1.7		1.2		
20	Air Launch Cruise Miss.			28.4		18.8		
21	Peacekeeper (M-X)			2.4		2.9		.2
22	AIM-9 Sidewinder			.9		11.7		4.7
23	MM II/III Mods			152.0		179.9		38.1
24	AGM-65D Maverick			3.4		1.5		.4
25	AGM-88A HARM			2.1		2.1		74.0
26	Mods Under $2M			.3		.3		.2
27	Advd Cruise Missile					4.9		
	TOTAL MODIFICATION OF INSERVICE MISSILES			**191.2**		**223.3**		**117.6**
BUDGET ACTIVITY 04: Spares and Repair Parts								
MISSILE SPARES AND REPAIR PARTS								
28	Spares & Rpr Prts.			65.3		51.8		54.2
	TOTAL SPARES AND REPAIR PARTS			**65.3**		**51.8**		**54.2**

(Continued on next page)

APPROPRIATION: 3020F MISSILE PROCUREMENT, AIR FORCE (Continued)

(Dollars) Millions of Dollars

No.	Item Nomenclature	FY 1994 Unit Cost	FY 1992 Quan.	FY 1992 Cost	FY 1993 Quan.	FY 1993 Cost	FY 1994 Quan.	FY 1994 Cost
BUDGET ACTIVITY 05: Other Support								
SPACE PROGRAMS								
29	Spaceborne Equip (Communications Security)			7.8		3.7		.2
30	Global Posit. (MYP)	45,272,250	4	(183.8)	4	(187.8)	4	(181.1)
	Less:Adv Pro(PY)			(-62.9)		(-62.3)		(-64.7)
				120.8		**125.6**		**116.4**
31	Global Posit. (MYP) Adv Pro (CY)			66.0		59.2		55.9
32	Space Shuttle Ops			31.9		87.0		74.9
33	Space Shuttle Ops Adv Pro (CY)			29.4				
34	Space Boosters(MYP)			290.5		375.8		470.6
35	Med. Lnc Veh.	84,348,000	4	(218.9)	4	(219.2)	2	(168.7)
	Less:Adv Pro(PY)			(-35.3)		(-38.9)		(-34.3)
				183.7		**180.3**		**134.4**
36	Med. Lnc Veh. Adv Pro (CY)			37.6		43.4		11.0
37	Def Meteor. Sat Prog (MYP)			(214.0)		(30.9)		(29.4)
	Less:Adv Pro(PY)			(-107.9)				
				106.1		**30.9**		**29.4**
38	Def Supp Prog (MYP)	367,358,000		(70.0)		(135.9)	1	(367.4)
	Less:Adv Pro(PY)			(-5.6)				(-101.6)
				64.4		**135.9**		**265.7**

(Continued on next page)

APPROPRIATION: 3020F MISSILE PROCUREMENT, AIR FORCE (Continued)

(Dollars) Millions of Dollars

No.	Item Nomenclature	FY 1994 Unit Cost	FY 1992		FY 1993		FY 1994	
			Quan.	Cost	Quan.	Cost	Quan.	Cost
39	Def Supp Prog (MYP) Adv Pro(CY)					107.1		193.4
40	Def Satell Comm System (MYP)			55.5		25.1		32.4
	SPECIAL PROGRAMS							
41	IONDS (MYP)	7,241,500	4	(23.1)	4	(33.6)	6	(43.4)
	Less:Adv Pro(PY)			(-4.0)		(-12.2)		(-11.7)
				19.2		**21.4**		**31.7**
42	IONDS (MYP) Adv Pro (CY)			12.9		11.2		10.1
43	Spec Updt Progs			72.6		59.9		141.1
44	Special Programs			2223.3		1586.3		1".9
	TOTAL OTHER SUPPORT			**3,321.8**		**2,852.7**		**3,181.2**
	TOTAL MISSILE PROCUREMENT, AIR FORCE			**4,745.2**		**4,334.1**		**4,361.1**

Section VIII.

U.S. Army

ARMY RDT&E REQUESTS FOR FISCAL YEAR 1994
(Research, Development, Test and Evaluation)
Dollar figures are in thousands.

Program Element Number	Item	FY 1992	FY 1993	FY 1994
0601101A	In-House Laboratory Indep. Research	8,808	11,729	10,954
0601102A	Def. Research Science	186,565	201,973	203,695
0601104A	Electromechanics and Hypervelocity Phys	2,922	3,686	3,712
0602104A	TRACTOR ROSE	2,758	3,382	5,720
0602105A	Materials Technology	14,440	17,436	11,288
0602120A	Elec. Survivability and Fuzing Tech	21,339	27,773	28,973
0602122A	TRACTOR HIP	10,822	23,227	11,921
0602123A	TRACTOR FIELD	1,221	3,873	
0602211A	Aviation Technology	50,518	44,781	34,150
0602270A	EW Technology	22,274	21,579	20,962
0602303A	Missile Technology	33,543	40,153	23,777
0602307A	Laser Weapons Tech	476	554	510
0602308A	Modeling & Simulation	7,900	4,727	
0602601A	Combat Vehicle and Automotive Tech	40,102	58,055	38,994
0602609A	Not Used			
0602618A	Ballistics Tech	61,890	56,904	29,547
0602622A	Chemical, Smoke & Eqpt. Defeating Tech	46,865	46,099	37,766
0602623A	Joint Service Small Arms Program	4,453	4,643	3,397
0602624A	Weapons and Munitions Technology	40,063	36,157	34,794
0602705A	Electronics and Elec. Devices	20,769	20,790	19,400
0602709A	Night Vision Tech	29,917	33,310	18,941
0602716A	Human Factors Engineering Tech	5,491	10,252	15,163
0602720A	Environmental Quality Technology	29,487	62,875	21,229
0602727A	Non-System Training Device Tech	3,889	8,021	4,413
0602782A	Command, Control, Comm. Tech	18,218	17,081	10,376
0602783A	Computer & Software Tech	5,282	2,938	5,743
0602784A	Military Eng Tech	44,478	40,551	41,183
0602785A	Manpower/Personnel/ Training Tech	15,789	15,325	13,319
0602786A	Logistics Tech	36,994	36,819	28,453
0602787A	Medical Tech	109,621	91,194	86,711
0602788A	TRACTOR FLOP	1,588	1,581	1,562
0602789A	ARMY Artificial Intelligence Tech	2,838	3,119	2,696
0602790A	Small Business Innovative Res/Small Bus Tech. Trns. Pilot Prog.		67,827	63,044
Technology Base		**881,311**	**1,018,414**	**832,213**
0602813A	TRACTOR DUMP		15,962	
0603001A	Logistics Advanced Tech	10,711	18,052	12,913
0603002A	Medical Advanced Tech	57,271	238,526	40,346
0603003A	Aviation Advanced Tech	34,048	38,323	53,073
0603004A	Weapons & Munitions Advanced Tech	59,854	59,068	17,291
0603005A	Combat Vehicle & Auto Advncd Tech	26,283	42,400	39,093

(Continued on next page)

ARMY RDT&E REQUESTS FOR FISCAL YEAR 1994 (Continued)

Program Element Number	Item	FY 1992	FY 1993	FY 1994
0603006A	Command, Control, Comm Advncd Tech	9,661	9,799	16,049
0603007A	Manpower, Persnl & Trng Advncd Tech	15,801	15,978	8,064
0603009A	TRACTOR HIKE	5,078	6,354	7,355
0603012A	TRACTOR HOLE	11,493	20,457	11,779
0603013A	TRACTOR DIRT	99	3,236	1,888
0603017A	TRACTOR RED	11,574	1,975	7,629
0603020A	TRACTOR ROSE			6,679
0603102A	Mtrls & Structures Advanced Tech	2,715	6,314	
0603105A	AIDS Research	27,339	53,790	3,410
0603238A	Global Surveillance/Air Def/Precision Strike		34,510	29,484
0603270A	EW Technology	6,093	40,817	28,533
0603313A	Missile & Rocket Advanced Tech	19,946	20,011	46,497
0603322A	TRACTOR CAGE	24,654	23,122	13,909
0603393A	TRACTOR TRAILER	11,059		
0603606A	Landmine Warfare & Barrier Advd Tech	17,197	18,091	9,995
0603607A	Joint Srvc Small Arms Prog	5,336	5,458	5,529
0603654A	Line-of-Sight, Antitank (LOSAT)	27,900	113,150	
0603710A	Night Vision Advd Tech	23,024	26,887	38,661
0603734A	Military Eng. Advd Tech	3,013	3,218	2,910
0603742A	Advd Elec Devices Dev.	7,337	6,283	
0603759A	Chemical Biological Def. Smoke Advd Tech	4,118	3,306	2,634
0603772A	Advanced Tactical Comp. Sci & Tech	11,371	18,599	30,946
Advanced Tech Development		**432,975**	**843,686**	**434,667**
0603392A	Anti-Satellite Weapon (ASAT)	34,022	18,437	
0102814A	Special Programs			
0303152A	World-Wide Military Command & Control Systems, Info Sys			
Strategic Programs		**57,208**	**30,498**	**8,647**
0603018A	TRACTOR TREAD		11,534	
0603019A	TRACTOR DUMP		4,037	19,010
0603053A	Advanced Command & Control Veh.	14,794	20,389	
0603303A	Surface-to-Surface Missile Rckt Sys	45,944		
0603604A	Nuclear Munitions - Adv Dev	3,315	3,053	2,006
0603612A	Adv Anti-Tank Weapon Sys	107,219		
0603617A	Non-Line of Sight (N-LOS)	8,800		34,702
0603619A	Landmine Warfare & Barrier - Adv Dev	7,367	12,761	21,685
0603627A	Smoke, Obscurant & Target Defeating Sys - Adv Dev	12,748	17,918	6,046
0603639A	Armament Enhancement Initiative			
0603640A	Artillery Propellant Dev.	7,890	16,127	12,033
0603645A	Armored System Mod. Adv Dev	297,719	314,146	148,342
0603647A	TRACTOR DIRT			265
0603649A	Engineer Mobility Eqpt. - Adv Dev		11,492	29,464
0603653A	Adv Tank Arm Sys			5,435
0603713A	Army Data Dist Sys	22,981	12,697	11,757

(Continued on next page)

ARMY RDT&E REQUESTS FOR FISCAL YEAR 1994 (Continued)

Program Element Number	Item	FY 1992	FY 1993	FY 1994
0603730A	Tact Surveillance System - Adv Dev	16,535	14,134	15,422
0603745A	Tact Elec Support Systems - Adv Dev	3,306	2,938	4,363
0603746A	Single Chnnl Ground & Air Radio System	1,670	4,930	
0603747A	Soldier Supp & Surv	9,438	13,161	13,193
0603757A	Forward Area Air Def (FAAD) Sys	54,924		
0603766A	Tact Elec Surv Sys	18,001	14,022	15,373
0603774A	Night Vision Sys	5,343	7,440	4,794
0603778A	MLRS Prod Imp Pro	20,213	23,669	40,915
0603801A	Aviation - Adv Dev	13,681	16,304	10,759
0603802A	Weapons & Muni.			764
0603804A	Log & Eng Eqpt	20,608	14,471	14,695
0603805A	Combat Srvc Supp Comp Sys Eval. & Analysis	30,148	18,746	20,502
0603806A	NBC Defense System	31,434	29,380	32,163
0603807A	Medical Systems - Adv Dev	30,153	27,459	27,628
0603808A	Classified Program	9,251		
0603811A	Meteorological Data	3,353	4,055	
0603813A	TRACTOR PULL	8,664		
0604201A	Aircraft Avionics			5,061
0604202A	Aircraft Weapons	4,013		
0604220A	Armed, Deployable OH-58D	9,171	7,659	
0604223A	Comanche	514,530	359,166	367,080
0604270A	EW Development	202,218	92,620	60,453
0604315A	Tri-Service Stndoff Attack Missile			89,682
0604321A	All Source Analysis Sys	105,172	51,476	971
0604328A	Not Used			
0604603A	Nuclear Munitions Eng Dev	1,730	4,539	
0604604A	Medium Tact Veh.	22,578	2,781	6,548
0604609A	Smoke, Obscurant & Target Defeating Sys - Eng Dev	13,295	10,692	17,118
0604611A	JAVELIN	118,297	95,929	44,937
0604619A	Landmine Warfare	34,616	22,869	21,322
0604622A	Heavy Tact Veh.	2,214	1,853	476
0604630A	Adv Tank Canon	35,995	25,571	
0604633A	Air Traffic Control	3,278	2,329	5,607
0604640A	Adv Command & Cntrl Vehicle (AC2V)			8,654
0604642A	Light Tact Whld Veh			2,064
0604645A	Armrd Sys Modern (ASM) - Eng Dev	42,348	72,027	89,504
0604649A	Eng Mob Eqpt Dev	7,900	2,116	13,304
0604710A	Night Vision Sys - Eng Dev	36,544	23,955	41,827
0604713A	Combat Feeding, Clothing & Eqpt	27,900	25,964	28,425
0604715A	Non-Sys Trng Dev	66,000	42,906	62,669
0604726A	Integrated Meteor Support System	4,411	905	949
0604740A	Tact Surveillance System - Eng Dev	21,398	18,944	38,815
0604741A	Air Defense Command, Control & Intell.	31,415	38,341	15,424
0604746A	Auto Test Eqpt Dev	17,928	16,521	14,472
0604766A	Tact Elec Surv. System - Eng Dev	20,451	32,275	52,547
0604767A	TRACTOR JEWEL	79,806	90,975	
0604768A	TRACTOR BAT	118,286	114,835	117,008
0604769A	TRACTOR HELM	31,387		

(Continued on next page)

ARMY RDT&E REQUESTS FOR FISCAL YEAR 1994 (Continued)

Program Element Number	Item	FY 1992	FY 1993	FY 1994
0604770A	Joint Surv/Target Attack Radar System	67,790	62,625	26,260
0604780A	Cmbnd Arms Tact Trainer (CATT)	52,988		
0604801A	Aviation - Eng Dev	13,800	15,349	5,733
0604802A	Weapons & Munitions	1,861	4,834	15,365
0604804A	Logistics & Engineer Eqpt. - Eng Dev	23,542	20,892	29,372
0604805A	Command, Control, Comm Sys-Eng Dev	5,612	7,159	9,244
0604806A	NBC Defense System	47,062	39,290	42,898
0604807A	Med Mat./Med Bio Def Eqpt-Eng Dev	23,290	19,109	21,128
0604808A	Landmine Warfare/ Barrier-Eng Dev	10,365	2,787	2,957
0604812A	Classified Program	45,875		
0604814A	Sense & Destroy Arm. Missile-Eng Dev	148,165	92,686	41,011
0604816A	LONGBOW - Eng Dev	248,626	290,083	277,954
0604817A	Non-Cooperative Target Recog.	28,152	26,081	34,547
0604818A	Army Tact Cmd & Cntrl Sys-Eng Dev	25,396	19,484	37,227
0604820A	Radar Devep.	39,378	17,656	25,834
0605710A	JTCB POC, Test/Assess, Smoke Assess, NBC Serv	10,765		
0102830A	Classified Program			
0603831A	Classified Program			
0203726A	Adv Field Artillery Tact Data System	49,143	39,381	46,285
0203735A	Combat Veh Improv. Programs	72,694	36,213	69,972
0203740A	Manuever Cntrl Sys.	36,793	26,951	29,702
0203744A	Aircraft Mod/Prod Improv. Programs	5,845	7,421	9,410
0203752A	Aircraft Eng Comp Improv. Program	6,285	6,318	6,567
0203755A	Field Artillery Ammo Supp. Vehicles	806		
0203801A	Missile/Air Defense Prod Improv Prog	58,848	67,939	59,782
0203802A	Other Missile Prod Improv Programs	54,420	4,729	66,438
0203806A	TRACTOR RIG	12,350	5,508	8,314
0203808A	TRACTOR CARD	8,537	6,629	7,615
0208010A	Joint Tact Comm Prog	4,966	7,178	16,529
Tactical Programs		**3,479,975**	**2,639,046**	**2,590,659**
0604716A	Terrain Info	14,274	11,507	9,929
0604778A	Pos. Sys Dev.	2,808	9,657	4,921
0301359A	Special Army Prog			
0303140A	Info Sys Sec. Prog	6,615	6,118	7,122
0303142A	SATCOM Grnd Env	102,148	110,148	153,931
0305127A	Foreign Counterintelligence Act.			
0305889A	Intelligence Support to OSD Counternarcotics	9,600		
Intelligence and Comm.		**140,354**	**147,828**	**184,561**
0604256A	Threat Sim Dev.	26,770	26,774	18,233
0604258A	Target Sys Dev.	12,141	9,866	18,945

(Continued on next page)

ARMY RDT&E REQUESTS FOR FISCAL YEAR 1994 (Continued)

Program Element Number	Item	FY 1992	FY 1993	FY 1994
0604759A	Major T&E Invest	18,427	20,738	28,893
0605103A	Rand Arroyo Ctr	16,737	19,111	15,492
0605301A	Army Kwajalein Atoll	195,257	181,899	171,380
0605502A	Small Bus. Innovative Research	62,948	54,543	
0605601A	Army Test Ranges & Facilities	174,335	154,817	145,415
0605602A	Army Tech Test Instr & Targets	76,593	62,511	25,540
0605604A	Suv/Lethal Analysis	51,496	38,485	33,179
0605605A	DOD High Energy Laser Test Fac.	27,902	26,405	4,808
0605702A	Meteor Supp to RDT&E Act.	20,371	17,999	17,970
0605706A	Mat. Sys Analysis	23,872	22,001	19,500
0605709A	Exploitation of Foreign Items	30,155	18,581	18,779
0605710A	JTCB POC, Test/Assess, Smoke Assess, NBC Srvc		10,082	7,404
0605712A	Supp of Optnl Test	64,812	57,535	58,433
0605801A	Programwide Acts	95,720	86,765	96,011
0605802A	Int'l Coop Res & Dev	1,485	1,820	1,861
0605803A	Tech Info Acts.	21,541	22,160	12,007
0605805A	Munitions Stnd., Effect. & Sfty	12,863	15,875	13,763
0605810A	RDT&E Supp for Nondev. Items	8,041	5,765	5,881
0605856A	Environ. Compliance	46,670	35,640	44,014
0605872A	Prod. Investments	21,284	11,108	
0605876A	Minor Const.-RDT&E	6,774	8,130	1,873
0605878A	Maint. & Rpr.-RDT&E	74,756	76,766	61,448
0605896A	Base Ops.-RDT&E	311,825	297,474	274,409
0605898A	Mgmt. Hdqs (R&D)	14,421	20,196	11,951
0708045A	End Item Indus. Prepdnss. Acts.	27,927	32,592	
0901600A	Contract Admin/Audit			92,012
Defensewide Mission Supp		**1,445,123**	**1,335,638**	**1,199,201**
Res Dev Test & Eval Army		**6,436,946**	**6,015,110**	**5,249,948**

ARMY AIRCRAFT AND MISSILE PROCUREMENT REQUESTS

| | | FY 1994 | Millions of Dollars
FY 1992 | | FY 1993 | | FY 1994 | |
No.	Item Nomenclature	Unit Cost	Quan.	Cost	Quan.	Cost	Quan.	Cost
BUDGET ACTIVITY 01: Aircraft								
FIXED WING								
1	Guardrail Common Sensor (TIARA)		6	185.6	5	111.4		4.9
2	Tractor Hall			0.5		0.2		
ROTARY								
3	TOTAL PACKAGE FIELDING			0.2		1.2		0.3
4	AH-64 Attck Hlcptr(APACHE)		4	204.0		146.6		17.6
5	UH-60 Blckhwk (MYP) Less:Adv Pro(PY)	6,545,700	60	-414.8 (-70.8) **343.9**	52	-339 (-143.5) **195.5**	60	(392.7) (-159.2) **233.6**
6	UH-60 Blckhwk(MYP) Adv Pro (CY)			163.5		153.8		174.7
7	Hlcptr New Trng	585,080	37	23.5	70	44.7	50	29.3
TOTAL AIRCRAFT				**921.2**		**653.3**		**460.2**
BUDGET ACTIVITY 02: Modification of Aircraft								
MODIFICATION OF AIRCRAFT								
8	Tractor Dew					0.2		
9	Guardrail Mods (TIARA)			35.8		92.5		111.5

(Continued on next page)

ARMY AIRCRAFT AND MISSILE PROCUREMENT REQUESTS (Continued)

Millions of Dollars

No.	Item Nomenclature	FY 1994 Unit Cost	FY 1992 Quan.	Cost	FY 1993 Quan.	Cost	FY 1994 Quan.	Cost
10	AH1F Mods			3.4		2.3		4.3
11	AH-64 Mods			75		64.9		46.4
12	CH-47 Cargo Hlcptr Mods(MYP)			-375.1		-14.9		-15.4
	Less:Adv Pro(PY)			(-92.3)				
				282.9		**14.9**		**15.4**
13	OH-58 Mods			21.9		3.7		7.7
14	C-20 Aircraft Mods							1
15	C-23 Mods			8				
16	Flght Data Recdr			0.5		9.7		2.4
17	Ext Fuel Tnks(UH-1)			5				
18	UH-1 Mods			10.3		5.2		14.2
19	UH-60A (Blckhwk) Mods			27.2		11.8		46.9
20	Kiowa Warrior			350.4		319.6		145.5
21	EH-60 Quickfix Mods			5.5		0.4		0.5
22	Airborne Avionics					1.0		4.8
23	ASE Mods			24.9		7.1		4.2
24	Mods < $2M							1.5
TOTAL MODIFICATION OF AIRCRAFT				**850.8**		**533.4**		**406.2**

(Continued on next page)

ARMY AIRCRAFT AND MISSILE PROCUREMENT REQUESTS (Continued)

Millions of Dollars

No.	Item Nomenclature	FY 1994 Unit Cost	FY 1992 Quan.	FY 1992 Cost	FY 1993 Quan.	FY 1993 Cost	FY 1994 Quan.	FY 1994 Cost
BUDGET ACTIVITY 03: Spares and Repair Parts								
25	Spares & Rpr Prts					98		86.3
TOTAL SPARES AND REPAIR PARTS						**98**		**86.3**
BUDGET ACTIVITY 04: Support Equipment and Facilities								
26	Aircraft Surv. Equip			43.5		51.9		37.6
OTHER SUPPORT								
27	Airborne Comd & Cntrl Consoles			10.5		6.8		11.4
28	Avionics Supp Equip			27.6		33.6		33.1
29	Common Ground Equip			48.5		34.7		27.6
30	Cntrct Audit/Mgmt							20.9
31	Aviation Life Supp Equip (ALS)					8.2		11.7
32	Air Traffic Cntrl			2		5.7		8.3
33	Indust Facilities			27.6		14.9		7.3
34	Launcher, 2.75 Rocket			1.3		1		
35	CLOSED ACCT ADJ			1.7				
TOTAL SUPPORT EQUIPMENT AND FACILITIES				**162.8**		**156.7**		**157.8**
TOTAL AIRCRAFT PROCUREMENT, ARMY				**1,934.80**		**1,441.40**		**1,110.40**

(Continued on next page)

ARMY AIRCRAFT AND MISSILE PROCUREMENT REQUESTS (Continued)

Millions of Dollars

No.	Item Nomenclature	FY 1994 Unit Cost	FY 1992 Quan.	FY 1992 Cost	FY 1993 Quan.	FY 1993 Cost	FY 1994 Quan.	FY 1994 Cost
	BUDGET ACTIVITY 02: Other Missiles							
	SURFACE-TO-AIR MISSILE SYSTEM							
1	Chaparral Sys Summary			6.8		6.6		
2	Hawk Sys Summary			1.6		1.7		2.8
3	Other Missile Supp			0.6		1.4		
4	Patriot System Summary (MYP)	97		163		24.9		40.6
5	Stinger Sys Summary			25.8		34.7		8.4
6	Avenger System Summary	1,164,798	144	(156.7)	144	(134.6)	144	(167.7)
	Less:Adv Pro(PY)			(-25.7)		(-32.6)		(-32.5)
				131		**102**		**135.2**
7	Avenger Sys Summary Adv Pro (CY)			52.7		44.9		
	AIR-TO-SURFACE MISSILE SYSTEM							
8	Hellfire Sys Summary	51,840	89	11.7	1781	82.9	1785	92.5
	ANTI-TANK/ASSAULT MISSILE SYSTEM							
9	Javeline (AAWS-M) Sys Summary	225,573					1000	(225.6)
	Less:Adv Pro(PY)							(-18.3)
								207.3
10	Javeline (AAWS-M) Sys Summary Adv Pro(CY) (FY93 for FY93)					18.2		
						-18.2		
11	TOW 2 Sys Summary	9550		200.6	8900	182		25.3

(Continued on next page)

ARMY AIRCRAFT AND MISSILE PROCUREMENT REQUESTS (Continued)

Millions of Dollars

No.	Item Nomenclature	FY 1994 Unit Cost	FY 1992 Quan.	FY 1992 Cost	FY 1993 Quan.	FY 1993 Cost	FY 1994 Quan.	FY 1994 Cost
12	MLRS Sys Summary			(66.0)	1002	-129.5		-9.8
	Less:Adv Pro(PY)			(-6.3)		(-19.8)		
				59.7		**109.8**		**9.8**
13	MLRS Launcher	6,371,058	44	-142.3	44	-163.4	34	(216.6)
	Less:Adv Pro(PY)			(-17.9)		(-18.6)		
				124.4		**144.8**		**216.6**
14	MLRS Launcher Adv Pro (CY)			0.7				
15	Army Tact MSL Sys (ATACMS)- Sys Su B	695,866	300	(165.2)	351	(187.3)	255	(177.4)
	Less:Adv Pro(PY)			(-14.4)		(-21.6)		(-24.9)
				150.8		**165.7**		**152.6**
16	Army Tact MSL Sys (ATACMS)-Sys Su Adv Pro (CY)			21.6		24.9		
17	Cntrct Admin/Audit							22
	TOTAL OTHER MISSILES			**950.8**		**944.3**		**913.1**

BUDGET ACTIVITY 03: Modification of Missiles

MODIFICATIONS

No.	Item Nomenclature	FY 1994 Unit Cost	FY 1992 Quan.	FY 1992 Cost	FY 1993 Quan.	FY 1993 Cost	FY 1994 Quan.	FY 1994 Cost
18	Patriot Mods			35.2		10		18.5
19	Hawk Mods			9.8		1.5		
20	Avenger Mods			0.1		4.1		9.3
21	Tow Mods			8.3		14.8		7.3

(Continued on next page)

ARMY AIRCRAFT AND MISSILE PROCUREMENT REQUESTS (Continued)

Millions of Dollars

No.	Item Nomenclature	FY 1994 Unit Cost	FY 1992 Quan.	FY 1992 Cost	FY 1993 Quan.	FY 1993 Cost	FY 1994 Quan.	FY 1994 Cost
22	MLRS Mods			44.3		12.1		23.2
23	Mods < $2M			1.4				1.9
24	Tractor Rig			8.2				
	TOTAL MODIFICATION OF MISSILES			**107.2**		**42.6**		**60.2**
	BUDGET ACTIVITY 04: Spares and Repair Parts							
25	Spares & Rpr Prts					39.6		50.6
	TOTAL SPARES AND REPAIR PARTS					**39.6**		**50.6**
	BUDGET ACTIVITY 05: Support Equipment & Facilities							
26	Air Def Targets			11.2		11.1		15
27	Items <$2M (Missiles)			1.6		0.8		1
28	Prod Base Support			7		10.1		3.8
29	Closed Acct Adj.			6				
	TOTAL SUPPORT EQUIPMENT AND FACILITIES			**25.7**		**22**		**19.7**
	TOTAL MISSILE PROCUREMENT, ARMY			**1,083.80**		**1,048.50**		**1,043.60**

Section IX.

U.S. Navy

NAVY RDT&E REQUESTS FOR FISCAL YEAR 1994
(Research, Development, Test and Evaluation)
Dollar figures are in thousands.

Program Element Number	Item	FY 1992	FY 1993	FY 1994
0601101A	In-House Laboratory Indep. Research	8,808	11,729	10,954
060 0601152N	In-House Indep Lab Research	14,000	16,746	16,985
0601153N	Defense Res. Sciences	377,934	408,844	416,922
0602111N	Surface/Aero Surveillance & Weapons Tech	68,677	70,942	67,305
0602121N	Surface Ship Tech	31,924	46,019	17,495
0602122N	Aircraft Technology	20,098	25,061	21,253
0602131M Technology	Marine Corps Landing Force	18,036	20,216	17,225
0602232N	Command, Control & Comm Tech	15,984	20,078	18,155
0602233N	Mission Support Tech	32,102	42,953	34,424
0602234N	Materials, Elec. & Comp Tech	78,977	94,645	71,063
0602270N	Elec. Warfare Technology	12,485	16,472	14,896
0602314N	Undersea Surveillance Weapon Technology	129,885	134,580	107,960
0602315N	Mine Countermeasures, Mining & Special Warfare Tech	21,430	41,154	21,944
0602323N	Submarine Technology	16,611	17,589	14,575
0602435N	Oceanographic & Atmospheric Technology	41,601	44,887	37,711
0602790N	Small Bus Innovative Res. Small Bus Tech Transfer Pilot Program		81,443	86,113
Technology Base		**879,744**	**1,081,629**	**964,026**
0603217N	Air Systems Advd Tech Dev	8,902	28,755	30,005
0603238N	Global Surv./Air Def/Prec. Strike Tech Demonstration		9,351	50,999
0603270N	Advd Elec Warfare Tech	4,865	27,913	12,983
0603508N	Ship Propulsion System	4,445	3,914	3,439
0603555N	Undersea Superiority Tech Dem.		61,029	95,438
0603563N	Ship Concept Adv. Design	3,710	8,033	18,820
0603640M	Marine Corps Advd Tech Dem.	15,989	25,348	35,815
0603706N	Medical Development	35,568	45,887	16,956
0603707N Adv Tech Dev	Manpower, Personnel & Trng	13,027	17,588	18,652
0603712N	Generic Logistics R&D Tech	17,838	28,549	13,720
0603747N	Advd Anti-Sub Warfare Tech	50,908	74,838	49,172
0603782N	Shallow Water MCM Demos	4,914	9,848	5,148
0603792N	Advd Tech Transition	61,645	86,599	63,394
0603794N	C3 Advd Technology	16,365	24,231	10,747
Advanced Technology Development		**238,176**	**451,883**	**425,288**
0603451N	Tactical Space Operations	4,075		2,018
0603735N	WWMCCS Arch Support	16	1,602	
0603741N	Satellite Laser Comm. (H)	10,000	14,226	
0605856N	Strategic Tech Support	10,645	4,553	3,781
0101221N	Stgic Sub & Weapons Sys Supp	80,602	75,264	54,295
0101224N	SSBN Security/Surv Program	69,667	76,543	27,835
0101226N	Sub Acoustic Warfare Dev.	36,347	38,823	16,800
0101402N	Navy Strategic Comm	14,046	19,553	36,184
0102427N	Naval Space Surveillance	855	863	735
0303152N	World-Wide Military Command & Cont Sys. Info System	4,268	3,687	
Strategic Programs		**230,521**	**235,114**	**141,648**

(Continued on next page)

Navy RDT&E requests for Fiscal Year 1994 (Continued)

Program Element Number	Item	FY 1992	FY 1993	FY 1994
0603109N	Integrated Aircraft Avionics	24,528		
0603207N	Air/Ocean Tactical Apps.	18,035	16,237	16,239
0603208N	Training System Aircraft	48,056	49,165	32,565
0603216N	Aviation Survivability	27,986	32,049	13,672
0603231N	Next Generation Fighter	1,975		
0603254N	ASW Systems Development	28,733	39,380	35,238
0603261N	Tactical Airborne Reconn.	13,776	14,444	30,358
0603320N	Low Cost Anti-Radiation Seeker	3,949		
0603321N	Advd Air-to-Air Missile	87,488		
0603382N	Advd Combat Sys Tech			3,750
0603502N	Undersea Warfare & MCM Dev	51,505	42,152	65,660
0603504N	Advd Sub Combat Sys Dev	39,697	32,137	20,341
0603506N	Surface Ship Torpedo Def	56,529	27,085	34,482
0603512N	Carrier Systems Dev	20,991	20,672	11,221
0603513N	Shipboard Sys Comp Dev	27,649	30,229	27,824
0603514N	Ship Combat Survivability	24,241	22,606	17,315
0603525N	PILOT FISH	54,215	34,947	26,884
0603528N	Non-Acoustic Anti-Sub Warfare			13,999
0603536N	RETRACT JUNIPER	50,199	60,676	32,560
0603542N	Radiological Control	2,857	3,599	3,291
0603553N	Surface ASW	65,913	46,606	21,150
0603561N	Advd Sub System Dev	57,978	128,093	142,068
0603562N	Sub Tactical Warfare Sys	11,470	10,946	9,518
0603564N	Ship Prelim Design & Feas. Studies	1,240	12,943	58,764
0603570N	Advd Nuclear Power Systems	88,695	92,126	136,651
0603573N	Advd Surface Mach Systems	38,997	73,610	92,328
0603576N	CHALK EAGLE	133,702	82,704	71,003
0603582N	Combat System Integration	10,322	9,626	6,842
0603591N	Joint Advd Systems	129,846	145,576	
0603601N	Mine Development	6,560	1	
0603609N	Conventional Munitions	42,653	60,518	42,632
0603610N	Advd Warhead Dev (MK-50)	6,587	8,824	
0603611M	Marine Corps Assault Vehicles	41,075	52,771	20,554
0603612M	Marine Corps Mine/Cntrmsrs. Systems - Adv Dev			2,743
0603634N	Electromagnetic Effects Protection Dev	3,962	7,549	5,104
0603635M	Marine Corps Ground Combat/ Support System	7,781	22,258	27,624
0603654N	Joint Srvc Explosive Ord Dev	8,171	9,108	9,359
0603691N	MK 48 ADCAP - Adv Dev	14,659	30,768	27,248
0603708N	ASW Signal Processing	3,262	3,325	
0603709N	Advd Marine Biological System	4,765	4,506	3,470
0603711N	Fleet Tact Dev & Eval Program	6,107	5,388	4,464
0603713N	Ocean Engineering Tech Dev	17,455	13,319	11,783
0603724N	Navy Energy Program	4,678	5,123	4,329
0603725N	Facilities Improvement	462	1,533	1,383
0603726N	Merchant Ship Naval Aug Prog	1		
0603734N	CHALK CORAL	66,708	61,986	71,969
0603737N	LINK HAZEL	13,025	23,369	
0603740N	LINK LAUREL	36,957		
0603746N	RETRACT MAPLE	141,517	143,987	124,408
0603748N	LINK PLUMERIA	20,625	14,717	40,109
0603750N	CHALK WEED	1,354	2,601	
0603751N	RETRACT ELM	147,293	107,225	62,997
0603752N	CHALK POINSETTIA	1,606	5,961	
0603755N	Ship Self Defense	220,407	210,329	237,204
0603763N	Warfare Sys Arch & Eng.	7,255	7,914	7,033
0603785N	Combat Sys Oceanographic Perf Assessment	26,273	25,699	19,850
0603787N	Special Processes	54,596	32,256	29,863
0603795N	Gun Weapon Sys Tech	7,570	11,122	17,247

(Continued on next page)

Navy RDT&E requests for Fiscal Year 1994 (Continued)

Program Element Number	Item	FY 1992	FY 1993	FY 1994
0604212N	ASW & Other Helo Dev	73,137	94,289	82,243
0604214N	AV-8B Aircraft - Eng Dev	9,084	11,735	18,284
0604215N	Standards Development	16,450	11,555	13,724
0604217N	S-3 Weapon System Improv		1,095	4,187
0604218N	Air/Ocean Eqpt Engineering	6,398	6,151	6,028
0604221N	P-3 Modernization Program	75,927	13,025	15,134
0604233N	AFX		155,884	399,218
0604261N	Acoustic Search Sensors	23,961	31,954	31,775
0604262N	V-22A	780,708	723,748	82,295
0604264N	Air Crew Systems Dev	21,149	20,694	11,126
0604265N	Air Launched Saturation Sys	16,792		
0604270N	EW Development	73,674	134,377	128,850
0604301N	MK 92 Fire Control Sys Upgrade	1,978	1,831	1,063
0604307N	AEGIS Combat System Eng	92,505	110,545	103,995
0604321N	Tri-Srvc Standoff Attck Mis.			75,430
0604314N	Advd Med Range Air-to-Air Mis.	2,599	2,649	15,159
0604354N	Air-to-Air Mis. Sys Eng.			7,098
0604366N	Standard Missile Improvements	70,313	50,106	63,022
0604372N	New Threat Upgrade	9,809	5,843	4,662
0604373N	Airborne MCM	28,803	30,398	33,155
0604503N	SSN-688 & Trident Modern.	56,624	72,151	56,549
0604504N	Air Control	7,300	13,359	9,993
0604507N	Enhanced Modular Signal Proc.	20,014	14,527	13,443
0604512N	Shipboard Aviation Systems			1,404
0604516N	Ship Survivability	10,583	10,900	10,292
0604518N	Combat Info Center Conversion	8,883	17,740	11,534
0604524N	Sub Combat System	261,050	52,474	87,481
0604558N	New Design SSN			240,222
0604561N	SSN-21 Developments	154,027	91,100	76,129
0604562N	Sub Tactical Warfare System	81,372	61,738	25,427
0604567N	Ship Cntrct Desgn/Live Fire	50,118	36,089	47,137
0604574N	Navy Tac Computer Resources	33,099	31,887	17,572
0604601N	Mine Development	8,673	8,044	5,666
0604602N	Naval Gunnery Improvements	4,409	5,014	
0604603N	Unguided Conv Air-Launched Weapons	12,228	10,162	29,972
0604610N	MK 50 Torpedo	5,965		
0604612M	Marine Corps Mine Cntrmsrs Systems - Eng Dev	1,239	2,845	1,298
0604618N	Joint Direct Attack Munition	10,306	24,697	10,352
0604654N	Joint Service Expl Ord Dev	5,665	5,737	6,266
0604656M	Marine Corps Assault Veh	19,100	13,995	
0604707N	Space & Elec Warfare Arch./ Eng Support		2,715	11,916
0604710N	Navy Energy Program	3,351	3,857	3,137
0604715N	Surface Warfare Trng Devices	1,224	4,734	
0604719M	Marine Corps Command/Cntrl/ Comms. Systems	17,614	11,527	26,223
0604727N	Joint Standoff Weapon Systems	47,668	63,343	80,503
0604755N	Ship Self Defense	64,388	58,081	116,760
0604761N	Intelligence	2,014	1,872	345
0604771N	Medical Developments	4,129	3,915	4,030
0604777N	Navigation/ID System	85,831	67,753	80,047
0604784N	Distributed Surveillance Sys	234,516	159,213	135,879
0605867N	SEW Surveillance/Reconaissance	13,726	10,508	17,863
0204134N	A-6 Squadrons	6,657	7,599	
0204136N	F/A-18 Squadrons	418,711	895,457	1,485,496
0204152N	E-2 Squadrons	6,268	6,352	48,930
0204163N	Fleet Telecomm. (Tactical)	35,830	42,892	34,435
0204229N	Tomahawk & Tomahawk Mission Plng Center (TMPC)	60,880	30,604	47,440
0204311N	Integrated Surveillance System 91,055	89,522	71,781	
0204413N	Amphibious Tact Support Units	3,344	3,378	2,823

(Continued on next page)

Navy RDT&E requests for Fiscal Year 1994 (Continued)

Program Element Number	Item	FY 1992	FY 1993	FY 1994
0204571N	Consolidated Trng Sys Dev	33,805	33,307	37,200
0205604N	Tactical Data Links	81,626	60,894	39,562
0205620N	Surface ASW Combat Sys Integ.	75,051	16,910	24,905
0205633N	Aviation Improvements	56,243	78,657	74,976
0205667N	F-14 Upgrade	115,068	120,064	71,995
0205675N	Operational Reactor Dev	60,986	59,274	57,784
0206313M	Marine Corps Communications	4,445	4,794	9,151
0206623M	Marine Corps Grnd Combat/ Supp Arms System	39,284	34,169	24,259
0206624M	Marine Corps Combat Srvc Supp	1,816	768	9,656
0206625M	Marine Corps Intelligence/Elec Warfare Systems	26,765	25,334	22,772
0206626M	Marine Corps Command/Cntrl/ Communications Systems	24,477	13,710	36,735
Tactical Programs		**5,704,648**	**5,689,306**	**6,018,884**
0604231N	Tactical Command System	41,259	26,034	30,617
0604721N	Battle Group Passive Horizon Extension System		10,781	24,735
0605866N	Navy Space & Elec Wrfr Supp	6,887	4,929	5,819
0301327N	Tech Reconnaissance & Surv			
0303109N	Satellite Communications	67,490	53,715	55,782
0303140N	Info Systems Security Program			
0304111N	Special Activities			
0305889N	Intelligence Support to OSD Counternarcotics	7,000		
Intelligence & Communications		**816,991**	**637,087**	**744,979**
0603721N	Environmental Protection	25,886	27,278	44,461
0604256N	Threat Simulator Dev	20,777	29,038	29,857
0604258N	Target Systems Dev	22,205	37,615	37,474
0604703N	Personnel, Trng, Simulation and Human Factor	1,775	1,087	1,069
0604759N	Major T&E Investment	35,729	41,088	52,496
0605152N	Studies & Analysis Supp-Navy	5,902	2,660	3,856
0605154N	Center for Naval Analyses	28,477	43,178	43,260
0605155N	Fleet Tact Dev & Eval	15,338	3,337	4,456
0605502N	Small Bus Innovative Res	43,406		
0605804N	Technical Info Services	11,609	14,018	10,273
0605853N	Mgmt., Tech & Int'l Supp	12,972	10,265	12,787
0605861N	RDT&E Science & Tech Mgmt	57,931	64,288	60,767
0605862N	RDT&E Instrumentation Modern.	14,924	22,285	39,419
0605863N	RDT&E Ship & Aircraft Supp	86,"	95,343	80,587
0605864N	Test & Eval Supp	291,063	307,131	293,422
0605865N	Operational Test & Eval Capa.	7,622	8,717	8,329
0605871M	Marine Corps Tact Exploit. of Nat'l Capabilities	1,247	1,191	1,314
0605872N	Productivity Investments	195	396	
0605873M	Long Range Planning Support	6,930	6,055	14,374
0205658N	Navy Science Asst. Program	4,723	7,464	6,668
0305160N	Defense Meteor. Satellite Prog	9,624	16,598	11,550
0708011N	Industrial Preparedness	68,756	99,485	
0901600N	Contract Admin./Audit	164,360		
0909999N	Financing for Cancelled Acct. Adjustments	110		
Defensewide Mission Support		**772,814**	**838,517**	**920,779**
Research Development Test & Eval Navy		**8,642,894**	**8,933,536**	**9,215,604**

U.S. NAVY PROCUREMENT REQUESTS
Dollar figures are in millions.

BUDGET ACTIVITY 01: Combat Aircraft

COMBAT AIRCRAFT

No.	Item Nomenclature	FY 1994 Unit Cost	FY 1992 Quan.	FY 1992 Cost	FY 1993 Quan.	FY 1993 Cost	FY 1994 Quan.	FY 1994 Cost
				(Dollars) Millions of Dollars				
1	EA-6B/Remanufacturing (Elec Wrfr)			(98.1)	3	(452.8)		(77.6)
	Less:Adv Pro. (PY).					(-17.0)		
				98.1		**435.8**		**77.6**
2	EA-6B/Remanufacturing (Elec Wrfr) Prowle							
	Adv Pro. (CY)			17.0		46.9		
3	AV-8B (V/STOL) Harrier	32,400,250	6	270.0		24.8	4	129.6
4	AV-8B (V/STOL) Harrier Adv							
	Proc. (CY)							15.0
5	F-14A/D (Fighter) Tomcat			162.1		141.1		
6	F/A-18C/D (Fighter) Hornet	44,504,277	48	(2057.3)	36	(1288.8)	36	(1602.2)
	Less:Adv Proc			(-151.8)		(-144.7)		(-109.4)
				1905.5		**1144.1**		**1492.7**
7	F/A-18C/D(Fighter) Hornet (MYP)							
	Adv Proc. (CY)			144.7		109.4		262.0
	(FY92 for FY92)			(144.7)				
	(FY93 for FY93)					(109.4)		
	(FY94 for FY94)							(9.4)
	(FY94 for FY95)							(181.3)
	(FY94 for FY96)							(42.4)
	(FY94 for FY97)							(28.9)
8	CH/MH-53E(HLCPTR) Super Stallion	27,509,750	16	(407.0)	20	(488.3)	12	(330.1)
	Less:Adv Pro(PY)			(-83.8)		(-41.9)		(-48.2)
				323.2		**446.4**		**281.9**

(Continued on next page)

U.S. NAVY PROCUREMENT REQUESTS (Continued)

(Dollars) Millions of Dollars

No.	Item Nomenclature	FY 1994 Unit Cost	FY 1992 Quan.	FY 1992 Cost	FY 1993 Quan.	FY 1993 Cost	FY 1994 Quan.	FY 1994 Cost
9	CH/MH-53E(HLCPTR) Super Stallion							
	Adv Pro (CY)			41.9		48.2		15.0
10	AH-1W (HLCPTR) Sea Cobra	11,939,500	14	140.3	12	122.2	12	143.3
11	SH-60B (ASW HLCPTR) Seahawk	29,995,571	13	(266.8)	12	(259.6)	7	(210.0)
	Less:Adv Pro(PY)			(-43.9)		(-45.8)		(-20.7)
				223.0		**213.8**		**189.3**
12	SH-60B (ASW HLCPTR) Seahawk							
	Adv Proc (CY)			43.7		20.7		27.2
	(FY92 for FY93)			(43.7)				
	(FY93 for FY94)					(20.7)		
	(FY94 for FY95)							(27.2)
13	SH-60F (CV ASW HLCPTR)	22,477,250	12	(241.2)	9	(217.6)	8	(179.8)
	Less:Adv Pro(PY)			(-51.8)		(-75.2)		(-30.0)
				189.3		**142.4**		**149.8**
14	SH-60F (CV ASW HLCPTR) Adv Pro(CY)			52.6		29.8		36.6
	(FY92 for FY93)			(52.4)				
	(FY92 for FY94)			(.2)				
	(FY93 for FY94)					(29.8)		
	(FY94 for FY95)							(36.6)
15	E-2C (Erly Wrng) Hawkeye		6	(536.9)		(94.8)		(27.9)
	Less:Adv Pro (PY)			(-37.6)				
				499.2		94.8		27.9
	TOTAL COMBAT AIRCRAFT			**4,110.7**		**3,020.5**		**2,838.4**

(Continued on next page)

U.S. NAVY PROCUREMENT REQUESTS (Continued)

(Dollars) Millions of Dollars

No.	Item Nomenclature	FY 1994 Unit Cost	FY 1992 Quan.	FY 1992 Cost	FY 1993 Quan.	FY 1993 Cost	FY 1994 Quan.	FY 1994 Cost
BUDGET ACTIVITY 02: Airlift Aircraft								
16	C-20				1	24.8		
	TOTAL AIRLIFT AIRCRAFT					**24.8**		
BUDGET ACTIVITY 03: Trainer Aircraft								
17	T-45TS(Trnr) Goshawk	23,728,833	12	(366.6)	12	(258.1)	12	(284.7)
	Less:Adv Pro(PY)			(-46.9)		(-21.0)		(-25.5)
				319.7		**237.1**		**259.2**
18	T-45TS(Trnr) Goshawk Adv Pro (CY)			21.0		25.5		30.8
	TOTAL TRAINER AIRCRAFT			**340.7**		**262.6**		**290.0**
BUDGET ACTIVITY 04: Other Aircraft								
19	HH-60H(HLCPTR) CSAR	16,016,222			7	(122.5)	9	(144.1)
	Less:Adv Pro(PY)					(-6.0)		
						116.5		**144.1**
20	HH-60J(HLCPTR) Coast Guard			30.0				
	TOTAL OTHER AIRCRAFT			**30.0**		**116.5**		**144.1**
BUDGET ACTIVITY 05: Modification of Aircraft								
21	A-4 Series			3.1				
22	A-6 Series			20.6		164.0		19.6

(Continued on next page)

U.S. NAVY PROCUREMENT REQUESTS (Continued)

(Dollars) Millions of Dollars

No.	Item Nomenclature	FY 1994 Unit Cost	FY 1992 Quan.	FY 1992 Cost	FY 1993 Quan.	FY 1993 Cost	FY 1994 Quan.	FY 1994 Cost
23	EA-6 Series			66.4		55.3		21.9
24	AV-8 Series			19.6		6.5		22.8
25	F-14 Series			228.1		194.0		116.2
26	ADVERSARY			3.9		.2		.2
27	ES-3 Series			10.9		5.7		10.7
28	OV-10 Series			.1				
29	F/A-18 Series			28.2		66.7		48.8
30	H-46 Series			67.4		132.3		74.3
31	H-53 Series			40.2		33.4		37.2
32	SH-60 Series			25.0		34.9		46.1
33	VH-60 Series			1.2				
34	H-1 Series			142.0		75.4		74.9
35	H-2 Series			15.4		30.5		
36	H-3 Series			36.1		2.8		2.8
37	EP-3 Series			24.7				34.2
38	P-3 Series			25.5		41.0		214.3
39	S-3 Series			78.2		13.3		12.9
40	E-2 Series			60.5		75.2		124.0

(Continued on next page)

U.S. NAVY PROCUREMENT REQUESTS (Continued)

(Dollars) Millions of Dollars

No.	Item Nomenclature	FY 1994 Unit Cost	FY 1992 Quan.	FY 1992 Cost	FY 1993 Quan.	FY 1993 Cost	FY 1994 Quan.	FY 1994 Cost
41	Trainer A/C Series			7.8		1.8		12.0
42	C-130 Series			20.3		9.5		13.6
43	Fleet Electronic Warfare Support Group			18.1		6.7		26.5
44	Cargo/Trnsprt A/C Series			1.2		1.6		15.0
45	E-6 Series			57.6		27.3		118.5
46	Exec. HLCPTR Series					27.5		52.3
47	Various			.1		.1		.1
48	Power Plant Chngs			25.7		26.0		9.5
49	Misc Flight Safety Chngs			.2		.2		.1
50	Common ECM Eqpt			119.7		105.8		65.8
51	Common Avionics Chngs			19.8		16.0		90.2
TOTAL MODIFICATION OF AIRCRAFT				**1,167.4**		**1,153.8**		**1,264.6**

BUDGET ACTIVITY 06: Aircraft Spares and Repair Parts

No.	Item Nomenclature	FY 1994 Unit Cost	FY 1992 Quan.	FY 1992 Cost	FY 1993 Quan.	FY 1993 Cost	FY 1994 Quan.	FY 1994 Cost
52	Spares & Rpr Parts			937.3		767.4		903.2
TOTAL AIRCRAFT SPARES AND REPAIR PARTS				**937.3**		**767.4**		**903.2**

(Continued on next page)

U.S. NAVY PROCUREMENT REQUESTS (Continued)

(Dollars) Millions of Dollars

No.	Item Nomenclature	FY 1994 Unit Cost	FY 1992 Quan.	FY 1992 Cost	FY 1993 Quan.	FY 1993 Cost	FY 1994 Quan.	FY 1994 Cost
BUDGET ACTIVITY 07; Aircraft Support Equipment and Facilities								
53	Common Grnd Eqpt			404.7		446.9		452.8
54	Aircraft Indus Facilities			37.6		30.3		37.9
55	War Consumables			15.1		15.1		18.1
56	Other Prod. Charges			45.5		69.5		41.5
57	Special Supp Eqpt			37.9		37.3		18.5
58	First Dest. Trans			6.1		5.9		4.7
59	Cntrct Admin/Audit							118.7
60	Cncld Acct Adj.			4.8				
	TOTAL AIRCRAFT SUPPORT EQUIPMENT AND FACILITIES			**551.7**		**605.1**		**692.3**
	TOTAL AIRCRAFT PROCUREMENT, NAVY			**7,137.8**		**5,950.7**		**6,132.6**

APPROPRIATION: 1507N WEAPONS PROCUREMENT, NAVY

No.	Item Nomenclature	FY 1994 Unit Cost	(Dollars) Millions of Dollars FY 1992		FY 1993		FY 1994	
			Quan.	Cost	Quan.	Cost	Quan.	Cost
	BUDGET ACTIVITY 01: Ballistic Missiles							
	Ballistic Missiles							
1	Trident I			6.8		1.1		7.6
2	Trident II	45,566,500	28	(1001.1)	21	(873.1)	24	(1093.6)
	Less:Adv Pro(PY)			(-22.9)		(-38.6)		(-110.3)
				978.1		**834.5**		**983.3**
3	Trident II							
	Avd Pro (CY)			117.2		146.8		145.3
	(FY92 for FY93)			(102.0)				
	(FY92 for FY94)			(.8)				
	(FY92 for FY95)			(5.2)				
	(FY92 for FY96)			(6.4)				
	(FY92 for FY97)			(1.6)				
	(FY92 for FY98)			(.6)				
	(FY92 for FY99)			(.6)				
	(FY93 for FY94)					(135.1)		
	(FY93 for FY95)					(9.2)		
	(FY93 for FY96)					(2.0)		
	(FY93 for FY97)					(.5)		
	(FY94 for FY95)							(142.6)
	(FY94 for FY96)							(2.1)
	(FY94 for FY97)							(.6)
	SUPPORT EQUIPMENT AND FACILITIES							
4	Missile Indus Facilities			2.0		1.5		2.2
	TOTAL BALLISTIC MISSILES			**1,104.2**		**984.0**		**1,138.4**

(Continued on next page)

APPROPRIATION: 1507N WEAPONS PROCUREMENT, NAVY (Continued)

No.	Item Nomenclature	FY 1994 Unit Cost	FY 1992 Quan.	FY 1992 Cost	FY 1993 Quan.	FY 1993 Cost	FY 1994 Quan.	FY 1994 Cost
				(Dollars)	Millions of Dollars			

BUDGET ACTIVITY 02: Other Missiles

STRATEGIC MISSILES

No.	Item Nomenclature	FY 1994 Unit Cost	FY 1992 Quan.	FY 1992 Cost	FY 1993 Quan.	FY 1993 Cost	FY 1994 Quan.	FY 1994 Cost
5	TOMAHAWK	1,149,481	176	411.2	200	411.9	216	248.3

TACTICAL MISSILES

No.	Item Nomenclature	FY 1994 Unit Cost	FY 1992 Quan.	FY 1992 Cost	FY 1993 Quan.	FY 1993 Cost	FY 1994 Quan.	FY 1994 Cost
6	AMRAAM	1,343,590	191	191.5	140	121.4	44	59.1
7	HARPOON	1,311,586	110	167.0	70	89.5	75	89.4
8	HARM		749	210.3		31.3		
9	STANDARD MISS	977,400	330	256.5	330	254.0	220	215.0
10	RAM	243,650		9.1		8.2	240	58.5
11	HELLFIRE	43,435			1000	49.9	1931	83.9
12	PENGUIN Less:Adv Pro(PY)	42		(48.1) (-3.7) **44.4**				
13	TWO IIA				938	23.6		
14	AERIAL TARGETS			173.3		164.3		114.4
15	DRONES & DECOYS			10.0		17.4		
16	OTHER MISS SUPP			17.6		11.0		9.8

MODIFICATION OF MISSILES

No.	Item Nomenclature	FY 1994 Unit Cost	FY 1992 Quan.	FY 1992 Cost	FY 1993 Quan.	FY 1993 Cost	FY 1994 Quan.	FY 1994 Cost
17	TOMAHAWK MODS			44.8		34.9		15.4

(Continued on next page)

APPROPRIATION: 1507N WEAPONS PROCUREMENT, NAVY (Continued)

(Dollars) Millions of Dollars

No.	Item Nomenclature	FY 1994 Unit Cost	FY 1992 Quan.	FY 1992 Cost	FY 1993 Quan.	FY 1993 Cost	FY 1994 Quan.	FY 1994 Cost
18	SPARROW MODS			29.8		21.2		35.9
19	SIDEWINDER MODS					15.1		18.2
20	PHOENIX MODS			18.2		9.1		
21	HARPOON MODS			21.2		32.0		2.8
22	HARM MODS							96.7
23	STANDARD MISS MODS			26.4		27.4		14.5

SUPPORT EQUIPMENT AND FACILITIES

No.	Item Nomenclature	FY 1994 Unit Cost	FY 1992 Quan.	FY 1992 Cost	FY 1993 Quan.	FY 1993 Cost	FY 1994 Quan.	FY 1994 Cost
24	Weapons Indust Facilities			44.8		41.2		22.1
25	Fleet Satellite Comm (MYP)		3	(369.4)		(265.6)		(159.8)
	Less:Adv Pro(PY)			(-86.4)		(-3.2)		
				283.1		**262.4**		**159.8**
26	Cntrct Admin/Audit							59.9

ORDNANCE SUPPORT EQUIPMENT

No.	Item Nomenclature	FY 1994 Unit Cost	FY 1992 Quan.	FY 1992 Cost	FY 1993 Quan.	FY 1993 Cost	FY 1994 Quan.	FY 1994 Cost
27	Ordnance Supp Eqpt			92.9		118.5		6.9
	TOTAL OTHER MISSILES			**2,052.2**		**1,744.3**		**1,319.5**

BUDGET ACTIVITY 03: Torpedoes and Related Equipment

TORPEDOES AND RELATED EQUIPMENT

No.	Item Nomenclature	FY 1994 Unit Cost	FY 1992 Quan.	FY 1992 Cost	FY 1993 Quan.	FY 1993 Cost	FY 1994 Quan.	FY 1994 Cost
28	MK-48 ADCAP Torpedo(MYP)	1,186,990	108	(146.4)	108	(195.6)	108	(128.2)
	Less:Adv Pro(PY)					(-29.7)		(-28.1)
				146.4		**165.9**		**100.1**

(Continued on next page)

APPROPRIATION: 1507N WEAPONS PROCUREMENT, NAVY (Continued)

(Dollars) Millions of Dollars

No.	Item Nomenclature	FY 1994 Unit Cost	FY 1992 Quan.	FY 1992 Cost	FY 1993 Quan.	FY 1993 Cost	FY 1994 Quan.	FY 1994 Cost
29	MK-48 ADCAP Torpedo(MYP)							
	Adv Pro(CY)			57.8				
	(FY92 for FY93)			(29.7)				
	(FY92 for FY94)			(28.1)				
30	MK-50 ALWT	218		261.2	212	241.0		21.4
31	ASW Targets			11.1		25.9		17.6
32	Anti-Submarine Rocket			2.8		2.1		
33	Vrtcl Launched Anti-Submarine Rocket (VLA)	815,675		(3.1)		(37.6)	40	(32.6)
	Less:Adv Pro(PY)							(-9.9)
				3.1		**37.6**		**22.7**
34	Vrtcl Launched Anti-Submarine Rocket (VLA)							
	Adv Pro(CY)					9.9		

MOD OF TORPEDOES AND RELATED EQUIP

No.	Item Nomenclature	FY 1994 Unit Cost	FY 1992 Quan.	FY 1992 Cost	FY 1993 Quan.	FY 1993 Cost	FY 1994 Quan.	FY 1994 Cost
35	MK-46 Torpedo MODS			12.4		48.8		24.1
36	Quickstrike Mine			8.5				3.5
37	MK-60 Captor MODS			1.3		1.3		

SUPPORT EQUIPMENT

No.	Item Nomenclature	FY 1994 Unit Cost	FY 1992 Quan.	FY 1992 Cost	FY 1993 Quan.	FY 1993 Cost	FY 1994 Quan.	FY 1994 Cost
38	Torpedo Supp Equip			45.4		43.3		37.6
39	ASW Range Supp			33.6		26.7		24.2

SUPPORT EQUIPMENT

No.	Item Nomenclature	FY 1994 Unit Cost	FY 1992 Quan.	FY 1992 Cost	FY 1993 Quan.	FY 1993 Cost	FY 1994 Quan.	FY 1994 Cost
40	First Dest Trans			9.0		8.9		7.1
	TOTAL TORPEDOES AND RELATED EQUIP			**592.7**		**611.3**		**258.4**

(Continued on next page)

APPROPRIATION: 1507N WEAPONS PROCUREMENT, NAVY (Continued)

No.	Item Nomenclature	FY 1994 Unit Cost	FY 1992 (Dollars) Millions of Dollars		FY 1993		FY 1994	
			Quan.	Cost	Quan.	Cost	Quan.	Cost
BUDGET ACTIVITY 04: Other Weapons								
GUNS AND GUN MOUNTS								
41	MK-15 Phalanx CIWS			2.0				
42	MK-19 40MM Machine Gun		568	11.1				
43	MK-38 25MM Gun Mount		55	10.0				
44	Small Arms & Wpns			12.4		14.4		.8
MODIFICATION OF GUNS AND GUN MOUNTS								
45	CIWS MODS			56.5		74.1		41.8
46	5/54 Gun Mount MODS			25.5		11.0		6.0
47	MK-75 76MM Gun Mount MODS			7.7		7.8		2.8
48	MODS Under $2 Mil			1.1		1.2		1.4
OTHER								
49	Cncld Acct Adj.			1.1				
TOTAL OTHER WEAPONS				**127.4**		**108.4**		**52.8**
BUDGET ACTIVITY 05: Other Ordnance								
AIR LAUNCHED ORDNANCE								
50	Gen Purpose Bombs			84.7		3.6		51.1
51	2.75 Inch Rockets			12.2		14.9		13.3

(Continued on next page)

APPROPRIATION: 1507N WEAPONS PROCUREMENT, NAVY (Continued)

(Dollars) Millions of Dollars

No.	Item Nomenclature	FY 1994 Unit Cost	FY 1992 Quan.	FY 1992 Cost	FY 1993 Quan.	FY 1993 Cost	FY 1994 Quan.	FY 1994 Cost
52	Machine Gun Ammo			31.6		1.0		7.4
53	Practice Bombs			9.1		13.8		10.9
54	Gator					9.6		
SHIP ORDNANCE								
55	5 Inch/54 Gun Ammo			36.3		82.3		55.2
56	CIWS Ammo			22.0		9.9		1.7
57	76MM Gun Ammo			9.5		8.7		15.6
58	Other Ship Gun Ammo			31.8		20.9		16.9
OTHER ORDNANCE								
59	Small Arms & Ldng Party Ammo			13.5		3.4		11.5
60	Pyrotechnic & Demo			14.7		19.7		13.4
61	Demilitarization							6.7
TOTAL OTHER ORDNANCE				**265.6**		**187.7**		**203.6**

BUDGET ACTIVITY 06: Spares and Repair Parts

No.	Item Nomenclature	FY 1994 Unit Cost	FY 1992 Quan.	FY 1992 Cost	FY 1993 Quan.	FY 1993 Cost	FY 1994 Quan.	FY 1994 Cost
SPARES AND REPAIR PARTS								
62	Spares & Rpr Parts			104.0		80.9		67.6
TOTAL SPARES AND REPAIR PARTS				**104.0**		**80.9**		**67.6**
TOTAL WEAPONS PROCUREMENT, NAVY				**4,246.0**		**3,716.6**		**3,040.3**

Section X.

National Aeronautics and Space Administration

(NASA)

NATIONAL AERONAUTICS AND SPACE ADMINISTRATION
MAJOR CONTRACTORS
Fiscal Years 1987-1991
By rank according to net value of NASA prime
contracts awarded during last fiscal year (Millions of Dollars)

Company	1987	1988	1989	1990	1991
TOTAL PROCUREMENTS	**$8,610**	**$9,545**	**$10,876**	**$12,565**	**$13,159**
Awards to Business Firms	**6,541**	**7,275**	**8,568**	**10,072**	**10,417**
% of TOTAL PROCUREMENTS	**76%**	**76%**	**79%**	**80%**	**79%**
Rockwell International Corp.	$1,610	$1,714	$ 1,692	$ 1,747	$ 1,560
McDonnell Douglas Corp.	285	299	506	851	1,089
Lockheed Space Operations Co.	323	474	553	583	591
Martin Marietta Corp.	326	341	355	507	572
The Boeing Co.	175	260	236	399	468
Lockheed Missiles & Space Co.	108	141	145	294	458
Thiokol Corp.	286	423	420	498	438
Rockwell Space Operations Inc.			287	309	343
General Electric Co.	225	211	300	402	308
Lockheed Engrg. & Science Co.	163	178	217	234	259
EG&G Florida Inc.	131	156	187	191	227
Computer Sciences Corp.	90	151	192	183	207
USBI Booster Production Co.	183	191	196	233	198
TRW Inc.	124	143	193	241	192
Loral Aerospace Co.	120	137	196	174	186
Bendix Field Engineering	142	152	156	156	176
Boeing Computer Support Services			158	165	159
United Technologies Corp.	166	91	133	136	133
Grumman Aerospace Corp.	23	74	80	86	100
Sverdrup Technology Inc.	27	38	65	79	97
Johnson Controls World Serv. Inc.					70
IBM Corp.	72	87	102	102	68
Teledyne Industries Inc.	38	40	52	73	65
BAMSI Inc.	31	40	30	38	52
Contel Corp.	81	76	51	65	50
Cray Research Inc.	11	31	48	43	47
Fairchild Industries Inc.	24	24	38	44	46
CAE Link Corp.			16	53	45
Harris Space Systems Corp.				25	45
Bionetics Corp.	7	11	29	36	41

Source: National Aeronautics and Space Administration, "Annual Procurement Report"
(Annually).

FEDERAL SPACE ACTIVITIES BUDGET AUTHORITY
Fiscal Years 1961 - 1991 (Millions of Current Dollars)

Year	TOTAL	NASA	DOD	Energy	Commerce	Other
1961	$ 1,808	$ 926	$ 814	$ 68	$ -	$ 1
1962	3,295	1,797	1,298	148	51	1
1963	5,435	3,626	1,550	214	43	2
1964	6,831	5,016	1,599	210	3	3
1965	6,956	5,138	1,574	229	12	3
1966	6,970	5,065	1,689	187	27	3
1967	6,710	4,830	1,664	184	29	3
1968	6,529	4,430	1,922	145	28	4
1969	5,976	3,822	2,013	118	20	3
1970	5,341	3,547	1,678	103	8	4
1971	4,741	3,101	1,512	95	27	5
1972	4,575	3,071	1,407	55	31	10
1973	4,825	3,093	1,623	54	40	15
1974	4,640	2,759	1,766	42	60	14
1975	4,914	2,915	1,892	30	64	13
1976	5,320	3,225	1,983	23	72	16
1977	5,983	3,440	2,412	22	91	18
1978	6,518	3,623	2,738	34	103	20
1979	7,244	4,030	3,036	59	98	21
1980	8,689	4,680	3,848	40	93	28
1981	9,978	4,992	4,828	41	87	30
1982	12,441	5,528	6,679	61	145	29
1983	15,589	6,328	9,019	39	178	25
1984	17,136	6,648	10,195	34	236	22
1985	20,167	6,925	12,768	34	423	17
1986	21,659	7,165	14,126	35	309	25
1987	26,448	9,809	16,287	48	278	27
1988	26,607	8,302	17,679	241	352	33
1989	28,443	10,098	17,906	97	301	42
1990	28,089	12,142	15,616	79	202	50
1991	27,592	13,036	14,181	108	211	56

Source: NASA, "Aeronautics and Space Report of the President" (Annually)
NOTE: Detail may not add to totals because of rounding.

FEDERAL SPACE ACTIVITIES BUDGET AUTHORITY
IN CONSTANT DOLLARS
Fiscal Years 1961 - 1991
(Millions of Constant Dollars, 1987 = 100)

Year	TOTAL	NASA	DOD	Energy	Commerce	Other
1961	$ 6,877	$ 3,522	$ 3,096	$259	$ -	$ 4
1962	12,299	6,708	4,845	552	190	4
1963	19,952	13,311	5,690	786	158	7
1964	24,714	18,148	5,785	760	11	11
1965	24,606	18,175	5,568	810	42	11
1966	23,977	17,423	5,810	643	93	10
1967	22,322	16,068	5,536	612	96	10
1968	20,926	14,199	6,160	465	90	13
1969	18,225	11,656	6,139	360	61	9
1970	15,450	10,260	4,854	298	23	12
1971	13,046	8,533	4,161	261	74	14
1972	11,967	8,033	3,680	144	81	26
1973	11,997	7,690	4,035	134	99	37
1974	10,723	6,376	4,081	97	139	32
1975	10,328	6,127	3,976	63	135	27
1976	10,387	6,296	3,872	45	141	31
1977	10,804	6,212	4,355	40	164	33
1978	10,942	6,082	4,596	57	173	34
1979	11,189	6,225	4,690	91	151	32
1980	12,311	6,631	5,452	57	132	40
1981	12,832	6,420	6,209	53	112	39
1982	14,890	6,616	7,994	73	174	35
1983	17,914	7,272	10,364	45	205	29
1984	18,862	7,318	11,222	37	260	24
1985	21,381	7,342	13,537	36	448	18
1986	22,301	7,377	14,545	36	318	26
1987	26,448	9,809	16,287	48	278	27
1988	25,675	8,011	17,060	233	340	32
1989	26,280	9,330	16,544	90	278	39
1990	24,997	10,805	13,897	70	180	44
1991	23,565	11,133	12,111	92	180	48

Source: AIA, derived from NASA, "Aeronautics and Space Report of the President" (Annually).

FEDERAL SPACE ACTIVITIES OUTLAYS
Fiscal Years 1961 - 1991 (Millions of Dollars)

Year	TOTAL	NASA	DOD	Energy	Commerce	Other
1961	$ 1,468	$ 694	$ 710	$ 64	$ -	$ -
1962	2,387	1,226	1,029	130	1	1
1963	4,079	2,517	1,368	181	12	1
1964	5,930	4,131	1,564	220	12	3
1965	6,886	5,035	1,592	232	24	3
1966	7,719	5,858	1,637	188	28	7
1967	7,237	5,337	1,673	184	39	5
1968	6,667	4,595	1,890	147	29	6
1969	6,326	4,078	2,095	118	31	5
1970	5,453	3,565	1,756	103	24	5
1971	4,999	3,171	1,693	97	30	8
1972	4,772	3,195	1,470	60	37	10
1973	4,719	3,069	1,557	51	29	13
1974	4,854	2,960	1,777	39	64	14
1975	4,891	2,951	1,831	34	64	11
1976	5,314	3,336	1,864	26	71	16
1977	5,559	3,600	1,833	22	87	18
1978	6,188	3,582	2,457	29	101	20
1979	6,808	3,744	2,892	55	97	21
1980	7,668	4,340	3,162	49	89	28
1981	9,166	4,877	4,131	47	81	30
1982	10,466	5,463	4,772	60	142	30
1983	12,590	6,101	6,247	40	178	25
1984	14,726	6,461	8,000	33	209	22
1985	17,255	6,607	10,441	34	155	17
1986	18,581	6,756	11,449	35	317	25
1987	21,844	7,254	14,264	37	262	26
1988	23,414	8,451	14,397	199	334	33
1989	25,143	10,195	14,504	97	306	41
1990	25,614	12,292	12,962	79	232	49
1991	28,170	13,351	14,432	108	223	56

Source: NASA, "Aeronautics and Space Report of the President" (Annually).
NOTE: Detail may not add to totals because of rounding.

FEDERAL SPACE ACTIVITIES OUTLAYS
IN CONSTANT DOLLARS
Fiscal Years 1961 - 1991 (Millions of Constant Dollars, 1987 = 100)

Year	TOTAL	NASA	DOD	Energy	Commerce	Other
1961	$ 5,584	$ 2,640	$ 2,701	$243	$ -	$ -
1962	8,909	4,576	3,840	485	4	3
1963	14,973	9,239	5,020	664	45	4
1964	21,454	14,947	5,657	796	45	9
1965	24,358	17,810	5,631	821	85	11
1966	26,551	20,151	5,633	648	97	23
1967	24,076	17,753	5,566	611	128	18
1968	21,368	14,729	6,058	470	93	18
1969	19,293	12,437	6,389	358	95	14
1970	15,774	10,313	5,080	297	69	15
1971	13,756	8,726	4,659	268	82	22
1972	12,482	8,357	3,845	156	98	26
1973	11,734	7,632	3,871	127	73	31
1974	11,218	6,842	4,107	90	148	32
1975	10,279	6,202	3,848	72	134	23
1976	10,375	6,514	3,640	50	139	32
1977	10,038	6,500	3,309	40	157	32
1978	10,388	6,014	4,125	48	169	33
1979	10,516	5,783	4,467	84	150	32
1980	10,864	6,149	4,480	69	126	39
1981	11,787	6,272	5,312	60	104	39
1982	12,527	6,539	5,711	71	170	35
1983	14,468	7,011	7,178	46	205	29
1984	16,209	7,112	8,806	37	230	25
1985	18,294	7,005	11,070	36	165	18
1986	19,132	6,956	11,788	36	326	26
1987	21,844	7,254	14,264	37	262	26
1988	22,594	8,154	13,893	192	322	32
1989	23,231	9,420	13,401	90	283	38
1990	22,794	10,939	11,535	70	206	44
1991	24,058	11,402	12,326	92	190	48

Source: AIA, derived from NASA, "Aeronautics and Space Report of the President" (Annually).

NOTE: Detail may not add to totals because of rounding.

FEDERAL AERONAUTICS RESEARCH AND DEVELOPMENT
IN CONSTANT DOLLARS
Fiscal Years 1973 - 1991 (Millions of Constant Dollars, 1987 = 100)

Year	TOTAL	NASA	DOD	DOT
BUDGET AUTHORITY				
1973	$5,438	$778	$4,473	$ 186
1974	4,691	642	3,878	171
1975	4,235	660	3,420	156
1976	4,590	635	3,790	166
1977	4,924	683	4,074	168
1978	5,603	734	4,712	158
1979	4,402	802	3,460	141
1980	4,238	793	3,310	135
1981	4,226	676	3,412	136
1982	4,286	618	3,572	97
1983	4,448	629	3,701	118
1984	4,499	660	3,549	289
1985	4,617	687	3,628	281
1986	6,857	619	5,073	1,166
1987	5,824	698	4,179	946
1988	6,730	698	4,814	1,218
1989	9,846	806	7,613	1,427
1990	9,513	829	7,001	1,683
1991	8,042	827	5,251	1,964
OUTLAYS				
1982	$3,961	$674	$3,180	$ 107
1983	4,084	647	3,356	82
1984	4,102	645	3,297	161
1985	4,251	682	3,288	282
1986	6,251	667	4,503	1,081
1987	5,866	622	4,182	1,062
1988	6,118	655	4,292	1,171
1989	7,845	790	5,932	1,124
1990	8,907	791	6,807	1,309
1991	8,266	869	5,801	1,597

Source: NASA, "Aeronautics and Space Report of the President" (Annually).

NATIONAL AERONAUTICS AND SPACE ADMINISTRATION OUTLAYS
Fiscal Years 1963 - 1993 (Millions of Current Dollars)

Year	TOTAL	Research and Development	Space Flight Control and Data Commun-cations	Construc-tion of Facilities	Research & Program Management
1963	$ 2,552	$1,912	$ -	$225	$ 416
1964	4,171	3,317	-	438	416
1965	5,093	3,984	-	531	578
1966	5,933	4,741	-	573	619
1967	5,426	4,487	-	289	650
1968	4,724	3,946	-	126	652
1969	4,251	3,530	-	65	656
1970	3,753	2,992	-	54	707
1971	3,382	2,630	-	44	708
1972	3,422	2,623	-	50	749
1973	3,315	2,541	-	45	729
1974	3,256	2,421	-	75	760
1975	3,266	2,420	-	85	761
1976	3,669	2,749	-	121	799
1977	3,945	2,980	-	105	860
1978	3,983	2,989	-	124	870
1979	4,196	3,139	-	133	925
1980	4,852	3,702	-	140	1,010
1981	5,426	4,228	-	147	1,050
1982	6,035	4,796	-	109	1,130
1983	6,664	5,316	-	108	1,240
1984	7,048	2,792	2,915	109	1,232
1985	7,251	2,118	3,707	170	1,322
1986	7,403	2,615	3,267	189	1,332
1987	7,591	2,436	3,597	149	1,409
1988	9,092	2,916	4,362	166	1,648
1989	11,051	3,922	5,030	190	1,908
1990	12,429	5,094	5,117	218	1,991
1991	13,878	5,765	5,590	326	2,185
1992 E	13,819	6,261	5,311	448	1,784
1993 E	14,088	6,768	5,150	492	1,661

Source: Office of Management and Budget, "Budget of the United States Government" (Annually).

NOTE: Detail may not add to totals because of rounding.
E Estimate. Latest year reflects Administration's budget proposal.

NATIONAL AERONAUTICS AND SPACE ADMINISTRATION OUTLAYS IN CONSTANT DOLLARS
Fiscal Years 1963 - 1993 (Millions of Constant Dollars, 1987 = 100)

Year	TOTAL	Research and Development	Space Flight Control and Data Commun- cations	Construc- tion of Facilities	Research & Program Management
1963	$ 9,369	$ 7,019	$ -	$ 826	$1,527
1964	15,090	12,001	-	1,585	1,505
1965	18,016	14,093	-	1,878	2,045
1966	20,409	16,309	-	1,971	2,129
1967	18,051	14,927	-	961	2,162
1968	15,141	12,647	-	404	2,090
1969	12,964	10,765	-	198	2,001
1970	10,856	8,655	-	156	2,045
1971	9,307	7,237	-	121	1,948
1972	8,951	6,861	-	131	1,959
1973	8,242	6,318	-	112	1,813
1974	7,525	5,595	-	173	1,756
1975	6,864	5,086	-	179	1,599
1976	7,163	5,367	-	236	1,560
1977	7,124	5,381	-	190	1,553
1978	6,686	5,018	-	208	1,460
1979	6,481	4,849	-	205	1,429
1980	6,874	5,245	-	198	1,431
1981	6,978	5,437	-	189	1,350
1982	7,223	5,740	-	130	1,352
1983	7,658	6,109	-	124	1,425
1984	7,758	3,073	3,209	120	1,356
1985	7,688	2,246	3,930	180	1,402
1986	7,623	2,693	3,364	195	1,371
1987	7,591	2,436	3,597	149	1,409
1988	8,774	2,814	4,209	160	1,590
1989	10,211	3,624	4,648	176	1,763
1990	11,031	4,521	4,542	193	1,767
1991	11,852	4,924	4,774	278	1,866
1992	E 11,447	5,186	4,399	371	1,478
1993	E 11,294	5,426	4,129	394	1,332

Source: Office of Management and Budget, "Budget of the United States Government" (Annually).
NOTE: Detail may not add to totals because of rounding.
E Estimate. Latest year reflects Administration's budget proposal.

**NATIONAL AERONAUTICS AND SPACE ADMINISTRATION
BUDGET AUTHORITY**
Fiscal Years 1963 - 1993 (Millions of Current Dollars)

Year	TOTAL	Research and Development	Space Flight Control and Data Communcations	Construction of Facilities	Research & Program Management
1963	$ 3,673	$2,929	$ -	$744	$
1964	5,099	3,890	-	713	496
1965	5,250	4,360	-	267	623
1966	5,175	4,502	-	61	602
1967	4,968	4,235	-	85	648
1968	4,589	3,912	-	38	639
1969	3,995	3,314	-	33	648
1970	3,749	2,993	-	53	703
1971	3,312	2,556	-	26	730
1972	3,308	2,523	-	53	732
1973	3,408	2,599	-	79	730
1974	3,040	2,194	-	101	745
1975	3,231	2,323	-	143	765
1976	3,552	2,678	-	82	792
1977	3,819	2,856	-	118	845
1978	4,064	3,012	-	162	890
1979	4,559	3,477	-	148	934
1980	5,243	4,088	-	159	996
1981	5,522	4,334	-	117	1,071
1982	6,020	4,772	-	114	1,134
1983	6,875	5,539	-	139	1,197
1984	7,316	2,064	3,772	223	1,256
1985	7,573	2,468	3,594	178	1,332
1986	7,807	2,619	3,670	176	1,342
1987	10,923	3,154	6,100	217	1,453
1988	9,062	3,280	3,806	213	1,763
1989	10,969	4,213	4,555	275	1,927
1990	12,324	5,225	4,645	218	2,023
1991	14,016	6,024	5,271	498	2,212
1992 E	14,321	6,851	5,352	525	1,578
1993 E	14,994	7,731	5,267	319	1,660

Source: Office of Management and Budget, "Budget of the United States Government" (Annually).
NOTE: Detail may not add to totals because of rounding.
E Estimate. Latest year reflects Administration's budget proposal.

NATIONAL AERONAUTICS AND SPACE ADMINISTRATION
BUDGET AUTHORITY IN CONSTANT DOLLARS
Fiscal Years 1963 - 1993 (Millions of Constant Dollars, 1987 = 100)

Year		TOTAL	Research and Development	Space Flight Control and Data Commun- cations	Construc- tion of Facilities	Research & Program Management
1963		$13,484	$10,753	$ -	$2,731	$ -
1964		18,448	14,074	-	2,580	1,795
1965		18,571	15,423	-	944	2,204
1966		17,802	15,487	-	210	2,071
1967		16,527	14,088	-	283	2,156
1968		14,708	12,538	-	122	2,048
1969		12,184	10,107	-	101	1,976
1970		10,845	8,658	-	153	2,034
1971		9,114	7,034	-	72	2,009
1972		8,653	6,600	-	139	1,915
1973		8,473	6,462	-	196	1,815
1974		7,026	5,070	-	233	1,722
1975		6,791	4,882	-	301	1,608
1976		6,935	5,228	-	160	1,546
1977		6,896	5,157	-	213	1,526
1978		6,822	5,056	-	272	1,494
1979		7,042	5,371	-	229	1,443
1980		7,428	5,792	-	225	1,411
1981		7,101	5,574	-	150	1,377
1982		7,205	5,712	-	136	1,357
1983		7,900	6,365	-	160	1,376
1984		8,053	2,272	4,152	245	1,382
1985		8,029	2,617	3,810	189	1,412
1986		8,039	2,697	3,779	181	1,382
1987		10,923	3,154	6,100	217	1,453
1988		8,745	3,165	3,673	206	1,701
1989		10,135	3,893	4,209	254	1,780
1990		10,938	4,637	4,123	193	1,796
1991		11,970	5,145	4,502	425	1,889
1992	E	11,863	5,675	4,433	435	1,307
1993	E	12,020	6,198	4,222	256	1,331

Source: Office of Management and Budget, "Budget of the United States Government" (Annually).

NOTE: Detail may not add to totals because of rounding.

**NATIONAL AERONAUTICS AND SPACE ADMINISTRATION
BUDGET AUTHORITY FOR RESEARCH AND DEVELOPMENT AND SPACE
FLIGHT, CONTROL, & DATA COMMUNICATIONS (Millions of Dollars)
Fiscal Years 1992 - 1993**

	1992	1993
RESEARCH AND DEVELOPMENT--TOTAL	**$6,851**	**$7,731**
Space Station--Total	$2,029	$2,250
Space Transport Capability Development--Total	720	864
Space Science & Applications--Total	2,729	2,985
Physics and Astronomy	1,047	1,114
Planetary Exploration	535	487
Life Sciences	149	177
Space Applications	998	1,207
Commercial Use of Space--Total	**148**	**172**
Aeronautics & Space Technology--Total	**1,103**	**1,302**
Aeronautical Research & Technology	784	890
Space Research & Technology	314	332
Transatmospheric Research & Technology	5	80
Safety, Reliability, & Quality Assurance--Total	**34**	**33**
Academic Programs--Total	67	71
Tracking & Data Advanced Systems--Total	**22**	**23**
SPACE FLIGHT, CONTROL, AND DATA COMMUNICATIONS--TOTAL	**$5,385**	**$5,267**
Space Shuttle Production & Capability Development--Total	**$1,328**	**$1,013**
Orbiter	246	306
Launch & Mission Support	262	211
Propulsion Systems	715	358
Assured Shuttle Availability	105	139
Space Shuttle Operations--Total	**2,943**	**3,115**
Flight Operations	790	740
Flight Hardware	1,297	1,455
Launch & Landing Operations	628	640
Research Operations Support	228	280
Expendable Launch Vehicles	195	218
Space and Ground Networks, Communications, & Data Systems--Total	**918**	**921**

Source: "NASA Budget Briefing Background Material" (Annually).
Note: Detail may not add to totals because of rounding.
E Estimate. Latest year reflects Administration's budget proposal.

ORDERS, SALES, AND BACKLOG
SPACE VEHICLE SYSTEMS
(Excluding Engines and Propulsion Units)
Calendar Years 1977 - 1991 (Millions of Dollars)

	SALES--Current Dollars			SALES--Constant Dollars		
Year	TOTAL	Military	Non-Military	TOTAL	Military	Non-Military
1977	$ 1,870	$ 814	$1,056	$ 3,425	$1,491	$1,934
1978	2,324	1,006	1,318	4,042	1,750	2,292
1979	2,539	1,105	1,434	3,998	1,740	2,258
1980	3,483	1,461	2,022	4,933	2,069	2,864
1981	3,856	1,736	2,120	4,850	2,184	2,667
1982	4,749	2,606	2,143	5,403	2,965	2,438
1983	4,940	2,420	2,520	5,358	2,625	2,733
1984	5,225	3,019	2,206	5,235	3,025	2,210
1985	6,300	4,241	2,059	6,383	4,297	2,086
1986	6,304	4,579	1,725	6,317	4,588	1,728
1987	8,051	5,248	2,803	8,051	5,248	2,803
1988	8,622	6,190	2,432	8,461	6,075	2,387
1989	9,758	6,457	3,301	9,197	6,086	3,111
1990	9,691	6,556	3,135	8,770	5,933	2,837
1991	10,953	6,898	4,055	9,558	6,019	3,538

	NEW NEW ORDERS			BACKLOG AS OF DEC 31		
Year	TOTAL	Military	Non-Military	TOTAL	Military	Non-Military
1977	$ 2,225	$1,175	$1,050	$ 1,589	$1,263	$ 326
1978	3,157	1,436	1,721	2,188	1,693	495
1979	2,698	1,018	1,680	1,448	909	539
1980	3,636	1,625	2,011	2,099	1,218	881
1981	5,062	2,878	2,184	3,163	2,166	997
1982	5,842	2,718	3,124	4,254	2,277	1,977
1983	5,399	3,016	2,383	4,865	2,733	2,132
1984	4,984	3,385	1,599	4,624	3,099	1,525
1985	8,383	6,083	2,300	6,707	4,941	1,766
1986	7,437	5,666	1,771	8,063	6,028	2,035
1987	11,455	9,000	2,455	12,393	9,460	2,933
1988	7,296	4,561	2,735	10,838	7,880	2,958
1989	11,709	8,107	3,602	13,356	9,192	4,164
1990	9,598	6,256	3,342	12,462	8,130	4,332
1991	11,721	5,575	6,146	12,450	6,539	5,911

Source: Bureau of the Census, "Aerospace Industry (Orders, Sales, and Backlog)"

ORDERS, SALES, AND BACKLOG
ENGINES AND PROPULSION UNITS FOR MISSILES AND SPACE VEHICLES
Calendar Years 1977 - 1991 (Millions of Dollars)

	SALES--Current Dollars			SALES--Constant Dollars		
Year	TOTAL	Military	Non-Military	TOTAL	Military	Non-Military
1977	$ 787	$ 757	$ 30	$1,441	$1,386	$ 55
1978	792	760	32	1,377	1,322	56
1979	952	915	37	1,499	1,441	58
1980	939	661	278	1,330	936	394
1981	1,204	786	418	1,514	989	526
1982	1,555	899	656	1,769	1,023	746
1983	1,814	951	863	1,967	1,031	936
1984	2,305	1,116	1,189	2,310	1,118	1,191
1985	2,466	1,256	1,210	2,498	1,273	1,226
1986	2,995	1,796	1,199	3,001	1,800	1,201
1987	2,993	1,563	1,430	2,993	1,563	1,430
1988	3,407	1,830	1,577	3,343	1,796	1,548
1989	3,602	1,771	1,831	3,395	1,669	1,726
1990	3,247	1,911	1,336	2,938	1,729	1,209
1991	3,851	1,843	2,008	3,360	1,608	1,752

	NET NEW ORDERS			BACKLOG AS OF DECEMBER 31		
Year	TOTAL	Military	Non-Military	TOTAL	Military	Non-Military
1977	$ 727	$ 693	$ 34	$ 613	$ 595	$ 18
1978	967	919	48	788	754	34
1979	1,187	1,141	46	1,024	980	44
1980	1,221	653	568	1,284	871	413
1981	1,284	746	538	1,343	828	515
1982	2,112	1,134	978	1,901	1,063	838
1983	1,618	942	676	1,691	1,052	639
1984	3,770	2,258	1,512	3,156	2,194	962
1985	3,823	1,323	2,500	4,513	2,261	2,252
1986	1,985	1,224	761	3,503	1,689	1,814
1987	3,335	1,995	1,340	3,849	2,121	1,728
1988	3,507	1,623	1,884	3,985	1,998	1,987
1989	6,113	2,475	3,638	6,410	2,595	3,815
1990	2,692	1,891	801	6,230	2,887	3,343
1991	5,719	1,092	4,627	8,491	2,324	6,167

Source: Bureau of the Census, "Aerospace Industry (Orders, Sales, and Backlog)."

NASA AWARDS BY TYPE OF CONTRACTOR
FISCAL YEARS 1988 - 1992

	FY 1988	FY 1989	FY 1990	FY 1991	FY 1992
AWARDS IN MILLIONS					
TOTAL	**$9,545**	**$10,876**	**$12,565**	**$13,159**	**$13,478**
BUSINESS FIRMS	7,275	8,568	10,071	10,417	10,717
EDUCATIONAL	370	464	514	592	659
NONPROFIT	129	180	201	244	298
JPL	980	1,058	1,107	1,140	1,230
GOV'T AGENCIES	735	543	610	693	498
OUTSIDE U.S.	56	63	62	73	76
PERCENT OF TOTAL					
TOTAL	**100**	**100**	**100**	**100**	**100**
BUSINESS FIRMS	76	79	80	79	79
EDUCATIONAL	4	4	4	4	5
NONPROFIT	1	2	2	2	2
JPL	10	10	9	9	9
GOV'T AGENCIES	8	5	5	5	4
OUTSIDE U.S.	1	*	*	1	1

* Less than .05 percent.

Appendix I shows distribution of NASA direct procurements by type of contractor for the period Fiscal Years 1961-1992.

DISTRIBUTION OF DIRECT NASA PROCUREMENTS
FISCAL YEARS 1986 - 1992

TYPE	FY 1986	FY 1987	FY 1988	FY 1989	FY 1990	FY 1991	FY 1992
NET VALUE OF AWARDS (MILLIONS)							
TOTAL	**$8,179.7**	**$8,609.8**	**$9,545.1**	**$10,876.4**	**$12,565.2**	**$13,159.0**	**$13,478.2**
Business Firms	6,356.0	6,540.5	7,274.9	8,567.6	10,071.5	10,417.3	10,716.7
Educational	276.6	315.4	370.3	464.2	513.6	592.0	659.3
Nonprofit	119.0	119.1	129.5	180.0	200.6	244.0	297.8
JPL	891.3	1,005.6	979.9	1,058.1	1,106.8	1,139.6	1,229.6
Government	489.7	594.9	734.6	543.2	610.4	693.4	498.6
Outside U.S.	47.1	34.3	55.9	63.3	62.3	72.7	76.2
PERCENT OF TOTAL							
TOTAL	**100**	**100**	**100**	**100**	**100**	**100**	**100**
Business Firms	78	76	76	79	80	79	79
Educational	3	4	4	4	4	4	5
Nonprofit	1	1	1	2	2	2	2
JPL	11	12	10	10	9	9	9
Government	6	7	8	5	5	5	4
Outside U.S.	1	*	1	*	*	1	1

* Less than 0.5 percent.

**TOP ONE HUNDRED CONTRACTORS (BUSINESS FIRMS) LISTED
ACCORDING TO TOTAL NASA AWARDS RECEIVED
FISCAL YEAR 1992
(S=Small Business/D=Disadvantaged Business)**

CONTRACTOR & PRINCIPAL PLACE OF CONTRACT PERFORMANCE	AWARDS (THOUSANDS)	PERCENT
TOTAL AWARDS TO BUSINESS FIRMS	**$10,716,743**	**100.00**
1. ROCKWELL INTERNATIONAL CORP Canoga Park, CA	1,449,346	13.52
2. MCDONNELL DOUGLAS CORP Huntington Beach, CA	1,045,418	9.75
3. LOCKHEED SPACE OPERATIONS CO Kennedy Space Center, FL	599,213	5.59
4. LOCKHEED MISSILES & SPACE CO Marshall Space Flight, AL	530,153	4.95
5. THIOKOL CORP Brigham City, UT	510,292	4.76
6. BOEING CO Marshall Space Flight, AL	500,115	4.67
7. MARTIN MARIETTA CORP New Orleans, LA	444,799	4.15
8. ROCKWELL SPACE OPERATIONS INC Houston, TX	345,886	3.23
9. GENERAL ELECTRIC CO King of Prussia, PA	299,400	2.79
10. LOCKHEED ENGRG & SCIENCE CO Houston, TX	269,905	2.52
11. COMPUTER SCIENCES CORP Greenbelt, MD	232,354	2.17
12. E G & G FLORIDA INC Kennedy Space Center, FL	212,843	1.99
13. U S B I BOOSTER PRODUCTION CO Huntsville, AL	207,274	1.93
14. T R W INC Redondo Beach, CA	194,369	1.81
15. BENDIX FIELD ENGINEERING CORP Greenbelt, MD	180,926	1.69
16. LORAL AEROSPACE CORP Houston, TX	140,521	1.31
17. BOEING COMPUTER SUPPORT SERVICES Marshall Space Flight, AL	139,816	1.30
18. UNITED TECHNOLOGIES CORP West Palm Beach, FL	135,840	1.27
19. SVERDRUP TECHNOLOGY INC Middleburgh Heights, OH	109,444	1.02
20. GRUMMAN AEROSPACE CORP Reston, VA	103,250	.96
21. SPACE SYSTEMS LORAL INC San Jose, CA	94,944	.89
22. JOHNSON CONTROLS WORLD SERV Stennis Space Center, MS	76,139	.71
23. INTERNATIONAL BUSINESS MACHINES Houston, TX	76,085	.71
24. CAE LINK CORP Houston, TX	61,467	.57
25. HARRIS SPACE SYSTEMS CORP Rockledge, FL	60,099	.56
26. BAMSI INC Marshall Space Flight, AL	(D) 58,739	.55
27. ORBITAL SCIENCES CORP(S) Denver, CO	55,631	.52
28. TELEDYNE INDUSTRIES INC Marshall Space Flight, AL	53,863	.50
29. G T E GOVERNMENT SYSTEMS CORP Gaithersburg, MD	49,687	.46

(Continued on next page)

TOP ONE HUNDRED CONTRACTORS (Continued)

CONTRACTOR & PRINCIPAL PLACE OF CONTRACT PERFORMANCE	AWARDS (THOUSANDS)	PERCENT
TOTAL AWARDS TO BUSINESS FIRMS	**$10,716,743**	**100.00**
30. BALL CORP Boulder, CO	49,345	.46
31. GENERAL DYNAMICS CORP San Diego, CA	49,058	.46
32. N S I TECHNOLOGY SERVICES CORP Greenbelt, MD	46,947	.44
33. STERLING FEDERAL SYSTEMS INC Moffett Field, CA	43,579	.41
34. BIONETICS CORP Marshall Space Flight, AL	43,174	.40
35. CRAY RESEARCH INC Chippewa Falls, WI	42,977	.40
36. P R C INC Washington, DC	41,267	.39
37. S T SYSTEMS CORP Greenbelt, MD	(D) 40,713	.38
38. SPACEHAB CORP Washington, DC	(S) 37,886	.35
39. METRIC CONSTRUCTORS INC Kennedy Space Center, FL	35,596	.33
40. RAYTHEON SERVICE CO Annapolis Junction, MD	33,847	.32
41. SANTA BARBARA RESEARCH CENTER Goleta, CA	32,367	.30
42. FAIRCHILD INDUSTRIES INC Germantown, MD	31,709	.30
43. CORTEZ III SERVICE CORP Cleveland, OH	(D) 31,283	.29
44. ANALEX CORP Fairview Park, OH	27,475	.26
45. AEROJET GENERAL CORP Azusa, CA	26,949	.25
46. SCIENCE APPLICATION INTL CORP San Diego,CA	26,658	.25
47. CALSPAN CORP Moffett Field, CA	26,286	.25
48. KRUG LIFE SCIENCES INC Houston, TX	24,892	.23
49. NORTHROP WORLDWIDE AIRCRAFT Houston, TX	22,208	.21
50. AIR PRODUCTS & CHEMICALS INC Allentown, PA	21,438	.20
51. PARAMAX SYSTEMS CORP Greenbelt, MD	21,082	.20
52. SWALES & ASSOCIATES INC Greenbelt, MD	(S) 20,690	.19
53. GRUMMAN DATA SYSTEMS CORP Houston, TX	19,013	.18
54. E E R SYSTEMS CORP Beltsville, MD	(S)(D) 18,382	.17
55. UNISYS GOVERNMENT SYSTEMS INC Hampton, VA	17,567	.16
56. BLAKE CONSTRUCTION CO INC Greenbelt, MD	17,501	.16
57. LOCKHEED CORP Burbank, CA	16,993	.16
58. OGDEN LOGISTICS SERVICES Greenbelt, MD	16,897	.16
59. SILICON GRAPHICS INC Mountain View, CA	16,381	.15
60. JACKSON & TULL INC Greenbelt, MD	(S)(D) 15,860	.15

(Continued on next page)

TOP ONE HUNDRED CONTRACTORS (Continued)

CONTRACTOR & PRINCIPAL PLACE OF CONTRACT PERFORMANCE	AWARDS (THOUSANDS)	PERCENT
TOTAL AWARDS TO BUSINESS FIRMS	**$10,716,743**	**100.00**
61. C B I SERVICES INC Moffett Field, CA	$15,238	.14
62. QUAD S CO Moffett Field, CA	(S) 15,162	.14
63. SPACE TRANSPORTATION PRO TEAM Huntsville, AL	14,760	.14
64. CLEVELAND ELECTRIC ILLUMINATING Cleveland, OH	14,627	.14
65. FERGUSON M K CO Cleveland, OH	14,559	.14
66. MICRO CRAFT INC Hampton, VA	(S) 14,555	.14
67. HERNANDEZ ENGINEERING INC Houston, TX	(S)(D) 14,109	.13
68. ENGINEERING DESIGN GROUP INC Cleveland, OH	(S) 13,856	.13
69. WYLE LABORATORIES Hampton, VA	13,148	.12
70. VIRGINIA ELECTRIC & POWER CO Hampton, VA	12,835	.12
71. DIGITAL EQUIPMENT CORP Moffett Field, CA	12,800	.12
72. R M S TECHNOLOGIES INC Cleveland, OH	(D) 12,730	.12
73. F D SERVICES INC Houston, TX	12,677	.12
74. JOHNSON ENGINEERING CORP Houston, TX	(S) 12,389	.12
75. PERKIN ELMER CORP Pomona, CA	12,304	.11
76. BOOZ ALLEN & HAMILTON INC Bethesda, MD	11,814	.11
77. COLEJON MECHANICAL CORP Cleveland, OH	(D) 11,750	.11
78. HUGHES DANBURY OPTICAL SYS Danbury, CT	11,695	.11
79. STERLING ZERO ONE INC Moffett Field, CA	(S) 11,640	.11
80. ADVANCED COMPUTER SYSTEMS INC Greenbelt, MD	(S)(D) 11,106	.10
81. B D M INTERNATIONAL INC Washington, DC	10,939	.10
82. GENERAL ELECTRIC U T C JV Evendale, OH	$10,924	.10
83. COMPUTER SCIENCES PAN AM SERV Slidell, LA	10,596	.10
84. ALLIED SIGNAL INC Tempe, AZ	9,942	.09
85. GOVERNMENT MICRO RESOURCES Chantilly, VA	(S)(D) 9,865	.09
86. FAIRCHILD SPACE & DEF CORP Greenbelt, MD	9,519	.09
87. ANALYTICAL SERVICES & MAT INC Hampton, VA	(S)(D) 9,293	.09
88. RECOM TECHNOLOGIES INC Moffett Field, CA	(S)(D) 9,180	.09
89. MASON & HANGER SERVICES INC Hampton, VA	9,166	.09
90. EDERER INC Seattle, WA	(S) 8,821	.08
91. TAFT BROADCASTING CO HOUSTON Houston, TX	(S) 8,716	.08

(Continued on next page)

TOP ONE HUNDRED CONTRACTORS (Continued)

CONTRACTOR & PRINCIPAL PLACE OF CONTRACT PERFORMANCE	AWARDS (THOUSANDS)	PERCENT
TOTAL AWARDS TO BUSINESS FIRMS	**$10,716,743**	**100.00**
92. VITRO CORP Washington, DC	8,633	.08
93. L T V AEROSPACE & DEFENSE CO Dallas, TX	8,424	.08
94. BOEING AEROSPACE OPERATIONS INC Moffett Field, CA	8,331	.08
95. I NET INC Kennedy Space Center, FL	(S)(D) 8,122	.08
96. HUGHES AIRCRAFT CO El Segundo, CA	7,869	.07
97. NYMA INC Greenbelt, MD	(S)(D) 7,747	.07
98. STANFORD TELECOMMUNICATIONS Reston, VA	(S) 7,734	.07
99. KELSEY SEYBOLD CLINIC Houston, TX	7,704	.07
100. CENTENNIAL CONTRACTORS ENTPR Greenbelt, MD	7,511	.07
OTHER*	1,122,115	10.47

* Includes other awards over $25,000 and smaller procurements of $25,000 or less.

**ONE HUNDRED EDUCATIONAL AND NONPROFIT INSTITUTIONS
LISTED ACCORDING TO TOTAL NASA AWARDS RECEIVED***
FISCAL YEAR 1992
(N=Nonprofit Institution)

CONTRACTOR & PRINCIPAL PLACE OF PERFORMANCE	AWARDS (THOUSANDS)	PERCENT
TOTAL AWARDS TO EDUCATIONAL & NONPROFIT INSTITUTIONS	**$957,085**	**100.00**
1. STANFORD UNIV Stanford, CA	53,963	5.64
2. ASSN UNIV RESEARCH & ASTRON Baltimore, MD	(N) 47,539	4.97
3. SMITHSONIAN INSTITUTION Cambridge, MA	(N) 38,293	4.00
4. MASS INSTITUTE TECHNOLOGY Cambridge, MA	37,085	3.88
5. UNIVERSITIES SPACE RESEARCH Greenbelt, MD	(N) 31,908	3.33
6. UNIV CALIF BERKELEY Berkeley, CA	24,497	2.56
7. C I E S I N Ann Arbor, MI	(N) 23,815	2.49
8. MITRE CORP Houston, TX	(N) 21,026	2.20
9. UNIV CALIF SAN DIEGO La Jolla, CA	20,950	2.19
10. UNIV MARYLAND COLLEGE PARK College Park, MD	20,935	2.19
11. UNIV ARIZONA Tucson, AZ	18,994	1.99
12. UNIV COLORADO BOULDER Boulder, CO	18,919	1.98
13. U T CALSPAN CENTER AEROSPACE RES Tullahoma, TN	(N) 18,750	1.96
14. NATIONAL ACADEMY SCIENCES Washington, DC	(N) 17,852	1.87
15. UNIV ALABAMA HUNTSVILLE Huntsville, AL	16,578	1.73
16. CHARLES STARK DRAPER LAB INC Cambridge, MA	(N) 16,561	1.73
17. NEW MEXICO STATE UNIV LAS CRUCES Palestine, TX	16,491	1.72
18. UNIV WISCONSIN MADISON Madison, WI	13,888	1.45
19. PENNSYLVANIA STATE UNIV UP University Park, PA	12,687	1.33
20. UNIV MICHIGAN ANN ARBOR Ann Arbor, MI	11,899	1.24
21. CALIF INSTITUTE TECHNOLOGY Pasadena, CA	11,477	1.20
22. UTAH STATE UNIV Logan, UT	11,437	1.20
23. UNIV NEW HAMPSHIRE Durham, NH	10,102	1.06
24. UNIV IOWA Iowa City, IA	9,381	.98
25. SOUTHWEST RESEARCH INSTITUTE San Antonio, TX	(N) 9,145	.96
26. UNIV WASHINGTON Seattle, WA	9,113	.95
27. CORNELL UNIV Ithaca, NY	8,726	.91
28. S E T I INSTITUTE Moffett Field, CA	(N) 8,573	.90

(Continued on next page)

ONE HUNDRED EDUCATIONAL AND NONPROFIT INSTITUTIONS (Continued)

CONTRACTOR & PRINCIPAL PLACE OF PERFORMANCE	AWARDS (THOUSANDS)	PERCENT
29. UNIV ALASKA FAIRBANKS Fairbanks, AK	8,552	.89
30. UNIV CALIF LOS ANGELES Los Angeles, CA	8,330	.87
31. UNIV TEXAS AUSTIN Austin, TX	8,127	.85
32. JOHNS HOPKINS UNIV Baltimore, MD	8,027	.84
33. SAN JOSE STATE UNIV Moffett Field, CA	7,752	.81
34. UNIV HAWAII Honolulu, HI	7,631	.80
35. UNIV VIRGINIA Charlottesville, VA	7,344	.77
36. CASE WESTERN RESERVE UNIV Cleveland, OH	7,081	.74
37. WHEELING JESUIT COLLEGE Wheeling, WV	6,956	.73
38. UNIV HOUSTON Houston, TX	6,918	.72
39. UNIV CHICAGO Chicago, IL	6,474	.68
40. COLUMBIA UNIV New York, NY	6,416	.67
41. UNIV HOUSTON CLEAR LAKE Houston, TX	6,307	.66
42. OKLAHOMA STATE UNIV Stillwater, OK	6,182	.65
43. BATTELLE MEMORIAL INSTITUTE Columbus, OH	(N) 5,980	.63
44. OHIO AEROSPACE INSTITUTE Brookpark, OH	(N) 5,747	.60
45. TEXAS A & M UNIV College Station, TX	5,656	.59
46. HARVARD UNIV Cambridge, MA	5,258	.55
47. PRINCETON UNIV Princeton, NJ	5,207	.54
48. AUBURN UNIV AUBURN Auburn, AL	5,104	.53
49. CARNEGIE MELLON UNIV Pittsburgh, PA	4,775	.50
50. UNIV CALIF SANTA BARBARA Santa Barbara, CA	4,685	.49
51. OREGON STATE UNIV Corvallis, OR	4,579	.48
52. ELORET INSTITUTE Moffett Field, CA	(N) 4,452	.47
53. OHIO STATE UNIV Columbus, OH	4,434	.46
54. UNIV ALABAMA BIRMINGHAM Birmingham, AL	4,255	.44
55. VIRGINIA POLYTECHNIC INSTITUTE Blacksburg, VA	4,084	.43
56. NORTH CAROLINA STATE UNIV Raleigh, NC	4,070	.43
57. OLD DOMINION UNIV Norfolk, VA	3,955	.41
58. UNIV FLORIDA Gainesville, FL	3,867	.40
59. WASHINGTON UNIV ST LOUIS St. Louis, MO	3,829	.40

(Continued on next page)

ONE HUNDRED EDUCATIONAL AND NONPROFIT INSTITUTIONS
(Continued)

CONTRACTOR & PRINCIPAL PLACE OF PERFORMANCE	AWARDS (THOUSANDS)	PERCENT
60. PURDUE UNIV West Lafayette, IN	3,751	.39
61. UNIV ILLINOIS URBANA Urbana, IL	3,653	.38
62. GEORGIA INSTITUTE TECHNOLOGY Atlanta, GA	3,610	.38
63. AMERICAN INSTIT AERO & ASTRO New York, NY	(N) $3,441	.36
64. CLEVELAND STATE UNIV Cleveland, OH	3,435	.36
65. WEST VIRGINIA UNIV Morgantown, WV	3,397	.36
66. UNIV SOUTHERN CALIF Los Angeles, CA	3,389	.35
67. HAMPTON CITY Hampton, VA	(N) 3,345	.35
68. RESEARCH TRIANGLE INSTITUTE Hampton, VA	(N) 3,333	.35
69. COLORADO STATE UNIV Fort Collins, CO	3,271	.34
70. MCAT INSTITUTE Moffett Field, CA	(N) 3,065	.32
71. UNIV CALIF IRVINE Irvine, CA	3,022	.32
72. GEORGE WASHINGTON UNIV Washington, DC	3,007	.31
73. UNIV MINNESOTA MINNPL ST PAUL Minneapolis, MN	2,924	.31
74. NORTH CAROLINA A & T STATE UNIV Greensboro, NC	2,883	.30
75. UNIV IDAHO Moscow, ID	2,785	.29
76. RENSSELAER POLY INST NEW YORK Troy, NY	2,621	.27
77. UNIV CORP ATMOSPHERIC RESEARCH Boulder, CO	(N) 2,597	.27
78. ARIZONA STATE UNIV Tempe, AZ	2,459	.26
79. S R I INTERNATIONAL CORP Menlo Park, CA	(N) 2,407	.25
80. UNIV TEXAS DALLAS Dallas, TX	2,359	.25
81. HOWARD UNIV Washington, DC	2,318	.24
82. RICE UNIV Houston, TX	2,294	.24
83. UNIV MIAMI Miami, FL	2,263	.24
84. UNIV CINCINNATI Cincinnati, OH	$2,213	.23
85. FLORIDA STATE UNIV Tallahassee, FL	2,206	.23
86. HAMPTON UNIV Hampton, VA	2,163	.23
87. FLORIDA A & M UNIV Tallahassee, FL	2,137	.22
88. AEROSPACE CORP Ann Arbor, MI	(N) 2,125	.22
89. ENVIRONMENTAL RES INSTIT MICH Ann Arbor, MI	(N) 2,057	.22
90. UNIV CALIF DAVIS Davis, CA	2,044	.21

(Continued on next page)

ONE HUNDRED EDUCATIONAL AND NONPROFIT INSTITUTIONS (Continued)

CONTRACTOR & PRINCIPAL PLACE OF PERFORMANCE	AWARDS (THOUSANDS)	PERCENT
91. BOSTON UNIV Boston, MA	2,022	.21
92. UNIV TOLEDO Toledo, OH	1,960	.20
93. CLARKSON UNIV Potsdam, NY	1,947	.20
94. UNIV PITTSBURGH Pittsburgh, PA	1,940	.20
95. VANDERBILT UNIV Irvine, CA	1,931	.20
96. FLORIDA ATLANTIC UNIV Boca Raton, FL	1,739	.18
97. UNIV CENTRAL FLORIDA Orlando, FL	1,684	.18
98. MOREHOUSE COLLEGE Atlanta, GA	1,673	.17
99. UNIV ROCHESTER Rochester, NY	1,650	.17
100. COLLEGE WILLIAM & MARY Williamsburg, VA	1,645	.17
OTHER**	109,702	11.46

*   Excludes JPL.
** Includes other awards over $25,000 and smaller procurements of $25,000 or less.

U.S. SPACECRAFT RECORD
Calendar Years 1957 - 1991

Year	Earth Orbit		Earth Escape		Year	Earth Orbit		Earth Escape	
	Success	Failure	Success	Failure		Success	Failure	Success	Failure
1957	-	1	-	-	1975	30	4	4	-
1958	5	8	-	4	1976	33	-	1	-
1959	9	9	1	2	1977	27	2	2	-
1960	16	12	1	2	1978	34	2	7	-
1961	35	12	-	2	1979	18	-	-	-
1962	55	12	4	1	1980	16	4	-	-
1963	62	11	-	-	1981	20	1	-	-
1964	69	8	4	-	1982	21	-	-	-
1965	93	7	4	1	1983	31	-	-	-
1966	94	12	7	1	1984	35	3	-	-
1967	78	4	10	-	1985	37	1	-	-
1968	61	15	3	-	1986	11	4	-	-
1969	58	1	8	1	1987	9	1	-	-
1970	36	1	3	-	1988	16	1	-	-
1971	45	2	8	1	1989	24	-	2	-
1972	33	2	8	-	1990	40	-	1	-
1973	23	2	3	-	1991	33	-	-	-
1974	27	2	1	-	**TOTAL**	**1,224**	**144**	**82**	**15**

Source: NASA, "Aeronautics and Space Report of the President" (Annually) and TRW Space & Defense Sector, "Space Log" (Annually).

WORLDWIDE SPACE LAUNCHINGS
WHICH ATTAINED EARTH ORBIT OR BEYOND
Calendar Years 1957 - 1991

Country	Total 1957 - 1991	1987	1988	1989	1990	1991
TOTAL	**3,397**	**110**	**116**	**100**	**116**	**89**
U.S.S.R.	2,315	95	90	74	75	59
United States	941	8	12	17	27	18
Japan	43	3	2	2	3	2
People's Republic of China	29	2	4	-	5	1
European Space Agency	43	2	7	7	5	8
Israel	2	-	1	-	1	-
Other	24	-	-	-	-	1

Source: NASA, "Aeronautics and Space Report of the President" (Annually) and TRW Space & Defense Sector, "Space Log" (Annually).

U.S. SPACE LAUNCH VEHICLES
As of 1991

Maximum Payload (Kg)

Vehicle and Initial Launch & First Launch of this Modification	Stages	Thrust (Kilo-newtons)	185-Km Orbit	Geo-synch.-Transfer Orbit	Circular Sun-Synch. Orbit
Scout (1960; 1979)	1. Algol IIIA* 2. Castor IIA* 3. Antares IIIA* 4. Altair IIIA*	431.1 285.2 83.1 25.6	255 205	-	155
Delta 3900 Series (Thor-Delta) (1960; 1982)	1. Thor plus 9 TX 526-2* 2. Delta	912.0 375.0 44.2	3,045 2,180	1,275	2,135
Delta II (1989)	1. Thor plus 9 TX 526-2* 2. Delta	920.8 432.0 43.0	-	1,819	-
Atlas E (1959; 1972)	1. Atlas booster & sustainer	1,722.0	2,090	-	1,500
Atlas-Centaur (1972; 1984)	1. Atlas booster & sustainer 2. Centaur	1,913.0 146.0	6,100	2,360	-
Titan IV (1989)	1. Two 7-segment, 3.05-m. dia* 2. LR-87 3. LR-91 4. IUS 1st stage* 5. IUS 2nd stage*	12,402.0 2,452.0 472.0 185.0 76.0	17,690	2,404	-

Maximum Payload (Kg)

Vehicle and Launch Date	Stages	Thrust (Kilo-newtons)	185-Km Orbit	24-Hour Polar Orbit	Sun-Synch. Transfer Orbit
Titan II (1962; 1988)	1. LR-87[2] 2. LR-91	2,108.4 444.8	2,200 1,905	-	-
Titan IIIB-Agena (1966)	1. LR-87 2. LR-91 3. Agena	2,341.0 455.1 71.2	3,600	-	3,060
Titan III(34)D/ IUS (1982)	1. Two 5 1/2-segment, 3.05-m. dia* 2. LR-87 3. LR-91 4. IUS 1st stage* 5. IUS 2nd stage*	11,564.8 2,366.3 449.3 275.8 115.7	14,920	1,850	-
Titan III(34)D/ Transtage (1984)	1. Two 5 1/2-segment, 3.05-m. dia* 2. LR-87 3. LR-91 4. Transtage	11,564.8 2,366.3 449.3 69.8	14,920	1,855	-

(Continued on next page)

U.S. SPACE LAUNCH VEHICLES (Continued)
As of 1991

| | | | Maximum Payload (Kg) | | |
	Vehicle and Launch Date	Stages	Thrust (Kilo- newtons)	185-Km Orbit	24-Hour Polar Orbit	Sun- Synch. Transfer Orbit
Space Shuttle (reusable) (1981)	1. Orbiter; 3 main engines (SSMEs) fire in parallel with SRBs	1,670	24,900	-	-
	2. Two solid-fueled rocket boosters (SRBs) mounted on external tank (ET) fire in parallel with SSMEs	11,790			

Source: NASA, "Aeronautics and Space Report of the President" (Annually) and NASA
Historian's office.
* Solid propellant; all others are liquid.

SUMMARY OF UNITED STATES MANNED SPACE FLIGHT

MISSION	CREW MEMBERS	MISSION DURATION (HR:MIN:SEC)	CREW HOURS (HR:MIN:SEC)
MERCURY REDSTONE (Suborbital)			
Freedom 7	Shepard	15:22	15:22:00
Liberty Bell 7	Grissom	15:37	15:37:00
Total Flights - 2		30:59	30:59
MERCURY ATLAS (Orbital)			
Friendship 7	Glenn	4:55:23	4:55:23
Aurora 7	Carpenter	4:56:05	4:56:05
Sigma 7	Schirra	9:13:11	295:13:38
Faith 7	Cooper	34:19:49	226:18:03
Total Flights - 4		53:24:28	53:24:28
TOTAL MERCURY FLIGHTS - 6		**53:55:27**	**53:55:27**
GEMINI TITAN			
Gemini 3	Grissom, Young	4:53:00	9:46:00
Gemini 4	McDivitt, White	97:56:11	195:52:22
Gemini 5	Cooper, Conrad	190:55:14	381:50:28
Gemini 6A	Schirra, Stafford	25:51:24	51:42:48
Gemini 7	Borman, Lovell	330:35:31	661:11:02
Gemini 8	Armstrong, Scott	10:41:26	21:22:52
Gemini 9A	Stafford, Ceman	72:21:00	144:42:00
Gemini 10	Young, Collins	70:46:39	141:33:18
Gemini 11	Conrad, Gordon	71:17:08	142:34:16
Gemini 12	Lovell, Aldrin	94:34:31	189:09:02
TOTAL GEMINI FLIGHTS - 10		**969:52:04**	**1939:44:08**
APOLLO SATURN I			
Apollo 7	Schirra, Eisele, Cunningham	260:09:03	780:27:09

(Continued on next page)

SUMMARY OF UNITED STATES MANNED SPACE FLIGHT (Continued)

MISSION	CREW MEMBERS	MISSION DURATION (HR:MIN:SEC)	CREW HOURS (HR:MIN:SEC)
APOLLO SATURN V			
Apollo 8	Borman, Lovell, Anders	147:00:42	441:02:06
Apollo 9	McDivitt, Scott, Schweickart	241:00:54	723:02:42
Apollo 10	Stafford, Young, Ceman	192:03:23	576:10:09
Apollo 11	Armstrong, Collins, Aldrin	195:18:35	585:55:45
Apollo 12	Conrad, Gordon, Bean	244:36:25	733:49:15
Apollo 13	Lovell, Swigert, Haise	142:54:41	428:44:03
Apollo 14	Shepard, Roosa, Mitchell	216:01:57	648:05:51
Apollo 15	Scott, Worden, Irwin	295:11:53	885:35:39
Apollo 16	Young, Mattingly, Duke	265:51:05	797:33:15
Apollo 17	Ceman, Evans, Schmitt	301:51:59	905:35:57
Total Flights - 10		2241:51:34	7506:01:51
TOTAL APOLLO - 11		**2502:00:37**	**7506:01:51**
SKYLAB SATURN IB			
Skylab 2	Conrad, Kerwin, Weitz	672:49:49	2018:29:27
Skylab 3	Bean, Garriott, Lousma	1427:09:04	4281:27:12
Skylab 4	Carr, E. Gibson, Pogue	2017:15:32	6051:46:36
TOTAL SKYLAB FLIGHTS - 3		**4117:14:25**	**12351:43:15**
APOLLO SATURN IB			
ASTP	Stafford, Brand, Slayton	217:28:23	652:25:09
STS-1 - Columbia	Young, Crippen	54:20:32	108:41:04
STS-2 - Columbia	Engly, Truly	54:13:13	108:26:26
STS-3 - Columbia	Lousma, Fullerton	192:04:45	384:09:30
STS-4 - Columbia	Mattingly, Hartsfield	169:09:40	338:19:20
STS-5 - Columbia	Brand, Overmyer, Allen, Lenoir	122:14:26	488:57:44
STS-6 - Challenger	Weitz, Bobko, Peterson, Musgrave	120:23:42	481:34:48
STS-7 - Challenger	Crippen, Hauch, Ride, Fabian, Thagard	146:23:59	731:59:55

(Continued on next page)

SUMMARY OF UNITED STATES MANNED SPACE FLIGHT (Continued)

MISSION	CREW MEMBERS	MISSION DURATION (HR:MIN:SEC)	CREW HOURS (HR:MIN:SEC)
STS-8 - Challenger	Truly, Brandenstein, D. Gardner, Bluford, W. Thornton	145:08:43	725:43:35
STS-9 - Columbia	Young, Shaw, Garriott, Parker, Lichtenberg, Merbold	247:47:24	1486:44:24
STS-41B - Challenger	Brand, Gibson, McCandless, McNair, Stewart	191:15:55	956:19:35
STS-41C - Challenger	Crippen, Scobee, van Hoften, G. Nelson, Hart	167:40:07	838:20:35
STS-41D - Discovery	Hartsfield, Coats, Resnik, Hawley, Mullane, C. Walker	144:56:04	869:36:24
STS-41G - Challenger	Crippen, McBride, Ride, Sullivan, Leetsma, Garneau, Scully-Power	197:23:33	1381:44:51
STS-51A - Discovery	Hauck, D. Walker, Gardner, A. Fisher, Allen	191:44:56	958:44:40
STS-51C - Discovery	Mattingly, Shriver, Onizuka, Buchli, Payton	73:33:23	367:46:55
STS-51D - Discoery	Bobko, Williams, Seddon, Hoffman, Griggs, C. Walker, Garn	167:55:23	1175:27:41
STS-51B - Challenger	Overmyer, Gregory, Lind, Thagard, W. Thornton, van den Berg, Wang	168:08:46	1177:01:22
STS-51G - Discovery	Brandenstein, Creighton, Lucid, Fabian, Nagel, Baudry, Al-Saud	169:38:52	1187:32:04
STS-51F - Challenger	Fullerton, Bridges, Musgrave, England, Henize, Acton, Bartoe	190:45:26	1335:18:02
STS-51I - Discovery	Engle, Covey, van Hoften, Lounge, W. Fisher	170:17:42	851:28:30
STS-51J - Atlantis	Bobko, Grabe, Hilmers, Stewart, Pailes	97:44:38	488:53:10
STS-61A - Challenger	Hartsfield, Nagel, Buchli, Bluford, Dunbar, Furrer, Messerschmid, Ockels	168:44:51	1349:58:48
STS-61B - Atlantis	Shaw, O'Connor, Cleave, Spring, Ross, Neri Vela, C. Walker	165:04:49	1155:33:43
STS-61C - Columbia	R. Gibson, Bolden, Chang-Diaz, Hawley, G. Nelson, Cenker, B. Nelson	146:03:51	1022:26:57
STS-51L - Challenger	Scobee, Smith, Resnik, Onizuka, McNair, Jarvis, McAuliffe	N/A	N/A
STS-26 - Discovery	Hauck, Covey, Lounge, Hilmers, G. Nelson	97:00:11	485:00:55
STS-27 - Atlantis	R. Gibson, Gardner, Mullane, Ross, Shepherd	105:05:37	525:28:05
STS-29 - Discovery	Coats, Blaha, Bagian, Buchi, Springer	119:38:52	598:14:20
STS-30 - Atlantis	Walker, Grabe, Thagard, Cleave, Lee	96:56:25	484:47:35
STS-28 - Columbia	Shaw, Richards, Leetsma, Adamson, Brown	121:00:09	605:00:45
STS-34 - Atlantis	Williams, McCully, Baker, Chang-Diaz, Lucid	119:39:24	596:17:00
STS-33 - Discovery	Gregory, Blaha, Musgrave, K. Thornton, Carter	120:06:49	600:34:05
STS-32 - Columbia	Brandenstein, Wetherbee, Dunbar, Ivins, Low	261:00:37	1305:03:05
STS-36 - Atlantis	Creighton, Casper, Hilmers, Mullane, Thuot	106:18:23	531:31:55
STS-31 - Discovery	Shriver, Bolden, McCandless, Hawley, Sullivan	121:16:05	606:20:25
STS-41 - Discovery	Richards, Cabana, Melnick, Shepard, Akers	98:11:00	490:45:00
STS-38 - Atlantis	Covey, Springer, Meade, Culbertson, Gemar	117:55:00	589:35:00
STS-35 - Columbia	Brand, Lounge, Hoffman, Parker, G. Gardner, Parise, Durrance	215:06:00	1505:42:00
STS-37 - Atlantis	Nagel, Cameron, Ross, Apt, Godwin	143:33:40	717:48:20
STS-39 - Discovery	Coats, Hammond, Harbaugh, Hieb, McMonagle, Bluford, Veach	199:23:16	1394:42:52
STS-40 - Columbia	Gutierrez, Seddon, Bagian, Jernigan, Gaffney, Hughes-Fullord, O'Connor	218:15:14	1527:46:38

(Continued on next page)

SUMMARY OF UNITED STATES MANNED SPACE FLIGHT (Continued)

MISSION	CREW MEMBERS	MISSION DURATION (HR:MIN:SEC)	CREW HOURS (HR:MIN:SEC)
STS-43 - Atlantis	Blaha, Baker, Lucid, Low, Adamson	213:22:26	1066:52:10
STS-48 - Discovery	Creighton, Reightler, Buchli, Brown, Gemar	128:28:17	642:21:25
STS-44 - Atlantis	Gregory, Henricks, Musgrave, Runco, Voss, Hennen	170:52:36	1025:15:36
TOTAL SHUTTLE FLIGHTS - 38		**6437:48:31**	**35271:47:14**

SUMMARY OF ANNOUNCED LAUNCHES

TOTAL		1957	1958	1959	1960	1961	1962	1963	1964	1965	1966	1967	1968	1969	1970	1971	1972	1973	1974
1	Australia	--	--	--	--	--	--	--	--	--	--	1	0	0	0	0	0	0	0
484	DOD	--	5	6	11	19	34	27	35	39	42	32	26	19	17	17	13	10	8
42	ESA	--	--	--	--	--	--	--	--	--	--	--	--	--	--	--	--	--	--
10	France	--	--	--	--	--	--	--	--	1	1	2	0	0	2	1	0	0	0
3	India	--	--	--	--	--	--	--	--	--	--	--	--	--	--	--	--	--	--
2	Israel	--	--	--	--	--	--	--	--	--	--	--	--	--	--	--	--	--	--
43	Japan	--	--	--	--	--	--	--	--	--	--	--	--	--	1	2	1	0	1
7	MDAC	--	--	--	--	--	--	--	--	--	--	--	--	--	--	--	--	--	--
3	MMarietta	--	--	--	--	--	--	--	--	--	--	--	--	--	--	--	--	--	--
457	NASA	--	2	5	5	10	18	11	22	24	31	26	19	21	12	15	18	13	16
1	Orbital Sciences	--	--	--	--	--	--	--	--	--	--	--	--	--	--	--	--	--	--
29	PRC	--	--	--	--	--	--	--	--	--	--	--	--	--	1	1	0	0	0
1	United Kingdom	--	--	--	--	--	--	--	--	--	--	--	--	--	--	1	0	0	0
2314	USSR	2	1	3	3	6	20	17	30	48	44	66	74	70	81	83	74	86	81
3397	TOTAL	2	8	14	19	35	72	55	87	112	118	127	119	110	114	120	106	109	106

NASA LAUNCHES

TOTAL		1957	1958	1959	1960	1961	1962	1963	1964	1965	1966	1967	1968	1969	1970	1971	1972	1973	1974
262	NASA	--	2	5	5	10	15	9	20	21	26	18	12	13	6	6	9	9	2
33	Cooperative	--	--	--	--	--	2	0	2	2	0	2	3	2	0	5	1	0	5
30	DOD	--	--	--	--	--	--	1	0	0	1	0	0	0	0	0	1	1	0
92	USA	--	--	--	--	--	1	1	0	1	4	6	3	4	4	3	3	2	4
39	Foreign	--	--	--	--	--	--	--	--	--	--	--	1	2	2	1	4	1	5
456	TOTAL	--	2	5	5	10	18	11	22	24	31	26	19	21	12	15	18	13	16

Summary of Announced Launches

TOTAL		1975	1976	1977	1978	1979	1980	1981	1982	1983	1984	1985	1986	1987	1988	1989	1990	1991	TOTAL
1	Australia	0	0	0	0	0	0	0	0	0	0	0	0	0	0	0	0	0	1
484	DOD	9	11	10	12	7	6	5	6	7	10	3	1	5	4	10	10	8	484
--	--	--	--	--	1	0	2	0	2	4	3	2	2	7	7	5	7	42	42 ESA
10	France	3	0	0	0	0	0	0	0	0	0	0	0	0	0	0	0	0	10
3	India	--	--	--	--	--	1	1	0	1	0	0	0	0	0	0	0	0	3

(Continued on next page)

SUMMARY OF ANNOUNCED LAUNCHES (Continued)

TOTAL		1975	1976	1977	1978	1979	1980	1981	1982	1983	1984	1985	1986	1987	1988	1989	1990	1991	TOTAL
2	Israel	--	--	--	--	--	--	--	--	--	--	--	--	--	1	0	1	0	2
43	Japan	2	1	2	3	2	2	3	1	3	3	2	2	3	2	2	3	2	43
7	MDAC	--	--	--	--	--	--	--	--	--	--	--	--	--	--	1	5	1	7
3	MMarietta	--	--	--	--	--	--	--	--	--	--	--	--	--	--	--	3	0	3
457	NASA	19	15	14	20	9	7	13	12	15	12	14	5	3	8	7	8	8	457
1	Orbital Sciences	--	--	--	--	--	--	--	--	--	--	--	--	--	--	--	1	0	1
29	PRC	3	2	0	1	0	0	1	1	1	3	1	2	2	4	0	5	1	29
1	United Kingdom	0	0	0	0	0	0	0	0	0	0	0	0	0	0	0	0	0	1
2314	USSR	89	99	98	88	87	89	98	101	98	97	97	919	5	90	74	75	59	2314
3397	**TOTAL**	**125**	**128**	**124**	**124**	**106**	**105**	**123**	**121**	**127**	**129**	**120**	**103**	**110**	**116**	**101**	**116**	**86**	**3397**

NASA LAUNCHES

TOTAL		1975	1976	1977	1978	1979	1980	1981	1982	1983	1984	1985	1986	1987	1988	1989	1990	1991	TOTAL
262	NASA	10	1	3	8	3	1	4	4	4	6	9	1	0	2	6	6	6	262
33	Cooperative	1	2	1	2	0	0	0	0	1	0	0	0	0	1	0	1	0	33
30	DOD	1	2	1	1	2	2	2	0	1	1	2	3	1	4	1	1	1	30
92	USA	4	8	2	4	3	4	7	6	8	4	3	1	1	1	0	0	1	92
39	Foreign	3	2	7	5	1	0	0	2	1	1	0	0	1	0	0	0	0	39
456	TOTAL	19	15	14	20	9	7	13	12	15	12	14	5	3	8	7	8	7	456

NASA ASTRONAUTS

NAME	SERVICE	MISSION	POSITION	FLIGHT TIME (HR:MIN:SEC)	EVA (HR:MIN)	TOTAL FLIGHT TIME (HR:MIN:SEC)
Acton, Loren W., PhD	Civ	STS-51F	PS	190:45:26		190:45:26
Adamson, James C. Lt. Col	USA	STS-28	MS	121:00:09		334:22:35
		STS-43	MS	213:22:26		
Akers, Thomas D. Maj	USAF	STS-41	MS	98:11:00		98:11:00
Aldrin, Edwin E., Jr., Col.	USAF Ret.	Gemini 12	Plt	68:34:31	05:37	289:53:06
		Apollo 11	LMP	195:18:35	02:15	
Allen, Joseph P. PhD	Civ	STS-5	MS	122:14:26		313:59:22
		STS-51A	MS	191:44:56	12:14	
Al-Saud, Salman	Civ	STS-51G	PS	169:38:52		169:38:52
Anders, William A., B. Gen.	USAF	Apollo 8	LMP	147:00:42		206:00:01
Apt, Jerome PhD	Civ	STS-37	MS	143:33:40	10:49	143:33:40
Armstrong, Neil	Civ	Gemini 8	Cdr	10:41:26		
		Apollo 11	Cdr	195:18:35	02:32	
Bagian, James P. MD	Civ	STS-29	MS	119:38:52		337:54:06
		STS-40	MS	218:15:14		
Baker, Ellen S., MD	Civ	STS-34	MS	119:39:24		119:39:24
Bartoe, John-David F., PhD	Civ	STS-51F	PS	190:45:26		190:45:26
Baudry, Patrick, Lt. Col.	FAF	STS-51G	PS	169:38:52		169:38:52
Bean, Alan F., Capt	USN Ret	Apollo 12	LMP	244:36:25	07:45	1671:45:29
		Skylab 3	Cdr	1427:09:04	2:45	
Blaha, John E., Col	USAF	STS-29	Plt	119:38:52		453:08:07
		STS-33	Plt	120:06:49		
		STS-43	Cdr	213:22:26		
Bluford, Guion S., Col	USAF	STS-8	MS	145:08:43		513:19:50
		STS-61A	MS	168:44:51		
		STS-39	MS	199:26:16		
Bobko, Karol J., Col	USAF	STS-6	Plt	120:23:42		386:03:43
		STS-51D	Cdr	167:55:23		
		STS-51J	Cdr	97:44:38		
Bolden, Charles F., Col	USMC	STS-61C	Plt	146:03:51		267:19:56
		STS-31	Plt	121:16:05		
Borman, Frank, Col.	USAF Ret	Gemini 7	Cdr	330:35:31		477:36:13
		Apollo 8	Cdr	147:00:42		

(Continued on next page)

NASA ASTRONAUTS (Continued)

NAME	SERVICE	MISSION	POSITION	FLIGHT TIME (HR:MIN:SEC)	EVA (HR:MIN)	TOTAL FLIGHT TIME (HR:MIN:SEC)
Brand, Vance D.	Civ	Apollo Soyuz	CMP	217:28:23		763:54:44
		STS-5	Cdr	122:14:26		
		STS-41B	Cdr	191:15:55		
		STS-35	Cdr	215:06:00		
Brandenstein, Daniel C., Capt	USN	STS-8	Plt	145:08:43		575:48:12
		STS-51G	Cdr	169:38:52		
		STS-32	Cdr	120:06:49		
Bridges, Roy D., Col	USAF	STS-51F	Plt	190:45:26		190:45:26
Brown, Mark ., Lt. Col	USAF	STS-28	MS	121:00:09		249:28:26
		STS-48	MS	128:28:17		
Buchli, James F., Col	USMC	STS-51C	MS	73:33:23		490:25:23
		STS-61A	MS	168:44:51		
		STS-29	MS	119:38:52		
		STS-48	MS	128:28:17		
Cabana, Robrt D., Lt. Col.	USMC	STS-41	Plt	98:11:00		98:11:00
Cameron, Kenneth D., Col.	USMC	STS-37	Plt	143:33:40		143:33:40
Carpenter, M. Scott, Cdr.	USN Ret	Aurora 7	Cdr	4:56:05		4:56:05
Carr, Gerald P., Col	USMC Ret	Skylab 4	Cdr	2017:15:32	15:48	2017:15:32
Carter, Manley, Cdr.	USN	STS-33	MS	120:06:49		120:06:49
Casper, John H., Col	USAF	STS-36	Plt	106:18:23		106:18:23
Cenker, Robert J.	Civ	STS-61C	PS	146:03:51		146:03:51
Cernan, Eugene A., Capt.	USN Ret	Gemini 9A	Plt	72:21:00	02:08	566:16:32
		Apollo 10	LMP	192:03:23		
		Apollo 17	Cdr	301:51:59	22:04	
Chang-Diaz, Franklin R., PhD.	Civ	STS-61C	MS	146:03:51		265:43:15
		STS-34	MS	119:39:24		
Cleave, Mary L., PhD	Civ	STS-61B	MS	165:04:49		262:02:20
		STS-30	MS	96:56:25		
Coats, Michael L., Capt.	USN	STS-41D	Plt	144:56:04		463:51:12
		STS-29	Cdr	119:38:52		
		STS-39	Cdr	199:26:16		
Collins, Michael, M. Gen	USAF	Gemini 10	Plt	70:46:39	1:30	266:11:14
		Apollo 11	CMP	195:18:35		

(Continued on next page)

NASA ASTRONAUTS (Continued)

NAME	SERVICE	MISSION	POSITION	FLIGHT TIME (HR:MIN:SEC)	EVA (HR:MIN)	TOTAL FLIGHT TIME (HR:MIN:SEC)
Conrad, Charles (Pete), Capt	USN Ret	Gemini 5	Plt	190:55:14		1179:28:36
		Gemini 11	Cdr	71:17:08		
		Apollo 12	Cdr	244:36:25	07:45	
		Skylab 2	Cdr	672:49:49	5:51	
Cooper, L. Gordon, Jr., Col.	USAF Ret	Faith 7	Plt	34:19:49		226:18:03
		Gemini 5	Cdr	190:55:14		
Covey, Richard O., Col	USAF	STS-51I	Plt	170:17:42		485:12:53
		STS-26	Plt	97:00:11		
		STS-38	Cdr	117:55:00		
Creighton, John O., Capt	USN	STS-51G	Plt	169:38:52		404:25:32
		STS-36	Cdr	106:18:23		
		STS-48	Cdr	128:28:17		
Crippen, Robert L., Capt.	USN	STS-1	Plt	54:20:32		565:48:11
		STS-7	Cdr	146:23:59		
		STS-41C	Cdr	167:40:07		
		STS-41G	Cdr	197:23:33		
Culbertson, Frank L.		STS-38	Plt	117:55:00		117:55:00
Cunningham, Walter	Civ	Apollo 7	LMP	260:09:03		260:09:03
Duke, Charles M., B. Gen.	USAF	Apollo 16	LMP	265:51:05	20:14	265:51:05
Dunbar, Bonnie J., PhD	Civ	STS-61A	MS	168:44:51		429:45:28
		STS-32	MS	261:00:37		
Durrance, Samuel T.		STS-35	PS	215:06:00		215:06:00
Eisele, Donn F., Col	USAF Ret	Apollo 7	CMP	260:09:03		260:09:03
England, Anthony W., PhD	Civ	STS-51F	MS	190:45:26		190:45:26
Engle, Joe H., Col	USAF	STS-2	Cdr	54:13:13		244:30:55
		STS-51I	Cdr	170:17:42		
Evans, Ronald R., Capt	USN Ret	Apollo 17	CMP	301:51:59	01:06	301:51:59
Fabian, John M. Col.	USAF	STS-7	MS	146:23:59		316:02:51
		STS-51G	MS	169:38:52		
Fisher, Anna L., MD	Civ	STS-51A	MS	191:44:56		191:44:56
Fisher, William F., MD	Civ	STS-51I	MS	170:17:42	11:51	170:17:42
Fullerton, C. Gordon, Col.	USAF	STS-3	Plt	192:04:45		382:50:11
		STS-51F	Cdr	190:45:26		

(Continued on next page)

NASA ASTRONAUTS (Continued)

NAME	SERVICE	MISSION	POSITION	FLIGHT TIME (HR:MIN:SEC)	EVA (HR:MIN)	TOTAL FLIGHT TIME (HR:MIN:SEC)
Furrer, Reinhard, PhD	Civ	STS-61A	PS	186:44:51		186:44:51
Gaffney, F. Drew Dr.	Civ	STS-40	PS	218:15:14		218:15:14
Gardner, Dale A.,	USN	STS-8	MS	145:08:43		336:53:39
		STS-51A	MS	191:44:56	12:14	
Gardner, Guy S., Lt. Col.	USAF	STS-27	Plt	105:05:37		320:11:37
		STS-35	Plt	215:06:00		
Garn, E. J. "Jake"	Civ	STS-51D	PS	167:55:23		167:55:23
Garneau, Marc, PhD	Civ	STS-41G	PS	197:23:33		197:23:33
Garriott, Owen K., PhD	Civ	Skylab 3	Plt	1427:09:04	13:44	1674:56:28
		STS-9	MS	247:47:24		
Gemar, Charles D.		STS-38	MS	117:55:00		246:23:17
		STS-48	MS	128:28:17		
Gibson, Edward G.,PhD	Civ	Skylab 4	Plt	2017:15:32	15:20	2017:15:32
Gibson, Robert L., Cdr	USN	STS-41B	Plt	191:15:55		442:25:23
		STS-61C	Cdr	146:03:51		
		STS-27	Cdr	105:05:37		
Glenn, John H., Jr. Col	USMC Ret	Friendship 7	Cdr	4:55:23		4:55:23
Godwin, Linda M., PhD	Civ	STS-37	MS	143:33:40		143:33:40
Gordon, Richard F., Jr., Capt.	USN Ret.	Gemini 11	Plt	71:17:08	1:57	315:53:33
		Apollo 12	CMP	244:36:25		
Grabe, Ronald J., Col	USAF	STS-51J	Plt	97:44:38		194:42:09
		STS-30	Plt	96:56:25		
Gregory, Frederick D., Col	USAF	STS-51B	Plt	168:08:46		459:06:11
		STS-33	Cdr	120:06:49		
		STS-44	Cdr	170:52:36		
Griggs, S. David	Civ	STS-51D	MS	167:55:23	3:10	167:55:23
Grissom, Virgil I., Lt. Col.	USAF	Liberty Bell	Plt	15:37		5:08:37
		Gemini 3	Cdr	4:53:00		
Gutierrez, Sidney M., Lt. Col.	USAF	STS-40	Plt	218:15:14		218:15:14
Haise, Fred W.	Civ	Apollo 13	LMP	142:54:41		142:54:41
Hammond, L. Blaine, Jr., Col	USAF	STS-39	Plt	199:26:16		199:26:16
Harbaugh, Gregory J.	Civ	STS-39	MS	199:26:16		199:26:16
Hart, Terry J.	Civ	STS-41C	MS	167:40:07		167:40:07

(Continued on next page)

NASA ASTRONAUTS (Continued)

NAME	SERVICE	MISSION	POSITION	FLIGHT TIME (HR:MIN:SEC)	EVA (HR:MIN)	TOTAL FLIGHT TIME (HR:MIN:SEC)
Hartsfield, Henry W.	USAF Ret	STS-4	Plt	169:09:40		482:50:35
		STS-41D	Cdr	144:56:04		
		STS-61A	Cdr	168:44:51		
Hauck, Frederick H., Capt	USN	STS-7	Plt	146:23:59		435:09:06
		STS-51A	Cdr	191:44:56		412:16:00
STS-26	Cdr	97:00:11				
Hawley, Steven A., PhD	Civ	STS-41D	MS	144:56:04		412:16:00
		STS-61C	MS	146:03:51		
		STS-31	MS	121:16:05		
Henize, Karl G., PhD	Civ	STS-51F	MS	190:45:26		190:45:26
Hennen, Thomas J.	USA	STS-44	PS	170:52:36		170:52:36
Henricks, Terence T., Col.	USAF	STS-44	Plt	170:52:36		170:52:36
Hieb, Richard J.	Civ	STS-39	MS	199:26:16		199:26:16
Hilmers, David C., Lt. Col.	USMC	STS-51J	MS	97:44:38		301:03:11
		STS-26	MS	97:00:11		
		STS-36	MS	106:18:23		
Hoffman, Jeffrey A., PhD	Civ	STS-51D	MS	167:55:23	3:10	383:01:23
		STS-35	MS	215:06:00		
Hughes-Fulford, Millie Dr.	Civ	STS-40	PS	218:15:14		218:15:14
Irwin, James B., Col	USAF Ret	Apollo 15	LMP	295:11:53	18:35*	295:11:53
Ivins, Marsha S.	Civ	STS-32	MS	261:00:37		261:00:37
Jarvis, Gregory B.	Civ	STS-51L	PS	N/A		N/A
Jernigan, Tamara E. PhD	Civ	STS-40	MS	218:15:14		218:15:14
Kerwin, Joseph P., Capt	USN Ret	Skylab 2	Plt	672:49:49	03:30	672:49:49
Lee, Mark C. Maj	USAF	STS-30	MS	96:56:25		96:56:25
Leetsma, David C., Cdr	USN	STS-41G	MS	197:23:33	3:29	318:23:42
		STS-28	MS	121:00:09		121:00:09
Lenoir, William B., PhD	Civ	STS-5	MS	122:14:26		122:14:26
Lichtenberg, Bryon K., PhD	Civ	STS-9	PS	247:47:24		247:47:24
Lind, Don Leslie, PhD	Civ	STS-51B	MS	168:08:46		168:08:46
Lounge, John M.	Civ	STS-51I	MS	170:17:42		482:23:53
		STS-26	MS	97:00:11		

(Continued on next page)

NASA ASTRONAUTS (Continued)

NAME	SERVICE	MISSION	POSITION	FLIGHT TIME (HR:MIN:SEC)	EVA (HR:MIN)	TOTAL FLIGHT TIME (HR:MIN:SEC)
Lousma, Jack R., Col	USMC	Skylab 3	Plt	1427:09:04	10:59	1619:13:49
		STS-3	Cdr	192:04:45		
Lovell, James A., Jr., Capt.	USN Ret	Gemini 7	Plt	330:35:31		715:05:25
		Gemini 12	Cdr	94:34:31		
		Apollo 8	CMP	147:00:42		
		Apollo 13	Cdr	142:54:41		
Low, G. David	Civ	STS-32	MS	261:00:37		474:23:03
		STS-43	MS	213:22:26		
Lucid, Shannon W., PhD	Civ	STS-51G	MS	169:38:52		502:40:42
		STS-34	MS	119:39:24		
		STS-43	MS	213:22:26		
Mattingly, Thomas K., Capt	USN	Apollo 16	CMP	265:51:05	01:24	508:34:08
		STS-4	Cdr	169:09:40		
		STS-51C	Cdr	73:33:23		
McAuliffe, S. Christa	Civ	STS-51L	PS	N/A	N/A	
McBride, Jon A., Cdr	USN	STS-41G	Plt	197:23:33		197:23:33
McCandless, Bruce, Capt.	USN	STS-41B	MS	121:16:05	11:37	121:16:05
McCulley, Michael, Cdr	USN	STS-34	Plt	119:39:24		119:39:24
McDivitt, James A., B. Gen	USAF Ret	Gemini 4	Cdr	97:56:11		338:57:05
		Apollo 9	Cdr	241:00:54		
McMonagle, Donald R. Lt. Col.	USAF	STS-39	MS	199:26:16		199:26:16
McNair, Ronald E., PhD	Civ	STS-41B	MS	191:15:55		191:15:55
		STS-51L	MS	N/A		
Meade, Carl J.		STS-38	MS	117:55:00		117:55:00
Melnick, Bruce E., Cdr	USCG	STS-41	MS	98:11:00		98:11:00
Merbold, Ulf, PhD	Civ	STS-9	PS	247:47:24		247:47:24
Messerschmid, Ernest, PhD	Civ	STS-61A	PS	168:44:51		168:44:51
Mitchell, Edger D., Capt	USN Ret	Apollo 15	LMP	216:01:57	09:23	216:01:57
Mullane, Richard M., Col	USAF	STS-41D	MS	144:56:04		356:20:04
		STS-27	MS	105:05:37		
		STS-36	MS	106:18:23		
		STS-35	MS	215:06:00		

(Continued on next page)

NASA ASTRONAUTS (Continued)

NAME	SERVICE	MISSION	POSITION	FLIGHT TIME (HR:MIN:SEC)	EVA (HR:MIN)	TOTAL FLIGHT TIME (HR:MIN:SEC)
Musgrave, F. Story, MD, PhD	Civ	STS-6	MS	120:23:42	03:54	602:08:33
		STS-51F	MS	190:45:26		
		STS-33	MS	120:06:49		
		STS-44	MS	170:52:36		
Nagel, Steven R., Col.	USAF	STS-51G	MS	169:38:52		481:57:23
		STS-61A	Plt	168:44:51		
		STS-37	Cdr	143:33:40		
Nelson, Bill	Civ	STS-61C	PS	146:03:51		146:03:51
Nelson, George D., PhD	Civ	STS-41C	MS	167:40:07	10:06	410:44:09
		STS-61C	MS	146:03:51		
		STS-26	MS	97:00:11		
Neri Vela, Rodolpho, PhD	Civ	STS-61B	PS	165:04:49		165:04:49
Ockels, Wubbo J., PhD	Civ	STS-61A	PS	168:44:51		168:44:51
O'Connor, Bryan O., Col	USMC	STS-61B	Plt	165:04:49		383:20:03
		STS-40	Cdr	218:15:14		
Onizuka, Ellison S., Lt. Col	USAF	STS-51C	MS	73:33:23		73:33:23
		STS-51L	MS	N/A		
Overmyer, Robert F., Col	USMC	STS-5	Plt	122:14:26		290:23:12
		STS-51B	Cdr	168:08:46		
Pailes, William A., Maj	USAF	STS-51J	PS	97:44:38		97:44:38
Parise, Ronald A.		STS-35	PS	215:06:00		215:06:00
Parker, Robert A., PhD	Civ	STS-9	MS	247:47:24		462:53:24
		STS-35	MS	215:06:00		
Payton, Gary E., Maj	USAF	STS-51C	PS	73:33:23		73:33:23
Peterson, Donald H.	USAF Ret	STS-6	MS	120:23:42	3:54	120:23:42
Pogue, William R., Col.	USAF Ret	Skylab 4	Plt	2017:15:32	13:34	2017:15:32
Reightler, Kenneth S., Jr. Cdr	USN	STS-48	Plt	128:28:17	128:28:17	
Resnik, Judith A., PhD	Civ	STS-41D	MS	144:56:04		144:56:04
		STS-51L	MS	N/A		
Richards, Richard N., Cdr	USN	STS-28	Plt	121:00:09		219:11:09
		STS-41	Cdr	98:11:00		
Ride, Sally K., PhD	Civ	STS-7	MS	146:23:59		343:47:32
		STS-41G	MS	197:23:33		

(Continued on next page)

NASA ASTRONAUTS (Continued)

NAME	SERVICE	MISSION	POSITION	FLIGHT TIME (HR:MIN:SEC)	EVA (HR:MIN)	TOTAL FLIGHT TIME (HR:MIN:SEC)
Roosa, Stuart A., Col	USAF Ret	Apollo 14	CMP	216:10:57		216:10:57
Ross, Jerry L., Lt. Col.	USAF	STS-61B	MS	165:04:49	12:20	413:44:06
		STS-27	MS	105:05:37		
		STS-37	MS	143:33:40		
Runco, Mario Jr., Lt. Cdr.	USN	STS-44	MS	170:52:36		170:52:36
Schirra, Walter M., Jr., Capt.	USN Ret	Sigma 7	Plt	9:13:11		295:13:38
		Gemini 6A	Cdr	25:51:24		
		Apollo 7	Cdr	260:09:03		
Schmitt, Harrison H., PhD	Civ	Apollo 17	LMP	301:51:59	22:04	301:51:59
Schweickart, Russell	Civ	Apollo 9	LMP	241:00:54	1:07	241:00:54
Scobe, Francis R. (Dick)	USAF Ret	STS-41C	Plt	167:40:07		167:40:07
		STS-51L	Cdr	N/A		
Scott, David R., Col	USAF Ret	Gemini 8	Plt	10:41:26		546:54:13
		Apollo 9	CMP	241:00:54	1:01	
		Apollo 15	Cdr	295:11:53	19:08	
Scully-Power, Paul D.	Civ	STS-41G	PS	197:23:33		197:23:33
Seddon, M. Rhea, MD	Civ	STS-51D	MS	167:55:23		386:10:37
		STS-40	MS	218:15:14		
Shaw, Brewster H., Col	USAF	STS-9	Plt	247:47:24		533:52:22
		STS-61B	Cdr	165:04:49		
		STS-28	Cdr	121:00:09		
Shepard, Alan B., Jr., R. Adm.	USN Ret	Freedom 7	Plt	15:22		216:17:19
		Apollo 14	Cdr	216:01:57	09:23	
Shepherd, William M., Capt	USN	STS-27	MS	105:05:37		203:16:37
		STS-41	MS	98:11:00		
Shriver, Loren J., Col	USAF	STS-51C	Plt	73:33:23		194:49:28
		STS-31	Cdr	121:16:05		
Slayton, Donald K., Maj	USAF Ret	Apollo Soyuz	CMP	217:28:23		217:28:23
Smith, Michael J., Cdr	USN	STS-51L	Plt	N/A		N/A
Spring, Sherwood C., Lt. Col	USA	STS-61B	MS	165:04:49	12:20	165:04:49
Springer, Robert C., Col	USMC	STS-29	MS	119:38:52		237:33:52
		STS-38	MS	117:55:00		

(Continued on next page)

NASA ASTRONAUTS (Continued)

NAME	SERVICE	MISSION	POSITION	FLIGHT TIME (HR:MIN:SEC)	EVA (HR:MIN)	TOTAL FLIGHT TIME (HR:MIN:SEC)
Stafford, Thomas P., Lt. Gen.	USAF Ret	Gemini 6A	Plt	25:51:24		507:44:10
		Gemini 9A	Cdr	72:21:00		
		Apollo 10	Cdr	192:03:23		
		Apollo Soyuz	Cdr	217:28:23		
Stewart, Robert L., Col	USA	STS-41B	MS	191:15:55	11:37	289:00:33
		STS-51J	MS	97:44:38		
Sullivan, Kathryn D., PhD	Civ	STS-41G	MS	197:23:33	03:29	318:39:38
		STS-31	MS	121:16:05		
Swigert, John L., Jr.	Civ	Apollo 13	CMP	142:54:41		142:54:41
Thagard, Norman E., MD	Civ	STS-7	MS	146:23:59		411:30:16
		STS-51B	MS	168:08:46		
		STS-30	MS	96:56:25		
Thornton, Kathryn	Civ	STS-33	MS	120:06:49		120:06:49
Thornton, William E., MD	Civ	STS-8	MS	145:08:43		313:17:29
		STS-51B	MS	168:08:46		
Thuot, Pierre J., Lt. Cdr	USG	STS-36	MS	106:18:23		106:18:23
Truly, Richard H., Capt	USN	STS-2	Plt	54:13:13		199:21:56
		STS-8	Cdr	145:08:43		
van den Berg, Lodewijk, PhD	Civ	STS-51B	PS	168:08:46		168:08:46
van Hoften, James D., PhD	Civ	STS-41C	MS	167:40:07	10:06	377:57:49
	STS-51I	MS	170:17:42	11:51		
Veach, Charles Lacy	USAF	STS-39	MS	199:26:16		199:26:16
Voss, James S. Lt. Col.	USA	STS-44	MS	170:52:36		170:52:36
Walker, Charles D.	Civ	STS-41D	PS	144:56:04		477:56:16
		STS-51D	PS	167:55:23		
		STS-61B	PS	165:04:49		
Walker, David M., Capt	USN	STS-51A	Plt	191:44:56		288:42:27
		STS-30	Cdr	96:56:25		
Wang, Taylor G., PhD	Civ	STS-51B	PS	168:08:46		168:08:46
Weitz, Paul J., Capt	USN Ret	Skylab 2	Plt	672:49:49	01:44	793:13:31
		STS-6	Cdr	120:23:42		
Wetherbee, James, Cdr	USN	STS-32	Plt	261:00:37		261:00:37
White, Edward H., Lt. Col	USAF	Gemini 4	Plt	97:56:11	0:23	97:56:11

(Continued on next page)

NASA ASTRONAUTS (Continued)

NAME	SERVICE	MISSION	POSITION	FLIGHT TIME (HR:MIN:SEC)	EVA (HR:MIN)	TOTAL FLIGHT TIME (HR:MIN:SEC)
Williams, Donald E., Capt.	USN	STS-51D	Plt	167:55:23		287:34:47
		STS-34	Cdr	119:39:24		
Worden, Alfred M., Col	USAF Ret	Apollo 15	CMP	295:11:53	00:39	295:11:53
Young, John W., Capt	USN Ret	Gemini 3	Plt	4:53:00		835:41:33
		Gemini 10	Cdr	70:46:39		
		Apollo 10	CMP	192:03:23		
		Apollo 16	Cdr	265:51:05	20:14	
		STS-1	Cdr	54:20:32		
		STS-9	Cdr	247:47:24		

PS	Payload Specialist
MS	Mission Specialist
Plt	Pilot
LMP	Lunar Module Pilot
Cdr	Commander
CMP	Command Module Pilot

Section XI.

Aerospace Export-Import Trade

U.S. TOTAL AND AEROSPACE FOREIGN TRADE
Calendar Years 1964 - 1991 (Millions of Dollars)

Year	Total U.S. Merchandise Trade			Aerospace		
	Trade Balance	Exports	Imports	Trade Balance	Exports	Imports
1964	7,006	$ 25,690	$ 18,684	$ 1,518	$ 1,608	$ 90
1965	5,334	26,699	21,366	1,459	1,618	159
1966	3,837	29,379	25,542	1,370	1,673	303
1967	4,122	30,934	26,812	1,961	2,248	287
1968	837	34,063	33,226	2,661	2,994	333
1969	1,289	37,332	36,043	2,831	3,138	307
1970	3,225	43,176	39,952	3,097	3,405	308
1971	(1,476)	44,087	45,563	3,830	4,203	373
1972	(5,729)	49,854	55,583	3,230	3,795	565
1973	2,390	71,865	69,476	4,360	5,142	782
1974	(3,884)	99,437	103,321	6,350	7,095	745
1975	9,551	108,856	99,305	7,045	7,792	747
1976	(7,820)	116,794	124,614	7,267	7,843	576
1977	(28,353)	123,182	151,534	6,850	7,581	731
1978	(30,205)	145,847	176,052	9,058	10,001	943
1979	(23,922)	186,363	210,285	10,123	11,747	1,624
1980	(19,696)	225,566	245,262	11,952	15,506	3,554
1981	(22,267)	238,715	260,982	13,134	17,634	4,500
1982	(27,510)	216,442	243,952	11,035	15,603	4,568
1983	(52,409)	205,639	258,048	12,619	16,065	3,446
1984	(106,703)	223,976	330,678	10,082	15,008	4,926
1985	(117,712)	218,815	336,526	12,593	18,725	6,132
1986	(138,279)	227,159	365,438	11,826	19,728	7,902
1987	(152,119)	254,122	406,241	14,575	22,480	7,905
1988	(118,526)	322,426	440,952	17,860	26,947	9,087
1989	(109,399)	363,812	473,211	22,083	32,111	10,028
1990	(101,010)	393,893	494,903	27,282	39,083	11,801
1991	(66,204)	421,851	488,055	30,785	43,788	13,003

Source: Bureau of the Census, Foreign Trade Division and Aerospace Industries Association, based on data from International Trade Administration.

NOTE: The Commerce Department began reporting international trade using the Harmonized Tariff Schedules of the United States in 1989. Previous years based on the Tariff Schedules of the United States Annotated.
Total U.S. and aerospace foreign trade are reported as (1) exports of domestic merchandise, including Department of Defense shipments and undocumented exports to Canada, f.a.s. (= free alongside ship) basis, (2) imports for consumption, customs value basis.

U.S. EXPORTS OF AEROSPACE PRODUCTS
BY MAJOR COUNTRIES OF DESTINATION
Calendar Years 1987 - 1991 (Millions of Dollars)

Major Countries of Destination

	1987	1988	1989	1990	1991
Australia	$1,036	$1,208	$1,271	$1,760	$1,596
Belgium/Luxembourg	373	348	538	681	825
Brazil	912	942	813	925	1,491
Canada	1,103	1,804	2,137	2,237	2,210
China	528	425	664	861	1,244
France	1,382	2,074	2,764	3,299	4,359
Germany, West	1,274	1,415	3,134	2,798	3,936
Hong Kong	351	166	381	587	759
Israel	487	454	453	503	738
Italy	455	578	625	737	1,051
Japan	2,313	2,710	2,700	4,185	3,907
Korea, South	343	823	1,257	1,113	1,715
Netherlands	565	744	1,448	1,613	1,458
Singapore	498	505	1,133	844	1,278
Spain	447	691	1,104	1,198	972
Sweden	307	627	815	952	1,081
Switzerland	334	294	458	283	1,226
Taiwan	153	164	460	732	1,324
Thailand	381	148	210	552	865
United Kingdom	2,297	2,908	3,520	4,966	3,961

Source: U.S. Department of Commerce, International Trade Administration.
NOTE: International trade reported using Harmonized Tariff Schedules after 1988.
Includes all civil products, free alongside ship basis; excludes military products
whose country of destination are not reported.

	1987	1988	1989	1990	1991
Bahrain	10	151	347	532	295
Malaysia	47	43	315	444	657
Turkey	216	358	292	468	580
Egypt	110	439	438	444	503
Mexico	196	178	432	462	608
Saudi Arabia	221	235	266	200	479

U.S. IMPORTS OF AEROSPACE PRODUCTS
BY MAJOR COUNTRIES OF ORIGIN
Calendar Years 1987 - 1991 (Millions of Dollars)

Major Countries of Destination

	1987	1988	1989	1990	1991
Brazil	$ 122	$ 183	$ 204	$ 360	$ 186
Canada	1,821	1,985	1,918	2,529	2,732
France	1,976	2,932	3,290	2,782	3,557
Germany, West	347	396	419	712	523
Israel	208	178	186	226	289
Italy	266	339	300	418	598
Japan	319	426	474	566	661
Netherlands	127	141	255	368	761
Sweden	278	246	257	317	332
United Kingdom	2,004	1,738	2,055	2,695	2,492

Source: U.S. Department of Commerce, International Trade Administration.

NOTE: International trade reported using Harmonized Tariff Schedules after 1988.
Includes civil and military products, c.i.f. (Cost, Insurance, and Freight) basis.

U.S. IMPORTS OF AEROSPACE PRODUCTS
Calendar Years 1987 - 1991 (Millions of Dollars)

Aerospace Imports	1987	1988	1989	1990	1991
TOTAL	**$7,905**	**$9,087**	**$10,028**	**$11,801**	**$13,003**
TOTAL CIVIL	**$6,409**	**$7,604**	**$ 7,200**	**$ 8,251**	**$ 9,268**
Complete Aircraft--TOTAL	**$2,038**	**$2,702**	**$ 2,788**	**$ 2,794**	**$ 3,413**
Transports	551	1,125	1,282	737	1,285
General Aviation	1,337	1,369	1,113	1,581	1,567
Helicopters	79	104	109	162	289
Other, Including Used Aircraft, & Gliders, Balloons, & Airships	70	103	285$Sa	314	272
Aircraft Engines--TOTAL	**1,117**	**951**	**999**	**1,234**	**1,226**
Turbine Engines	1,110	951	961	1,204	1,185
Piston Engines	7	-	38	31	42
Aircraft & Engine Parts--TOTAL	**3,254**	**3,951**	**3,414**	**4,222**	**4,629**
Aircraft Parts and Accessories	659	2,585	2,305	2,751	3,166
Turbine Engine Parts	1,058	1,323	924	1,147	1,279
Piston Engine Parts	19	14	136	57	43
Spacecraft, Other Parts & Accessories$Sc	1,519	29	50	267	141
TOTAL MILITARY	**$1,496**	**$1,483**	**$ 2,828**	**$ 3,550**	**$ 3,735**
Complete Aircraft--TOTAL	**$ 33**	**$ 2**	**$ 17**	**$ 44**	**$ 26**
Aircraft Engines--TOTAL	**199**	**106**	**971**	**1,217**	**1,203**
Turbine Engines	196	101	961	1,204	1,185
Piston Engines Including Parts	3	5	10	13	18
Aircraft & Engine Parts--TOTAL	**1,265**	**1,376**	**1,841**	**2,290**	**2,507**
Aircraft Parts	699	869	797	858	1,033
Turbine Engine Parts	370	480	881	1,088	1,238
Spacecraft, Missiles, Rockets, Other Parts, & Accessories	196	27	162	343	236

Source: Aerospace Industries Association, based on data from International Trade Administration.

NOTE: International trade reported using Harmonized Tariff Schedules after 1988.
Products within this category are not designated civil or military by the Harmonized Tariff Schedules. Historically, these products have been predominantly civil.
Category contains products whose use (civil or military) is unspecified by the Harmonized Tariff Schedules. Figures for those products distributed equally between civil and military.
Includes satellites, propulsion engines, and parachutes.

U.S. IMPORTS OF COMPLETE AIRCRAFT
Calendar Years 1987 - 1991

Aircraft Imports	1987	1988	1989	1990	1991
TOTAL NUMBER OF AIRCRAFT	**816**	**737**	**702**	**848**	**1,036**
Civil Aircraft--TOTAL	**630**	**706**	**673**	**820**	**955**
New Complete Aircraft:					
Helicopters	98	114	124	167	244
General Aviation:					
Single-Engine	41	40	53	80	72
Multi-Engine, Under 4,400 lbs	1	3	1	5	1
Multi-Engine, 4,400-10,000 lbs	81	74	32	53	41
Multi-Engine, Turbojet/Turbofan, 10,000-33,000 lbs	76	74	39	63	45
Multi-Engine, Other, Including Turboshaft, 10,000-33,000 lbs	79	78	87	100	95
Transports, Multi-Engine, Over 33,000 lbs	22	18	36	30	44
Other Civil Aircraft:					
Used or Rebuilt	115	194	210	130	246
Aircraft Previously Exported from U.S.	NA	NA	NA	NA	NA
Gliders	117	111	76	184	140
Balloons & Airships	NA	NA	15	8	27
Military Aircraft--TOTAL	**186**	**31**	**29**	**28**	**81**
New Complete Aircraft	123	27	25	28	8
Gliders	63	4	-	-	-
Balloons & Airships	NA	NA	-	-	-
VALUE (Millions of Dollars)	**$2,070.4**	**$2,703.3**	**$2,804.5**	**$2,838.3**	**$3,438.1**
Civil Aircraft--TOTAL	**$2,037.7**	**$2,701.5**	**$2,788.1**	**$2,794.2**	**$3,412.7**
New Complete Aircraft:					
Helicopters	79.3	103.9	108.7	162.4	288.8
General Aviation:					
Single-Engine	3.1	4.5	6.7	9.0	23.4
Multi-Engine, Under 4,400 lbs	0.3	6.5	0.1	1.3	0.0
Multi-Engine, 4,400-10,000 lbs	206.7	163.6	119.1	217.3	176.3
Multi-Engine, Turbojet/Turbofan, 10,000-33,000 lbs	677.3	729.1	372.0	643.6	526.9
Multi-Engine, Other, Including Turboshaft, 10,000-33,000 lbs	449.8	465.3	614.9	709.9	840.3
Transports, Multi-Engine, Over 33,000 lbs	551.1	1,125.4	1,281.8	737.0	1,285.3
Other Civil Aircraft:					
Used or Rebuilt	60.7	92.0	236.7	292.4	269.5
Aircraft Previously Exported from U.S.	8.0	0.1	48.8	0.4	-
Gliders	0.6	0.5	0.3	0.8	0.9
Balloons & Airships	0.9	10.8	0.6	2.3	1.3
Military Aircraft--TOTAL	**$ 32.7**	**$ 1.8**	**$ 16.5**	**$ 44.2**	**$ 25.5**
New Aircraft	29.8	1.6	16.4	44.2	21.0
Gliders	1.3	0.1			
Balloons & Airships	1.6	0.1			

Source: Aerospace Industries Association, based on data from International Trade Administration.

NOTE: International trade reported using Harmonized Tariff Schedules after 1988. NA Not Available

TOTAL U.S. EXPORTS AND EXPORTS OF AEROSPACE PRODUCTS
Calendar Years 1964 - 1991 (Millions of Dollars)

Exports of Aerospace Products

Year	TOTAL Exports of U.S. Merchandise	TOTAL	Percent of Total U.S. Exports	Total	Trans-ports	Military
1964	$ 25,690	$ 1,608	6.3%	$ 764	$ 211	$ 844
1965	26,699	1,618	6.1	854	353	764
1966	29,379	1,673	5.7	1,035	421	638
1967	30,934	2,248	7.3	1,380	611	868
1968	34,063	2,994	8.8	2,289	1,200	705
1969	37,332	3,138	8.4	2,027	947	1,111
1970	43,176	3,405	7.9	2,516	1,283	889
1971	44,087	4,203	9.5	3,080	1,567	1,123
1972	49,854	3,795	7.6	2,954	1,119	841
1973	71,865	5,142	7.2	3,788	1,664	1,354
1974	99,437	7,095	7.1	5,273	2,655	1,822
1975	108,856	7,792	7.2	5,324	2,397	2,468
1976	116,794	7,843	6.7	5,677	2,468	2,166
1977	123,182	7,581	6.2	5,049	1,936	2,532
1978	145,847	10,001	6.9	6,018	2,558	3,983
1979	186,363	11,747	6.3	9,772	4,998	1,975
1980	225,566	15,506	6.9	13,248	6,727	2,258
1981	238,715	17,634	7.4	13,312	7,180	4,322
1982	216,442	15,603	7.2	9,608	3,834	5,995
1983	205,639	16,065	7.8	10,595	4,683	5,470
1984	223,976	15,008	6.7	9,659	3,195	5,350
1985	218,815	18,725	8.6	12,942	5,518	5,783
1986	227,159	19,728	8.7	14,851	6,276	4,875
1987	254,122	22,480	8.8	15,768	6,377	6,714
1988	322,426	26,947	8.4	20,298	8,766	6,651
1989	363,812	32,111	8.8	25,619	12,313	6,492
1990	393,893	39,083	9.9	31,517	16,691	7,566
1991	421,851	43,788	10.4	35,548	20,881	8,239

Source: Bureau of the Census, Foreign Trade Division and Aerospace Industries Association, based on data from International Trade Administration.

NOTE: International trade reported using Harmonized Tariff Schedules after 1988. Includes DOD shipments and undocumented exports to Canada, free alongside ship basis.

U.S. EXPORTS OF AEROSPACE PRODUCTS
Calendar Years 1987 - 1991 (Millions of Dollars)

Aerospace Exports	1987	1988	1989	1990	1991
TOTAL	**$22,480**	**$26,947**	**$32,111**	**$39,083**	**$43,788**
TOTAL CIVIL	**$15,765**	**$20,297**	**$25,619**	**$31,517**	**$35,548**
Complete Aircraft--TOTAL	**$ 7,517**	**$10,294**	**$13,447**	**$18,150**	**$22,385**
Transports	6,377	8,766	12,313	16,691	20,881
General Aviation	295	348	413	555	576
Helicopters	201	219	156	161	168
Used Aircraft	503	639	533	712	738
Other, Incl. Spacecraft	141	323	217	360	176
Aircraft Engines--TOTAL	**1,206**	**1,570**	**1,948**	**1,754**	**2,127**
Turbine Engines	1,154	1,492	1,856	1,679	2,050
Piston Engines	53	78	93	75	77
Aircraft and Engine Parts Incl. Spares--TOTAL	**7,042**	**8,432**	**10,019**	**11,257**	**10,878**
Aircraft Parts & Accessories	4,650	5,442	6,258	6,964	6,859
Aircraft Engine Parts	2,393	2,990	3,761	4,293	4,018
TOTAL MILITARY	**$ 6,714**	**$ 6,651**	**$ 6,492**	**$ 7,566**	**$8,239**
Complete Aircraft--TOTAL	**$ 2,628**	**$ 2,157**	**$ 892**	**$ 1,481**	**$1,788**
Fighters & Fighter Bombers	1,986	1,469	368	533	323
Transports	363	212	234	432	633
Helicopters	81	198	180	381	587
Used Aircraft	4	59	56	75	146
Other, Incl. Spacecraft	194	219	247	391	253
Aircraft Engines--TOTAL	**157**	**223**	**236**	**203**	**206**
Turbine Engines	154	213	198	168	171
Piston Engines	4	10	38	35	35
Aircraft and Engine Parts Incl. Spares--TOTAL	**3,085**	**3,214**	**4,134**	**4,261**	**4,891**
Aircraft Parts & Accessories	2,457	2,546	3,450	3,640	4,202
Aircraft Engine Parts	628	668	684	622	689
Guided Missiles, Rockets, & Parts--TOTAL	**845**	**1,056**	**1,037**	**1,290**	**1,200**
Guided Missiles & Rockets	353	383	375	551	298
Missile & Rocket Parts	456	622	656	724	899
Missile & Rocket Engines	21	30	6	15	3
Missile & Rocket Engine Parts	15	21	-	-	-

Source: Aerospace Industries Association, based on data from International Trade Administration.

NOTE: International trade reported using Harmonized Tariff Schedules after 1988.
All fixed-wing aircraft under 33,000 pounds.
Products within this category are not designated civil or military by the Harmonized Tariff Schedules. Historically, aircraft herein have been predominantly civil. Also, spacecraft not included in "Complete Aircraft--Total."
Includes aircraft exported under Military Assistance Programs and Foreign Military Sales.

U.S. EXPORTS OF CIVIL AIRCRAFT
Calendar Years 1987 - 1991

Civil Aircraft Exports	1987	1988	1989	1990	1991
TOTAL NUMBER OF AIRCRAFT	**1,811**	**2,784**	**3,564**	**3,375**	**3,071**
Helicopters--TOTAL	**242**	**280**	**294**	**349**	**318**
Under 2,200 lbs	115	161	186	266	246
Over 2,200 lbs	127	119	108	83	72
General Aviation--TOTAL	**487**	**643**	**1,310**	**809**	**534**
Single-Engine	295	459	1,119	561	345
Multi-Engine, Under 4,400 lbs	51	51	39	33	22
Multi-Engine, 4,400-10,000 lbs	126	109	104	136	98
Multi-Engine, 10,000-33,000 lbs	15	24	48	79	69
Transports--TOTAL	**170**	**217**	**260**	**306**	**385**
Passenger Aircraft, Over 33,000 lbs	160	205	256	294	371
Cargo Aircraft, Over 33,000 lbs	4	8	1	3	5
Other, Over 33,000 lbs, Incl. Pass./Cargo Combi	6	4	3	9	9
Other Aircraft--TOTAL	**912**	**1,644**	**1,700**	**1,911**	**1,834**
Used or Rebuilt Aircraft	912	1,644	1,700	1,911	1,834
Other Aircraft, Including Balloons, Gliders, & Kites	NA	NA	2,888	1,448	1,133
TOTAL VALUE (Millions of Dollars)	**$7,518**	**$10,296**	**$13,447**	**$18,150**	**$22,385**
Helicopters--TOTAL	**$ 201**	**$ 219**	**$ 156**	**$ 161**	**$ 168**
Under 2,200 lbs	27	30	29	39	40
Over 2,200 lbs	174	189	127	123	129
General Aviation--TOTAL	**295**	**348**	**413**	**555**	**576**
Single-Engine	26	47	56	44	40
Multi-Engine, Under 4,400 lbs	8	12	9	10	8
Multi-Engine, 4,400-10,000 lbs	219	239	184	256	249
Multi-Engine, 10,000-33,000 lbs	42	49	164	245	279
Transports--TOTAL	**6,377**	**8,766**	**12,313**	**16,691**	**20,881**
Passenger Aircraft, Over 33,000 lbs	5,635	7,770	11,859	15,307	19,349
Cargo Aircraft, Over 33,000 lbs	208	599	90	264	405
Other, Over 33,000 lbs, Incl. Pass./Cargo Combi	534	396	364	1,121	1,127
Other Aircraft--TOTAL	**645**	**963**	**560**	**742**	**760**
Used or Rebuilt Aircraft	503	639	533	712	738
Other Aircraft, Including Balloons, Gliders, & Kites	141	323	33	30	23

Source: Aerospace Industries Association, based on data from International Trade Administration.

NOTE: International trade reported using Harmonized Tariff Schedules after 1988. Included spacecraft until 1989.

U.S. EXPORTS OF COMMERCIAL TRANSPORT AIRCRAFT
Calendar Years 1987 - 1991

Region of Destination	1987	1988	1989	1990	1991
TOTAL NUMBER EXPORTED	**170**	**217**	**260**	**306**	**385**
Canada & Greenland	-	10	9	4	3
Latin America & Caribbean	20	15	28	25	32
Europe	88	127	151	172	228
Middle East	7	4	8	9	16
Asia	40	41	47	70	83
Oceania	8	11	8	16	14
Africa	7	9	9	10	9
TOTAL VALUE (Millions of Dollars)	**$6,377**	**$8,766**	**$12,313**	**$16,691**	**$20,881**
Canada & Greenland	$ -	$ 547	$ 535	$ 309	$ 221
Latin America & Caribbean	725	669	726	1,001	1,472
Europe	2,753	3,944	6,335	8,166	10,461
Middle East	185	227	631	440	648
Asia	2,263	2,404	2,951	5,010	6,382
Oceania	289	503	640	1,256	1,177
Africa	162	471	496	509	520

Source: Aerospace Industries Association, based on data from the International Trade Administration.

NOTE: International trade reported using Harmonized Tariff Schedules after 1988. Airframe weight exceeding 33,000 pounds.

U.S. EXPORTS OF CIVIL HELICOPTERS
Calendar Years 1987 - 1991

Region of Destination	1987	1988	1989	1990	1991
TOTAL NUMBER EXPORTED	**242**	**280**	**294**	**349**	**318**
Canada & Greenland	13	17	11	11	20
Latin America & Caribbean	42	25	54	46	45
Europe	97	131	170	140	125
Middle East	10	15	6	1	2
Asia	46	52	51	65	66
Oceania	27	31	33	68	38
Africa	7	9	9	18	22
TOTAL VALUE (Millions of Dollars)	**$200.5**	**$218.6**	**$155.5**	**$161.2**	**$168.4**
Canada & Greenland	$ 4.9	$ 5.2	$ 2.6	$ 5.1	$ 7.9
Latin America & Caribbean	47.8	24.5	39.7	20.1	19.6
Europe	37.7	36.0	37.1	46.8	56.3
Middle East	53.1	70.6	5.4	3.6	16.5
Asia	47.0	68.1	60.0	71.3	59.2
Oceania	6.1	10.3	9.2	8.7	5.7
Africa	4.0	3.9	1.6	5.6	3.1

Source: Aerospace Industries Association, based on data from the International Trade Administration.

NOTE: International trade reported using Harmonized Tariff Schedules after 1988. Excludes used helicopters.

U.S. IMPORTS OF CIVIL HELICOPTERS
Calendar Years 1987 - 1991

Country of Origin	1987	1988	1989	1990	1991
TOTAL NUMBER IMPORTED	**98**	**114**	**124**	**167**	**243**
Canada	32	33	52	82	146
France	29	30	45	49	57
Germany	33	43	25	25	30
Italy	4	7	2	11	10
United Kingdom	-	1	-	-	-
TOTAL VALUE (Millions of Dollars)	**$79.3**	**$103.9**	**$108.7**	**$162.4**	**$288.2**
Canada	$18.9	$ 21.5	$ 44.5	$ 86.3	$182.1
France	24.0	21.6	32.0	29.9	53.6
Germany	31.2	50.1	28.9	34.9	35.6
Italy	5.2	10.5	3.3	11.3	16.9
United Kingdom	-	0.2	-	-	-

Source: Aerospace Industries Association, based on data from the International Trade Administration.

NOTE: International trade reported using Harmonized Tariff Schedules after 1988. Excludes used helicopters.

U.S. EXPORTS OF GENERAL AVIATION AIRCRAFT
Calendar Years 1987 - 1991

Region of Destination	1987	1988	1989	1990	1991
TOTAL NUMBER EXPORTED	**487**	**643**	**1,310**	**809**	**534**
Canada & Greenland	25	14	35	34	9
Latin America & Caribbean	93	100	155	133	80
Europe	213	322	634	379	317
Middle East	27	2	7	15	11
Asia	67	50	154	55	54
Oceania	33	125	164	72	18
Africa	29	30	161	121	45
TOTAL VALUE (Millions of Dollars)	**$295.1**	**$347.7**	**$413.1**	**$554.9**	**$576.0**
Canada & Greenland	$ 12.0	$ 12.8	$ 11.7	$ 41.7	$ 31.2
Latin America & Caribbean	51.4	114.0	120.4	152.8	142.9
Europe	148.6	126.7	168.0	197.1	253.1
Middle East	1.6	0.1	4.7	18.1	21.7
Asia	49.8	38.7	43.0	47.9	95.0
Oceania	3.4	35.8	18.0	22.0	6.9
Africa	28.4	19.6	47.4	75.3	25.2

Source: Aerospace Industries Association, based on data from the International Trade Administration.

NOTE: International trade reported using Harmonized Tariff Schedules after 1988. All fixed-wing aircraft under 33,000 pounds.

U.S. IMPORTS OF GENERAL AVIATION AIRCRAFT
Calendar Years 1987 - 1991

Country of Origin	1987	1988	1989	1990	1991
TOTAL NUMBER IMPORTED	**278**	**269**	**212**	**301**	**254**
Brazil	20	30	30	51	24
Canada	34	40	31	32	42
France	76	60	65	93	92
Israel	8	5	8	12	8
Japan	12	29	-	-	-
United Kingdom	80	64	49	77	48
Other	48	41	38	36	40
TOTAL VALUE (Millions of Dollars)	**$1,337.0**	**$1,369.0**	**$1,112.8**	**$1,581.2**	**$1,566.8**
Brazil	$ 97.8	$ 163.8	$ 175.6	$ 306.9	$ 152.2
Canada	209.6	268.6	275.2	354.7	469.8
France	510.5	532.7	335.0	336.2	469.9
Israel	30.7	24.6	41.5	70.6	51.7
Japan	12.6	23.9	-	-	-
United Kingdom	301.9	271.7	212.7	414.6	276.9
Other	173.9	83.7	72.8	98.1	146.3

Source: Aerospace Industries Association, based on data from the International Trade Administration.

NOTE: International trade reported using Harmonized Tariff Schedules after 1988. All fixed-wing aircraft under 33,000 pounds.

U.S. EXPORTS OF MILITARY AIRCRAFT
Calendar Years 1987 - 1991

	1987	1988	1989	1990	1991
TOTAL NUMBER OF AIRCRAFT	**492**	**743**	**846**	**445**	**490**
Fighters and Fighter Bombers	122	87	32	39	16
Transports	99	14	74	43	40
Helicopters	39	53	36	47	72
New Aircraft, NEC	218	464	505	259	235
Used or Rebuilt Aircraft	14	125	199	57	127
Airships, Balloons, Gliders, etc.	NA	NA	NA	NA	NA
TOTAL VALUE (Millions of Dollars)	**$2,628**	**$2,157**	**$892**	**$1,481**	**$1,783**
Fighters and Fighter Bombers	$1,986	$1,469	$368	$ 533	$ 323
Transports	363	212	234	432	633
Helicopters	81	198	180	381	587
New Aircraft, NEC	135	173	53	61	98
Used or Rebuilt Aircraft	4	59	56	75	142
Airships, Balloons, Gliders, etc.	59	46	-	-	-

Source: Aerospace Industries Association, based on data from the International Trade Administration.

NOTE: International trade reported using Harmonized Tariff Schedules after 1988.
Includes aircraft exported under Military Assistance Programs and Foreign Military Sales.
Includes spacecraft until 1989.
Products within this category are not designated civil or military by the Harmonized Tariff Schedules. Historically, these products have been predominantly civil.
NA Not available.
NEC Not elsewhere classified.

DIRECT EXPORT SHIPMENTS OF MILITARY HELICOPTERS
Calendar Years 1987 - 1991

Manufacturer and Model	1987	1988	1989	1990	1991
DIRECT MILITARY EXPORT SHIPMENTS	21	66	46	48	45
Value (Millions of Dollars)	$57	$352	$278	$337	$489
Bell AH-1S	-	24	26	-	-
Boeing Vertol CH-47/414/352	4	1	-	11	9
McDonnell Douglas 500MD (TOW)/ 500 Scout	11	19	-	-	-
Schweizer 300C	6	-	-	-	-
Sikorsky S-70C	-	13	17	35	36
Sikorsky S-80M	-	-	3	2	-
Sikorsky MH53	-	9	-	-	-

Source: Aerospace Industries Association, company reports. Shipments of helicopters in military configuration exported directly from U.S. manufacturers to foreign governments.

U.S. EXPORTS OF AIRCRAFT ENGINES
Calendar Years 1989 - 1991 (Values in Millions of Dollars)

	1989		1990		1991	
	Number	**Value**	**Number**	**Value**	**Number**	**Value**
TOTAL	**12,570**	**$2,184**	**9,419**	**$1,957**	**10,651**	**$2,333**
Turbine Engines	3,917	$2,053	3,008	$1,846	3,199	$2,221
Civil	3,031	1,856	2,277	1,679	2,114	2,050
Military	886	198	731	168	1,085	171
Piston Engines	8,653	131	6,411	110	7,452	112
Civil, New, Under 500 HP	1,964	19	1,108	15	1,168	17
Civil, New, Over 500 HP	423	13	256	10	76	4
Civil, Used	4,036	70	3,183	50	3,486	56
Military	2,230	38	1,864	35	2,722	35

Source: Aerospace Industries Association, based on data from the International Trade Administration.
NOTE: International trade reported using Harmonized Tariff Schedules after 1988.

U.S. IMPORTS OF AIRCRAFT ENGINES
Calendar Years 1989 - 1991 (Values in Millions of Dollars)

	1989		1990		1991	
	Number	**Value**	**Number**	**Value**	**Number**	**Value**
Turbine Engines	2,283	$1,921	5,007	$2,408	2,032	$2,370
Piston Engines	3,562	43	3,152	36	9,379	53
Military	1,079	6	251	5	6,648	12
Civil, New, S	1,500	3	2,070	5	2,085	3
Civil, New, L	729	18	136	15	29	29
Civil, Used	254	17	695	11	617	9

Source: Aerospace Industries Association, based on data from the International Trade Administration.
NOTE: International trade reported using Harmonized Tariff Schedules after 1988. New and used.

EXPORT-IMPORT BANK
TOTAL AUTHORIZATIONS OF LOANS AND GUARANTEES
AND AUTHORIZATIONS IN SUPPORT OF AIRCRAFT EXPORTS
Fiscal Years 1979 - 1991 (Millions of Dollars

Authorizations in Support of Aircraft Exports

Year	TOTAL AUTHORI- ZATIONS	TOTAL	Percent of TOTAL Authori- zations	Commercial Jet Aircraft	Other Aircraft
LOANS					
1979	$4,475	$1,469.4	32.8%	$1,399.4	$ 70.0
1980	4,578	1,743.3	38.1	1,692.6	50.7
1981	5,431	2,576.6	47.4	2,550.3	26.3
1982	3,516	263.9	7.5	199.1	64.8
1983	845	396.7	46.9	383.8	12.9
1984	1,465	608.0	41.5	531.8	76.2
1985	659	39.7	6.0	12.6	27.1
1986	578	54.6	9.4	46.4	8.2
1987	599	17.0	2.8	13.3	3.7
1988	685	-	-	-	-
1989	695	166.4	23.9	158.0	8.4
1990	614	5.0	0.8	-	5.0
1991	604	-	-	-	-
GUARANTEES					
1979	$ 908	$ 261.4	28.8%	$ 239.3	$ 22.1
1980	2,510	1,131.9	45.1	1,088.1	43.8
1981	1,506	562.6	37.4	533.4	29.2
1982	727	104.2	14.3	78.4	25.8
1983	1,741	629.6	36.2	601.3	28.3
1984	1,333	355.5	26.7	293.5	62.0
1985	1,320	322.4	24.4	288.9	33.5
1986	1,128	329.2	29.2	277.4	51.8
1987	1,506	808.3	53.4	808.3	-
1988	601	89.2	14.8	73.4	15.8
1989	1,292	496.4	38.4	390.4	106.0
1990	3,333	1,666.3	50.0	224.7	1,441.6
1991	6,016	606.0	10.1	566.0	40.0

Source: Export-Import Bank of the United States.
Includes complete aircraft, engines, parts, and retrofits
Includes business aircraft, general aviation aircraft, helicopters, and related goods
and services.
Loans are commitments for financing by the Export-Import Bank to foreign buyers of
U.S. equipment and services, including Direct Credits, loans authorized under the
Cooperative Financing Facilty (CFF), (until the termination of the CFF program in
1981), and Discount Loans, which are made by the Export-Import Bank to commercial banks and which subsequently may be guaranteed by the Export-Import Bank,
in which case the value of the loans is also included with Guarantees.
Guarantees by the Export-Import Bank provide assurances of repayment of principal
and interest on loans made by private lending institutions, such as commercial
banks, for major export transactions. Excludes insurance.

EXPORT-IMPORT BANK
SUMMARY OF COMMERCIAL JET AIRCRAFT AUTHORIZATIONS
FOR LOANS AND GUARANTEES
Fiscal Years 1957 - 1991 (Values in Millions of Dollars)

Year	No. of Jet Aircraft Loans	No. of Jet Aircraft Guar-antees	Export Value Loans	Export Value Guar-antees	No. of New Commitments Loans	No. of New Commitments Guar-antees	Gross Authorizations Loans	Gross Authorizations Guar-antees
New Authorizations:								
1957-1973	919	92	9,374	615	321	215	3,961	1,201
1974	189	-	2,195	-	79	22	895	133
1975	136	1	2,070	5	64	10	691	64
1976	77	6	1,017	139	34	11	398	87
1977	31	25	330	902	16	14	138	294
1978	29	5	479	253	18	5	189	77
1979	118	7	2,938	317	35	10	1,399	239
1980	136	21	3,975	901	36	24	1,693	1,088
1981	121	18	4,568	637	26	17	2,550	533
1982	11	6	441	113	5	2	199	78
1983	21	9	779	619	3	4	384	601
1984	37	8	1,023	327	7	4	532	294
1985	-	14	19	481	1	5	13	289
1986	3	13	74	451	1	9	46	277
1987	-	27	22	1,449	1	14	13	808
1988	-	2	-	97	-	2	-	73
1989	3	5	253	459	1	2	158	390
1990	-	6	-	264	-	2	-	225
1991	-	12	-	657	-	3	-	566
Cumulative New Authorizations	1,846	282	29,776	8,210	654	378	13,353	$7,376
Transfers, Reversals, & Participation	-	-	(8)	8	4	-	(140)	(20)
Cumulative Gross Authorizations (net of Adjustments)	1,846	282	29,767	8,875	658	378	13,213	$7,357

Source: Export-Import Bank of the United States.

Loans are commitments for direct financing by the Export-Import Bank to foreign buyers of U.S. equipment and services, including Direct Credits and loans authorized under the Cooperative Financing Facility (CFF) until the termination of the CFF program in 1981, but excluding Discount Loans, which are made by the Export-Import Bank to commercial banks and which subsequently may be guaranteed by the Export-Import Bank in which case the value of the loans is included with Guarantees. Guarantees by the Export-Import Bank provide assurances of repayment of principal and interest on loans made by private lending institutions, such as commercial banks, for major export transactions.

For Export-Import Bank commitments including both loan and guarantee authorization, number of aircraft and export value reported under "Loans."

First year of commercial jet aircraft authorizations.

EXPORT-IMPORT BANK
AUTHORIZATIONS OF LOANS AND GUARANTEES
IN SUPPORT OF EXPORTS OF COMMERCIAL JET AIRCRAFT
Fiscal Years 1987 - 1991 (Values in Millions of Dollars)

Customer (Country/Airline)	Number and Aircraft Model or Related Product	AUTHORIZATION Loans (Direct Credits) Export Value	Amt	Percent Cover-age	Int-erest Rate	Repay-ment Terms	Guar-antees Amt
FY 1991							
TOTALS	**12 aircraft**	**$ 657**	-	-	-	-	**$566**
Bahrain/Gulf Air Co.	6 x 767	427	-	-	-	-	366
Greece/Olympic Airways	6 x 737	230	-	-	-	-	200
FY 1990							
TOTALS	**6 aircraft**	**$ 264**	-	-	-	-	**$225**
Columbia/Avianca	2 x 767	150	-	-	-	-	128
Morocco/Royal Air Maroc	4 x 737	114	-	-	-	-	97
FY 1989							
TOTALS	**8 aircraft**	**$ 712**	**$293**	-	-	-	**$605**
Algeria/Algerie Air	3 x 737	253	158	62.5	8.95%	24-S	215
Yugoslavia/Jugoslovenski Aerotransport	3 x MD-11	301	-	-	-	-	255
Zimbabwe/Government of	2 x 747	159	135	85.0	9.68%	24-S	135
FY 1988							
TOTALS	**2 aircraft**	**$ 94**	-	-	-	-	**$ 76**
Bangladesh/Bangladesh Biman Corp.	1 x DC-10-30	67	-	-	-	-	50
Israel/El Al	1 x 757	27	-	-	-	-	22
Uganda/Uganda Airlines	707 Hushkit	3	-	-	-	-	3
FY 1987							
TOTALS	**32 aircraft**	**$1,411**	**$ 7**	-	-	-	**$768**
Brazil/VARIG	6 x 767	324	-	-	-	-	275
Israel/El Al	2 x 757	59	-	-	-	-	50
Japan/All Nippon	15 x 767	857	-	-	-	-	300
Mauritius/Air Mauritius	2 x 767	5	-	-	-	-	5
Mauritius/ Air Mauritius	spare parts for 2 x 767	16	7	42.5	9.10%	20-S	7
Nepal/Royal Nepal Airlines	2 x 757	76	-	-	-	-	64
Yugoslavia/Jugoslovenski Aerotransport	2 x 737	53	-	-	-	-	45
Yugoslavia/Aviogenex	2 x 737	18	-	-	-	-	14
Yugoslavia/Aviogenex	1 x 737	19	-	-	-	-	15

Source: Aerospace Industries Assocation, based on data from the Export-Import Bank of the United States.

Section XII.

Aerospace Employment

ANNUAL AVERAGE EMPLOYMENT IN ALL MANUFACTURING, DURABLE GOODS, AND AEROSPACE INDUSTRIES
Calendar years 1979 - 1991 (Thousands of Employees)

Year	All Manu-facturing Industries	Durable Goods Industries	TOTAL	Aerospace Industry As Percent of All Manufac-turing	Durable Goods
1979	21,040	12,730	1,007	4.8%	7.9%
1980	20,285	12,159	1,080	5.3	8.9
1981	20,170	12,082	1,087	5.4	9.0
1982	18,781	11,014	1,038	5.5	9.4
1983	18,434	10,707	1,019	5.5	9.5
1984	19,378	11,479	1,058	5.5	9.2
1985	19,260	11,464	1,152	6.0	10.0
1986	18,965	11,203	1,242	6.5	11.1
1987	19,024	11,167	1,283	6.7	11.5
1988	19,350	11,381	1,294	6.7	11.4
1989	19,442	11,420	1,314	6.8	11.5
1990	19,111	11,115	1,303	6.8	11.7
1991	18,426	10,556	1,216	6.6	11.5

Source: Bureau of Labor Statistics, "Employment and Earnings" (Monthly) and Aerospace Industries Association estimates.

ANNUAL PAYROLL
AEROSPACE INDUSTRY AND ALL MANUFACTURING INDUSTRIES
Calendar Years 1979 - 1991 (Millions of Dollars)

Year	All Manufacturing Industries	Aerospace Industry			Aerospace As Percent of All Manufacturing
		TOTAL	Production Workers	Other Workers	
1979	$334,800	$15,150	$ 6,465	$ 8,685	4.5%
1980	355,600	18,026	7,658	10,368	5.1
1981	386,700	19,910	8,150	11,760	5.1
1982	384,000	20,750	8,043	12,707	5.4
1983	397,400	21,643	8,074	13,569	5.4
1984	439,100	23,780	8,746	15,034	5.4
1985	460,900	26,753	9,837	16,915	5.8
1986	473,200	29,554	11,038	18,517	6.2
1987	490,300	31,108	11,703	19,405	6.3
1988	524,000	32,576	11,750	20,826	6.2
1989	541,800	34,164	12,446	21,718	6.3
1990	555,800	35,598	13,023	22,574	6.6
1991	556,500	34,571	12,561	22,011	6.4

AEROSPACE -- INCLUDING LUMP-SUM PAYMENTS

Year	TOTAL	Production Workers	Other Workers	Aerospace As Percent of All Manufacturing
1984	$23,825	$ 8,791	$15,034	5.4%
1985	26,789	9,873	16,915	5.8
1986	29,624	11,108	18,517	6.3
1987	31,269	11,865	19,405	6.4
1988	32,767	11,941	20,826	6.3
1989	34,406	12,688	21,718	6.4
1990	35,873	13,298	22,574	6.6
1991	34,745	12,734	22,011	6.4

Source: Bureau of Economic Analysis, "Survey of Current Business" (Monthly) and Aerospace Industries Association estimates based on Bureau of Labor Statistics, "Employment and Earnings" (Monthly).

Based on combined annual average employment and average weeklyearnings for SICs 372 and 376.

Many aerospace manufacturers have included lump-sum payments in labor settlements since late 1983 in lieu of general wage increases and/or cost of living adjustments. These payments are reported by BLS in separate wage series for SICs 3721 & 3761 and are included by AIA in the totals for production workers and all aerospace.

EMPLOYMENT IN THE AEROSPACE INDUSTRY
Calendar Years 1979 - 1991 (Annual Average, Thousands of Employees)

Year	TOTAL	Aircraft, Engines, & Parts (SIC 372)	Missiles & Space Vehicles (SIC 376)	Other
TOTAL EMPLOYMENT				
1979	1,007	593	102	313
1980	1,080	633	111	336
1981	1,087	627	123	338
1982	1,038	584	131	323
1983	1,019	562	141	317
1984	1,058	575	154	329
1985	1,152	616	177	358
1986	1,242	656	200	386
1987	1,283	678	206	399
1988	1,294	684	208	402
1989	1,314	711	194	409
1990	1,303	712	185	405
1991	1,216	671	167	378
PRODUCTION WORKERS				
1979	378	322	33	24
1980	404	344	35	25
1981	395	333	37	25
1982	361	296	40	24
1983	343	274	46	24
1984	353	276	52	25
1985	384	295	62	27
1986	419	323	67	29
1987	435	339	67	30
1988	424	332	63	30
1989	434	344	60	31
1990	432	345	57	30
1991	401	324	48	28

Source: Bureau of Labor Statistics, "Employment and Earnings" (Monthly) and Aerospace Industries Association estimates.
Communications, navigation, flight control, and displays (aerospace- related portions of SICs 366, 381, & 382).

EMPLOYMENT IN THE AIRCRAFT, ENGINES, AND PARTS INDUSTRY
Calendar Years 1979 - 1991 (Annual Average, Thousands of Employees)

Year	TOTAL (SIC 372)	Airframes (SIC 3721)	Engines and Parts (SIC 3724)	Other Parts & Equipment (SIC 3728)
TOTAL EMPLOYMENT				
1979	592.5	333.2	151.6	107.8
1980	633.1	349.3	162.9	120.9
1981	626.5	344.2	162.5	119.8
1982	584.0	319.9	148.8	115.3
1983	561.6	304.7	140.1	116.9
1984	575.1	306.1	140.2	128.8
1985	616.3	325.6	147.5	143.3
1986	656.0	338.9	153.6	163.4
1987	678.2	356.4	158.2	163.6
1988	683.8	368.5	155.8	159.5
1989	711.3	382.2	153.5	175.6
1990	712.5	381.0	151.7	179.7
1991	671.1	357.3	143.2	170.6
PRODUCTION WORKERS				
1979	322.1	165.9	86.4	70.2
1980	343.9	173.7	93.0	77.4
1981	332.6	167.0	92.4	73.5
1982	296.2	144.7	84.2	67.4
1983	274.0	131.5	74.7	67.1
1984	276.0	128.2	73.0	73.5
1985	294.6	135.5	74.8	82.3
1986	322.5	146.6	78.7	94.5
1987	338.6	159.1	80.5	96.4
1988	331.5	162.1	77.1	92.3
1989	343.9	167.4	76.8	99.7
1990	344.7	164.1	77.2	103.4
1991	324.3	152.4	73.1	98.9

Source: Bureau of Labor Statistics, "Employment and Earnings" (Monthly).

AEROSPACE INDUSTRY EMPLOYMENT
BY OCCUPATIONAL CLASSIFICATION
As of December 1982 - 1992 (Thousands of Employees)

Year	TOTAL	Production Workers	Scientists & Engineers	Technicians	Others
1982	765	353	134	54	224
1983	765	344	135	55	231
1984	817	365	147	60	245
1985	898	405	163	66	264
1986	948	436	168	67	277
1987	968	436	175	69	288
1988	977	431	184	66	296
1989	992	439	198	68	287
1990	946	422	205	68	251
1991	868	381	193	63	231
1992 E	809	NA	NA	NA	NA

Source: Aerospace Industries Association, company reports and Bureau of Labor Statistics, "Employment and Earnings" (Monthly).

NOTE: Employment estimates have been revised for 1982 and after to reflect recently released March 1988 benchmarks. Consequently, figures for years prior to 1982 are not directly comparable.

Totals for employment by occupational classification reflect only establishments in SICs 372, 376, 366, 381, and 382. As a result, they do not match the totals for aerospace employment by product group which include other industries with employment related to aerospace.

End-of-year figures often differ from annual averages appearing in other tables.

E Estimate.

NA Not available.

GEOGRAPHIC DISTRIBUTION OF AEROSPACE EMPLOYMENT BY OCCUPATIONAL CLASSIFICATION AND PRODUCT GROUP
As of December 1991

PERCENT DISTRIBUTION BY OCCUPATION

Region TOTAL	TOTAL 100.0%	Production Workers 100.0%	Scientists & Engineers 100.0%	Technicians 100.0%	All Others 100.0%
New England	10.9%	14.4%	8.1%	7.0%	10.0%
Middle Atlantic	7.3	6.3	11.6	2.6	6.3
East North Central	7.6	11.0	6.7	7.4	4.0
West North Central	10.2	13.1	7.1	9.2	9.4
South Atlantic	9.2	5.1	10.9	11.1	12.4
South Central	8.5	7.9	8.5	6.4	9.7
Mountain	5.8	4.7	5.9	6.6	6.9
Pacific	40.5	37.5	41.2	49.7	41.3

PERCENT DISTRIBUTION BY PRODUCT GROUP

Region TOTAL	Aircraft			Missiles 100.0%	Space 100.0%	Other	
	Total 100.0%	Civil 100.0%	Military 100.0%			Aero 100.0%	Non-Aero 100.0%
New England 31.8%	10.9%	3.4%	19.0%	− 25.1%	− 6.6%	8.7%	−
Middle Atlantic	7.3	1.4	8.2	/	/	16.6	/
East North Central	7.6	14.6	9.9	− 8.3	− 1.6	− 15.7	6.0
West North Central	10.2	11.0	12.4	/	/	/	2.9
South Atlantic	9.2	\	8.6	15.5	11.5	12.9	− 48.0
South Central	8.5	− 8.7	13.7	2.0	9.0	5.9	/
Mountain	5.8	4.9	1.2	5.5	20.6	5.6	− 11.3
Pacific	40.5	56.0	27.0	43.6	50.7	34.6	/

Source: Aerospace Industries Association, company reports.
NOTE: Data for two regions are combined where employment for one region within a product group represented three or fewer companies. Employment in 37 surveyed aerospace manufacturing corporations accounted for approximately two-thirds of total industry employment.

TOTAL EMPLOYMENT AND SCIENTISTS & ENGINEERS
IN COMMERCIAL TRANSPORT AIRCRAFT
& HELICOPTER MANUFACTURING ESTABLISHMENTS
As of December 1977 - 1991

Commercial Transport Aircraft

Year	Total	Scientists & Engineers	Total	Scientists & Engineers
1977	55,900	8,100	21,100	3,500
1978	58,700	8,700	24,200	3,300
1979	99,800	12,900	27,500	3,000
1980	106,500	13,700	29,800	3,200
1981	84,000	12,000	28,000	3,000
1982	69,000	11,100	26,500	3,100
1983	49,400	8,600	27,600	3,500
1984	59,100	9,400	31,300	3,800
1985	69,200	11,000	37,900	5,000
1986	79,100	12,900	37,400	4,000
1987	88,100	14,400	39,000	4,300
1988	99,100	15,900	36,600	4,200
1989	115,900	13,400	34,200	4,900
1990	116,800	15,500	30,600	4,500
1991	120,200	16,800	30,200	4,600

Source: Aerospace Industries Association, company reports and AIA estimates.
Includes only establishments identified as prime manufacturers of commercial transport aircraft and of civil and military helicopters.
Excludes subcontractors and propulsion manufacturers.
Industry strike during this period.

FEDERAL CIVILIAN EMPLOYMENT IN THE DEPARTMENT OF DEFENSE
Fiscal Years 1967 - 1993

Year	TOTAL	Civil Functions	Military Functions
1967	1,225,637	31,980	1,193,657
1968	1,288,130	32,062	1,256,068
1969	1,257,091	31,214	1,225,877
1970	1,159,935	30,293	1,129,642
1971	1,092,804	30,063	1,062,741
1972	1,040,147	30,585	1,009,562
1973	987,281	29,971	957,310
1974	1,002,850	29,072	973,778
1975	983,790	29,069	954,721
1976	951,034	28,648	922,386
1977	940,549	28,912	911,637
1978	933,071	28,962	904,109
1979	914,582	28,592	885,990
1980	907,700	27,700	880,000
1981	981,400	34,400	947,000
1982	1,009,344	31,263	978,081
1983	1,015,779	30,973	984,806
1984	1,040,213	28,681	1,011,532
1985	1,065,551	28,681	1,036,870
1986	1,069,863	28,511	1,041,352
1987	1,059,516	28,199	1,031,317
1988	1,052,848	28,267	1,024,581
1989	1,051,019	27,934	1,023,085
1990	1,049,422	28,259	1,021,163
1991	996,300	27,241	969,059
1992 E	966,394	27,725	938,669
1993 E	925,216	27,444	897,772

Source: Office of Management and Budget, "The Budget of the United States Government" (Annually).
Full-time equivalent civilian employment.
Data are estimated for portions of Civil Functions.
The Department of Defense is exempt from full-time equivalent controls.
Data shown are estimated civilian employment for military functions and military assistance.
E Estimate.

EMPLOYMENT IN NATIONAL AERONAUTICS AND SPACE ADMINISTRATION PROGRAMS
End of Fiscal Years 1961 - 1993

Year	TOTAL	NASA Employees	Contractor Employees
1961	74,577	17,077	57,500
1962	137,656	22,156	115,500
1963	246,304	27,904	218,400
1964	379,084	31,984	347,100
1965	409,900	33,200	376,700
1966	393,924	33,924	360,000
1967	306,926	33,726	273,200
1968	267,871	32,471	235,400
1969	218,345	31,745	186,600
1970	160,850	31,350	129,500
1971	143,578	29,478	114,100
1972	138,800	27,500	111,300
1973	134,850	26,850	108,000
1974	125,220	25,020	100,200
1975	127,733	24,333	103,400
1976	130,739	24,039	108,000
1977	124,136	23,636	100,500
1978	124,637	23,237	101,400
1979	131,931	22,831	109,100
1980	135,613	22,613	113,000
1981	133,473	21,873	111,600
1982	127,952	21,652	106,300
1983	129,246	22,246	107,000
1984	162,080	22,080	140,000
1985	131,991	21,991	110,000
1986	154,660	21,660	133,000
1987	165,001	22,001	143,000
1988	172,326	22,326	150,000
1989	213,054	23,054	190,000
1990	221,829	23,829	198,000
1991	223,149	24,149	199,000
1992 E	230,737	24,737	206,000
1993 E	230,947	24,947	206,000

Source: Office of Management and Budget, "Budget of the United States Government" (Annually) and NASA Headquarters.
Includes estimates of manpower for hardware and related contracts, as well as actual work-years for support service contracts. Increase in FY 1984 caused by change in estimating methodology to reflect more accurately the mix of support and development contractors.
E Estimate.

EMPLOYMENT AND COST OF R&D SCIENTISTS AND ENGINEERS
ALL INDUSTRIES AND AEROSPACE INDUSTRY
Calendar Years 1979 - 1991

| | Employment | | | Cost Per R&D Scientist and Engineer | |
| | | | | | |
Year	All Industries (Thousands)	Aerospace (Thousands)	Aerospace as a Percent of All Industries	All Industries	Aerospace
1979	423.9	86.5	20.4%	$ 87,400	$ 93,300
1980	450.6	85.9	19.1	94,900	101,600
1981	487.8	95.2	19.5	103,900	128,500
1982	509.8	91.1	17.9	111,600	148,800
1983	540.9	103.1	19.1	116,000	143,600
1984	584.1	111.5	19.1	124,000	156,000
1985	622.5	130.2	20.9	130,200	161,700
1986	671.0	144.8	21.6	128,500	149,800
1987	695.8	136.3	19.6	131,200	179,400
1988	708.6	136.4	19.2	137,000	185,900
1989	720.2	142.3	19.8	140,400	189,400
1990	730.9	128.5	17.6	144,900	205,500
1991	709.7	118.3	16.7	NA	NA

Source: National Science Foundation.
Employment as of January. Scientists and engineers working less than full time have been included in terms of their full time equivalent number.
All manufacturing industries and those non-manufacturing industries known to conduct or finance research and development.
Standard Industrial Classification codes 372 and 376.
The arithmetic mean of the numbers of R&D scientists and engineers reported for January in two consecutive years, divided into the total R&D expenditures of each industry during the earlier year.
NA Not available.

AVERAGE HOURLY EARNINGS IN THE AEROSPACE INDUSTRY
Production Workers Only
Calendar Years 1974 - 199

Year	TOTAL	TOTAL	Airframes (SIC 3721)	Engines & Parts (SIC 3724)	Other Parts & Equipment (SIC 3728)	TOTAL	Guided Missiles & Space Vehicles (SIC 3761)
		Aircraft (SIC 372)				Guided Missiles, Space Vehicles & Parts (SIC 376)	
AVERAGE HOURLY EARNINGS							
1974	$ 5.43	$ 5.42	$ 5.58	$ 5.41	$ 5.05	$ 5.48	$ 5.44
1975	6.00	6.00	6.21	6.04	5.47	6.02	5.99
1976	6.44	6.44	6.63	6.46	5.95	6.48	6.49
1977	6.93	6.92	7.07	7.05	6.44	7.04	7.15
1978	7.54	7.54	7.70	7.80	6.93	7.56	7.72
1979	8.26	8.26	8.50	8.53	7.48	8.25	8.38
1980	9.27	9.28	9.66	9.42	8.40	9.22	9.33
1981	10.29	10.31	10.74	10.41	9.35	10.06	10.34
1982	11.20	11.23	11.85	11.16	10.17	10.95	11.21
1983	11.79	11.82	12.58	11.61	10.73	11.59	11.84
1984	12.24	12.32	12.91	12.40	11.37	11.82	12.01
1985	12.54	12.62	13.18	12.85	11.66	12.14	12.36
1986	12.75	12.86	13.48	13.08	11.90	12.20	12.48
1987	13.10	13.17	13.74	13.33	12.23	12.73	13.09
1988	13.48	13.55	14.18	13.80	12.28	13.13	13.53
1989	14.10	14.17	14.89	14.42	12.81	13.70	14.20
1990	14.73	14.79	15.66	14.84	13.37	14.39	14.82
1991	15.52	15.61	16.72	15.38	14.07	14.90	15.21
AVERAGE HOURLY EARNINGS INCLUDING LUMP-SUM WAGE PAYMENTS							
1984	$12.37	$12.46	$13.11	$12.40	$11.37	$11.92	$12.14
1985	12.69	12.77	13.40	12.85	11.66	12.29	12.56
1986	12.94	13.06	13.80	13.08	11.90	12.33	12.66
1987	13.37	13.48	14.32	13.33	12.23	12.80	13.19
1988	13.72	13.79	14.65	13.80	12.28	13.36	13.87
1989	14.37	14.44	15.41	14.42	12.81	13.98	14.63
1990	15.04	15.10	16.32	14.84	13.37	14.68	15.26
1991	15.72	15.82	17.16	15.38	14.07	15.09	15.49

Source: Bureau of Labor Statistics, "Employment and Earnings" (Monthly) and Aerospace Industries Association estimates.
TOTAL columns are employment-based weighted averages.
Includes overtime premiums.
Many aerospace manufacturers have included lump-sum payments in labor settlements since late 1983 in lieu of general wage increases and/or cost of living adjustments. These payments are reported by BLS in separate wage series for SICs 3721 & 3761 and are included by AIA in totals.

AVERAGE WEEKLY EARNINGS IN THE AEROSPACE INDUSTRY
Production Workers Only
Calendar Years 1979 - 1991

	Aircraft (SIC 372)					Guided Missiles, Space Vehicles & Parts (SIC 376)	
Year	TOTAL	TOTAL	Airframes (SIC 3721)	Engines & Parts (SIC 3724)	Other Parts & Equipment (SIC 3728)	TOTAL	Guided Missiles & Space Vehicles (SIC 3761)
AVERAGE WEEKLY EARNINGS							
1979	$351	$351	$360	$361	$322	$347	$348
1980	389	390	404	394	358	378	383
1981	424	426	444	422	396	410	420
1982	460	462	485	454	426	447	461
1983	486	487	513	476	453	480	494
1984	513	516	532	523	486	496	508
1985	531	534	547	542	506	515	527
1986	545	550	568	561	520	517	533
1987	556	558	578	567	523	541	556
1988	573	575	596	582	529	567	585
1989	593	594	616	616	542	589	611
1990	624	626	656	637	570	612	634
1991	648	651	694	654	584	632	649
AVERAGE WEEKLY EARNINGS INCLUDING LUMP-SUM PAYMENTS							
1984	$516	$519	$540	$523	$486	$501	$514
1985	532	535	556	542	506	521	535
1986	548	553	581	561	520	523	541
1987	563	567	603	567	523	544	561
1988	583	584	615	582	529	577	599
1989	605	605	638	616	542	601	629
1990	637	639	684	637	570	624	653
1991	657	660	712	654	584	640	661

Source: Bureau of Labor Statistics, "Employment and Earnings" (Monthly) and Aerospace Industries Association estimates.
TOTAL columns are employment-based weighted averages.
Includes overtime premiums.
Many aerospace manufacturers have included lump-sum payments in labor settlements since late 1983 in lieu of general wage increases and/or cost of living adjustments. These payments are reported by BLS in separate wage series for SICs 3721 & 3761 and are included by AIA in totals.

AVERAGE HOURS IN THE AEROSPACE INDUSTRY
Production Workers Only
Calendar Years 1977 - 1991

Year	Aircraft (SIC 372) TOTAL	TOTAL	Airframes (SIC 3721)	Engines & Parts (SIC 3724)	Other Parts & Equipment (SIC 3728)	Guided Missiles, Space Vehicles & Parts (SIC 376)	Complete Guided Missiles, & Space Vehicles (SIC 3761)
AVERAGE WEEKLY HOURS							
1977	41.8	41.9	41.9	41.4	42.5	40.9	40.5
1978	42.2	42.2	42.1	41.7	43.1	41.9	40.9
1979	42.5	42.5	42.3	42.3	43.1	42.0	41.5
1980	41.9	42.0	41.8	41.8	42.6	41.0	41.1
1981	41.3	41.3	41.3	40.5	42.4	40.8	40.6
1982	41.1	41.1	40.9	40.7	41.9	40.8	41.1
1983	41.2	41.2	40.8	41.0	42.2	41.4	41.7
1984	41.9	41.9	41.2	42.2	42.7	42.0	42.3
1985	42.3	42.3	41.5	42.2	43.4	42.4	42.6
1986	42.7	42.8	42.1	42.9	43.7	42.4	42.7
1987	42.4	42.4	42.1	42.5	42.8	42.5	42.5
1988	42.5	42.4	42.0	42.2	43.1	43.2	43.2
1989	42.1	41.9	41.4	42.7	42.3	43.0	43.0
1990	42.3	42.3	41.9	42.9	42.6	42.5	42.8
1991	41.8	41.7	41.5	42.5	41.5	42.4	42.7
AVERAGE WEEKLY OVERTIME HOURS							
1977	3.5	3.5	2.8	3.9	4.5	3.2	2.8
1978	4.4	4.4	3.6	5.0	5.3	4.1	3.4
1979	4.7	4.7	4.1	5.1	5.3	4.4	3.8
1980	4.1	4.2	3.5	5.0	5.0	3.6	3.2
1981	3.5	3.5	3.1	3.5	4.4	3.2	2.9
1982	3.2	3.2	2.7	3.6	3.7	3.1	3.1
1983	3.1	3.1	2.5	3.7	3.7	3.3	3.5
1984	3.9	4.0	3.0	5.1	4.6	3.3	3.4
1985	4.6	4.6	3.5	5.4	5.3	4.6	5.0
1986	4.8	4.9	4.2	5.5	5.5	4.4	4.7
1987	4.8	4.9	4.4	5.0	5.4	4.2	4.3
1988	4.6	4.6	4.3	4.6	5.1	4.5	4.6
1989	5.0	5.1	5.0	5.4	5.0	4.4	4.5
1990	4.5	4.6	4.3	5.3	4.5	3.8	4.1
1991	4.0	4.0	4.1	4.5	3.5	3.9	4.5

Source: Bureau of Labor Statistics, "Employment and Earnings" (Monthly) and Aerospace Industries Association estimates.
TOTAL columns are employment-based weighted averages.

OCCUPATIONAL INJURY AND ILLNESS INCIDENCE RATES
ALL MANUFACTURING AND AEROSPACE INDUSTRIES
Calendar Years 1986 - 1990

	1986	1987	1988	1989	1990
All Manufacturing:					
Total Cases	10.6	11.9	13.0	13.1	13.2
Lost Workday Cases	4.7	5.3	5.7	5.8	5.8
Nonfatal Cases without Lost Workdays	5.9	6.7	7.3	7.3	7.3
Lost Workdays	85.2	95.5	107.3	113.0	120.7
Aircraft and Parts (SIC 372):					
Total Cases	7.0	8.3	9.9	10.1	10.4
Lost Workday Cases	2.6	3.1	3.6	3.7	4.0
Nonfatal Cases without Lost Workdays	4.4	5.2	6.3	6.4	6.4
Lost Workdays	43.8	55.7	67.9	70.2	90.3
Aircraft (SIC 3721):					
Total Cases	6.6	7.4	10.1	10.2	10.0
Lost Workday Cases	2.1	2.6	3.3	3.5	3.9
Nonfatal Cases without Lost Workdays	4.5	4.8	6.7	6.7	6.1
Lost Workdays	38.3	48.0	66.1	70.5	95.3
Aircraft Engines and Parts (SIC 3724):					
Total Cases	5.4	7.1	8.7	7.9	9.3
Lost Workday Cases	2.8	3.4	3.7	3.7	4.2
Nonfatal Cases without Lost Workdays	2.6	3.7	5.0	4.2	5.1
Lost Workdays	48.0	67.4	81.9	72.5	89.5
Aircraft Parts (SIC 3728):					
Total Cases	9.0	10.8	10.5	12.0	11.9
Lost Workday Cases	3.3	3.9	3.9	4.1	3.9
Nonfatal Cases without Lost Workdays	5.7	6.9	6.6	7.8	8.0
Lost Workdays	50.1	60.4	59.1	67.7	80.5
Guided Missiles, Space Vehicles & Parts (SIC 376):					
Total Cases	3.1	4.4	4.6	4.8	4.0
Lost Workday Cases	1.5	2.0	2.2	2.2	1.9
Nonfatal Cases without Lost Workdays	1.6	2.4	2.4	2.6	2.1
Lost Workdays	28.3	34.0	41.3	39.7	39.5
Guided Missiles & Space Vehicles (SIC 3761):					
Total Cases	2.8	4.3	4.6	4.6	4.0
Lost Workday Cases	1.4	2.2	2.3	2.2	1.9
Nonfatal Cases without Lost Workdays	1.4	2.2	2.3	2.5	2.1
Lost Workdays	29.5	37.4	44.6	41.4	37.3
Space Propulsion Units & Parts (SIC 3764):					
Total Cases	4.8	4.5	4.5	4.6	4.4
Lost Workday Cases	1.7	1.8	1.9	2.1	2.2
Nonfatal Cases without Lost Workdays	3.1	2.7	2.6	2.5	2.2
Lost Workdays	29.2	34.3	32.6	33.5	48.7
Other Space Vehicle Equipment (SIC 3769):					
Total Cases	3.1	4.2	NA	5.6	3.8
Lost Workday Cases	1.3	1.2	NA	2.3	1.6
Nonfatal Cases without Lost Workdays	1.8	3.0	NA	3.3	2.3
Lost Workdays	21.0	16.3	NA	41.5	38.4

Source: Bureau of Labor Statistics, "Occupational Injuries and Illnesses in the United States by Industry" (Annually).
Defined as the number of injuries and illnesses per 100 full-time workers.
Separate incidence rates also available for occupational injuries only.
NA Not available.

AEROSPACE INDUSTRY WORK STOPPAGES
Calendar Years 1979 - 1991

Year	Number of Strikes	Number of Workers Involved	Work-Days Idle in Year
1979	12	6,600	103,400
1980	17	4,400	92,900
1981	12	6,100	188,900
1982	4	11,900	45,200
1983	2	8,700	404,100
1984	4	14,600	188,200
1985	4	19,700	289,800
1986	-	-	-
1987	-	-	-
1988	3	10,600	415,800
1989	2	58,500	1,848,000
1990	1	2,300	56,700
1991	1	1,500	-

Source: Bureau of Labor Statistics, "Compensation and Working Conditions" (Monthly).
Based on SIC 372 of the 1967 Code, which includes missile and space propulsion units and parts and missile and space vehicle equipment not elsewhere classified, but which excludes complete guided missiles and space vehicles.
Effective 1982, data not available for work stoppages involving fewer than 1,000 employees.

Section XIII.

Directory
of
Selected
Aviation and Aerospace
Committees in the
103rd Congress

House addresses are Washington, D.C. 20515

Senate addresses are Washington, D.C. 20510

Note: Where available, one staff member is listed for each majority and minority staff.

U.S. HOUSE OF REPRESENTATIVES

House Appropriations Committee

Democrats	Republicans
William Natcher (Ky.), Chair	Joseph McDade (Pa.)
Jamie Whitten (Miss.)	John Myers (Ind.)
Neal Smith (Iowa)	C.W. Bill Young (Fla.)
Sidney Yates (Ill.)	Ralph Regula (Ohio)
David Obey (Wis.)	Bob Livingston (La.)
Louis Stokes (Ohio)	Jerry Lewis (Calif.)
Tom Bevill (Ala.)	John Porter (Ill.)
John Murtha (Pa.)	Harold Rogers (Ky.)
Charles Wilson (Texas)	Joe Skeen (N.M.)
Norman Dicks (Wash.)	Frank Wolf (Va.)
Martin Sabo (Minn.)	Tom DeLay (Texas)
Julian Dixon (Calif.)	Jim Kolbe (Ariz.)
Vic Fazio (Calif.)	Dean Gallo (N.J.)
W.G. (Bill) Hefner (N.C.)	Barbara Vucanovich (Nev.)
Steny Hoyer (Md.)	Jim Lightfoot (Iowa)
Bob Carr (Mich.)	Ron Packard (Calif.)
Richard Durbin (Ill.)	Sonny Callahan (Ala.)
Ronald Coleman (Texas)	Helen Bentley (Md.)
Alan Mollohan (W.Va.)	Jim Walsh (N.Y.)
Jim Chapman (Texas)	Charles Taylor (N.C.)
Marcy Kaptur (Ohio)	Dave Hobson (Ohio)
David Skaggs (Colo.)	Ernest Istook (Okla.)
David Price (N.C.)	Henry Bonilla (Texas)
Nancy Pelosi (Calif.)	
Peter Visclosky (Ind.)	
Tom Foglietta (Pa.)	
Esteban Torres (Calif.)	
Buddy Darden (Ga.)	
Nita Lowey (N.Y.)	
Ray Thornton (Ark.)	
Jose Serrano (N.Y.)	
Rosa DeLauro (Conn.)	
Jim Moran (Va.)	
Pete Peterson (Fla.)	
John Olver (Mass.)	
Ed Pastor (Ariz.)	
Carrie Meek (Fla.)	

Staff	Room	Telephone
Frederick Mohrman, clerk and staff director	H-218 Capitol	(202) 225-2771
James Kulikowski, minority staff director	1016 Longworth	(202) 225-3481

Appropriations
Subcommittee on Commerce, Justice, State, the Judiciary and Related Agencies

Democrats	Republicans
Neal Smith (Iowa), Chair	Harold Rogers (Ky.)
Bob Carr (Mich.)	Jim Kolbe (Ariz.)
Alan Mollohan (W.Va.)	Charles Taylor (N.C.)
William Natcher (Ky.)	
Jim Moran (Va.)	
David Skaggs (Colo.)	
David Price (N.C.)	

Staff	Room	Telephone
John Osthaus, staff assistant	H-309 Capitol	(202) 225-3351
Elizabeth Dawson, minority staff assistant	1016 Longworth	(202) 225-3481

Appropriations
Subcommittee on Defense

Democrats	Republicans
John Murtha (Pa.), Chair	Joseph McDade (Pa.)
Norman Dicks (Wash.)	C.W. Bill Young (Fla.)
Charles Wilson (Texas)	Bob Livingston (La.)
W.G. (Bill) Hefner (N.C.)	Jerry Lewis (Calif.)
Martin Sabo (Minn.)	Joe Skeen (N.M.)
Julian Dixon (Calif.)	
William Natcher (Ky.)	
Peter Visclosky (Ind.)	
Buddy Darden (Ga.)	

Staff	Room	Telephone
Richbourg, Donald, staff assistant	H-144 Capitol	(202) 225-2847
Kevin Roper, minority staff assistant	1016 Longworth	(202) 225-3481

Appropriations
Subcommittee on Foreign Operations, Export Financing and Related Programs

Democrats	Republicans
David Obey (Wis.), Chair	Bob Livingston (La.)
Sidney Yates (Ill.)	John Porter (Ill.)
Charles Wilson (Texas)	Jim Lightfoot (Iowa)
William Natcher (Ky.)	Sonny Callahan (Ala.)
John Olver (Mass.)	
Nancy Pelosi (Calif.)	
Esteban Torres (Calif.)	
Nita Lowey (N.Y.)	
Jose Serrano (N.Y.)	

Staff	Room	Telephone
Terry Peel, staff assistant	H-307 Capitol	(202) 225-2041
James Kulikowski, minority staff assistant	1016 Longworth	(202) 225-3481

Appropriations
Subcommittee on Military Construction

Democrats	**Republicans**
W.G. (Bill) Hefner (N.C.), Chair	Barbara Vucanovich (Nev.)
Tom Foglietta (Pa.)	Sonny Callahan (Ala.)
Carrie Meek (Fla.)	Helen Bentley (Md.)
Norman Dicks (Wash.)	Dave Hobson (Ohio)
Julian Dixon (Calif.)	
Vic Fazio (Calif.)	
Steny Hoyer (Md.)	
William Natcher (Ky.)	
Ronald Coleman (Texas)	

Staff	**Room**	**Telephone**
William Marinelli, staff assistant	B-300 Rayburn	(202) 225-3047
Elizabeth Dawson, minority staff assistant	1016 Longworth	(202) 225-3481

Appropriations
Subcommittee on Transportation

Democrats	**Republicans**
Bob Carr (Mich.), Chair	Frank Wolf (Va.)
Richard Durbin (Ill.)	Tom DeLay (Texas)
Martin Sabo (Minn.)	Ralph Regula (Ohio)
David Price (N.C.)	
Ronald Coleman (Texas)	
William Natcher (Ky.)	
Tom Foglietta (Pa.)	

Staff	**Room**	**Telephone**
Richard Efford, staff assistant	2358 Rayburn	(202) 225-2141
John Blazey, minority staff assistant	1016 Longworth	(202) 225-3481

Appropriations
Subcommittee on VA, HUD and Independent Agencies (including NASA)

Democrats	**Republicans**
Louis Stokes (Ohio), Chair	Jerry Lewis (Calif.)
Alan Mollohan (W.Va.)	Tom DeLay (Texas)
Jim Chapman (Texas)	Dean Gallo (N.J.)
Marcy Kaptur (Ohio)	
William Natcher (Ky.)	
Esteban Torres (Calif.)	
Ray Thornton (Ark.)	

Staff	**Room**	**Telephone**
Michelle Burkett, staff assistant	H-143 Capitol	(202) 225-3241
William Warfield, minority staff assistant	1016 Longworth	(202) 225-3481

House Armed Services Committee

Democrats	**Republicans**
Ronald Dellums (Calif.), Chair	Floyd Spence (S.C.)
G.V. Montgomery (Miss.)	Bob Stump (Ariz.)
Patricia Schroeder (Colo.)	Duncan Hunter (Calif.)

Earl Hutto (Fla.)
Ike Skelton (Mo.)
Dave McCurdy (Okla.)
Marilyn Lloyd (Tenn.)
Norman Sisisky (Va.)
John Spratt (S.C.)
Frank McCloskey (Ind.)
Solomon Ortiz (Texas)
George Hochbrueckner (N.Y.)
Owen Pickett (Va.)
Martin Lancaster (N.C.)
Lane Evans (Ill.)
James Bilbray (Nev.)
John Tanner (Tenn.)
Glen Browder (Ala.)
Gene Taylor (Miss.)
Neil Abercrombie (Hawaii)
Thomas Andrews (Maine)
Chet Edwards (Texas)
Don Johnson (Ga.)
Frank Tejeda (Texas)
David Mann (Ohio)
Bart Stupak (Mich.)
Martin Meehan (Mass.)
Robert Underwood (Guam)
Jane Harman (Calif.)
Paul McHale (Pa.)
Tim Holden (Pa.)
Pete Geren (Texas)
Elizabeth Furse (Ore.)

John Kasich (Ohio)
Herbert Bateman (Va.)
James Hansen (Utah)
Curt Weldon (Pa.)
Jon Kyl (Ariz.)
Arthur Ravenel (S.C.)
Robert Dornan (Calif.)
Joel Hefley (Colo.)
Ronald Machtley (R.I.)
Jim Saxton (N.J.)
Randy Cunningham (Calif.)
James Inhofe (Okla.)
Stephen Buyer (Ind.)
Peter Torkildsen (Mass.)
Tillie Fowler (Fla.)
John McHugh (N.Y.)
James Talent (Mo.)
Terry Everett (Ala.)
Roscoe Bartlett (Md.)

Staff

	Room	Telephone
Marilyn Elrod, staff director	2120 Rayburn	(202) 225-4158

Armed Services
Subcommittee on Military Acquisition

Democrats
Ronald Dellums (Calif.), Chair
Marilyn Lloyd (Tenn.)
John Spratt (S.C.)
Frank McCloskey (Ind.)
Lane Evans (Ill.)
John Tanner (Tenn.)
Gene Taylor (Miss.)
Neil Abercrombie (Hawaii)
Thomas Andrews (Maine)
David Mann (Ohio)
Bart Stupak (Mich.)
Paul McHale (Pa.)
Tim Holden (Pa.)
Pete Geren (Texas)

Republicans
Floyd Spence (S.C.)
Herbert Bateman (Va.)
Curt Weldon (Pa.)
Arthur Ravenel (S.C.)
Robert Dornan (Calif.)
Joel Hefley (Colo.)
Ronald Machtley (R.I.)
Jim Saxton (N.J.)
Randy Cunningham (Calif.)
James Inhofe (Okla.)

Norman Sisisky (Va.)

Armed Services
Subcommittee on Military Forces and Personnel

Democrats	**Republicans**
Ike Skelton (Mo.), Chair	Jon Kyl (Ariz.)
G.V. Montgomery (Miss.)	Arthur Ravenel (S.C.)
Owen Pickett (Va.)	Stephen Buyer (Ind.)
Martin Lancaster (N.C.)	Tillie Fowler (Fla.)
James Bilbray (Nev.)	James Talent (Mo.)
Bart Stupak (Mich.)	Roscoe Bartlett (Md.)
Martin Meehan (Mass.)	
Robert Underwood (Guam)	
Jane Harman (Calif.)	

Armed Services
Subcommittee on Military Installations and Facilities

Democrats	**Republicans**
Dave McCurdy (Okla.), Chair	Duncan Hunter (Calif.)
G.V. Montgomery (Miss.)	Tillie Fowler (Fla.)
Frank McCloskey (Ind.)	John McHugh (N.Y.)
Solomon Ortiz (Texas)	Terry Everett (Ala.)
George Hochbrueckner (N.Y.)	Bob Stump (Ariz.)
James Bilbray (Nev.)	Ronald Machtley (R.I.)
Glen Browder (Ala.)	Jim Saxton (N.J.)
Gene Taylor (Miss.)	Peter Torkildsen (Mass.)
Neil Abercrombie (Hawaii)	
Chet Edwards (Texas)	
Don Johnson (Ga.)	
Frank Tejeda (Texas)	
Robert Underwood (Guam)	

Armed Services
Subcommittee on Oversight and Investigations

Democrats	**Republicans**
Norman Sisisky (Va.), Chair	James Hansen (Utah)
John Spratt (S.C.)	Jon Kyl (Ariz.)
John Tanner (Tenn.)	Joel Hefley (Colo.)
Glen Browder (Ala.)	John McHugh (N.Y.)
Chet Edwards (Texas)	Terry Everett (Ala.)
Don Johnson (Ga.)	Robert Dornan (Calif.)
Frank Tejeda (Texas)	
David Mann (Ohio)	
Jane Harman (Calif.)	
Tim Holden (Pa.)	

Armed Services
Subcommittee on Readiness

Democrats	Republicans
Earl Hutto (Fla.), Chair	John Kasich (Ohio)
Solomon Ortiz (Texas)	Herbert Bateman (Va.)
Owen Pickett (Va.)	Curt Weldon (Pa.)
Martin Lancaster (N.C.)	Robert Dornan (Calif.)
Lane Evans (Ill.)	Randy Cunningham (Calif.)
Glen Browder (Ala.)	James Inhofe (Okla.)
Martin Meehan (Mass.)	
Robert Underwood (Guam)	
Paul McHale (Pa.)	
Dave McCurdy (Okla.)	

Armed Services
Subcommittee on Research and Technology

Democrats	Republicans
Patricia Schroeder (Colo.), Chair	Bob Stump (Ariz.)
George Hochbrueckner (N.Y.)	Stephen Buyer (Ind.)
Owen Pickett (Va.)	Peter Torkildsen (Mass.)
Martin Lancaster (N.C.)	James Talent (Mo.)
James Bilbray (Nev.)	Roscoe Bartlett (Md.)
Chet Edwards (Texas)	Duncan Hunter (Calif.)
Don Johnson (Ga.)	John Kasich (Ohio)
Frank Tejeda (Texas)	James Hansen (Utah)
Martin Meehan (Mass.)	
Jane Harman (Calif.)	
Elizabeth Furse (Ore.)	
Earl Hutto (Fla.)	
Dave McCurdy (Okla.)	

House Government Operations Committee
Subcommittee on Employment, Housing and Aviation

Democrats	Republicans
Colin Peterson (Minn.), Chair	Ronald Machtley (R.I.)
Tom Lantos (Calif.)	Christopher Shays (Conn.)
Bobby Rush (Ill.)	John McHugh (N.Y.)
Floyd Flake (N.Y.)	
Karen Thurman (Fla.)	
Barbara-Rose Collins (Mich.)	

Staff	Room	Telephone
Andrea Nelson, staff director	B-349A Rayburn	(202) 225-6751
Michael Nannini, minority professional staff	2158 Rayburn	(202) 225-2738

House Public Works and Transportation Committee

Democrats	Republicans
Norman Mineta (Calif.), Chair	Bud Shuster (Pa.)
James Oberstar (Minn.)	William Clinger (Pa.)

Nick Joe Rahall (W.Va.)
Douglas Applegate (Ohio)
Ron de Lugo (Virgin Islands)
Robert Borski (Pa.)
Tim Valentine (N.C.)
William Lipinski (Ill.)
Robert Wise (W.Va.)
James Traficant (Ohio)
Peter DeFazio (Ore.)
James Hayes (La.)
Bob Clement (Tenn.)
Jerry Costello (Ill.)
Mike Parker (Miss.)
Greg Laughlin (Texas)
Pete Geren (Texas)
George Sangmeister (Ill.)
Glenn Poshard (Ill.)
Dick Swett (N.H.)
Bud Cramer (Ala.)
Barbara-Rose Collins (Mich.)
Eleanor Holmes Norton (D.C.)
Lucien Blackwell (Pa.)
Jerrold Nadler (N.Y.)
Sam Coppersmith (Ariz.)
Leslie Byrne (Va.)
Maria Cantwell (Wash.)
Pat Danner (Mo.)
Karen Shepherd (Utah)
Robert Menendez (N.J.)
Jim Clyburn (S.C.)
Corrine Brown (Fla.)
Nathan Deal (Ga.)
James Barcia (Mich.)
Dan Hamburg (Calif.)
Bob Filner (Calif.)
Walter Tucker (Calif.)
Eddie-Bernice Johnson (Texas)

Thomas Petri (Wis.)
Sherwood Boehlert (N.Y.)
James Inhofe (Okla.)
Bill Emerson (Mo.)
John Duncan (Tenn.)
Susan Molinari (N.Y.)
William Zeliff (N.H.)
Thomas Ewing (Ill.)
Wayne Gilchrest (Md.)
Jennifer Dunn (Wash.)
Tim Hutchinson (Ark.)
Bill Baker (Calif.)
Mac Collins (Ga.)
Jay Kim (Calif.)
David Levy (N.Y.)
Steve Horn (Calif.)
Bob Franks (N.J.)
Peter Blute (Mass.)
Buck McKeon (Calif.)
John Mica (Fla.)
Peter Hoekstra (Mich.)
Jack Quinn (N.Y.)

Staff	Room	Telephone
Paul Schoellhamer, chief of staff	2165 Rayburn	(202) 225-4472
Jack Schenendorf, minority chief counsel and staff director	2163 Rayburn	(202) 225-9446

House Public Works and Transportation Committee
Subcommittee on Aviation

Democrats
James Oberstar (Minn.), Chair
Del. Ron de Lugo (V.I.), Vice Chair
William Lipinski (Ill.)
Pete Geren (Texas)
George Sangmeister (Ill.)

Republicans
William Clinger (Pa.)
Sherwood Boehlert (N.Y.)
James Inhofe (Okla.)
John Duncan (Tenn.)
Thomas Ewing (Ill.)

Barbara-Rose Collins (Mich.)
Sam Coppersmith (Ariz.)
Robert Borski (Pa.)
Tim Valentine (N.C.
Peter DeFazio (Ore.)
Jimmy Hayes (La.)
Bob Clement (Tenn.)
Jerry Costello (Ill.)
Mike Parker (Miss.)
Greg Laughlin (Texas)
Dick Swett (N.H.)
Bud Cramer (Ala.)
Lucien Blackwell (Pa.)
Maria Cantwell (Wash.)
Pat (Patsy Ann) Danner (Mo.)
Karen Shepherd (Utah)
Corrine Brown (Fla.)
Norman Mineta (Calif.)
(ex officio)

Wayne Gilchrest (Md.)
Jennifer Dunn (Wash.)
Michael "Mac" Collins (Ga.)
)Jay Kim (Calif.)
David Levy (N.Y.)
Stephen Horn (Calif.)
Howard "Buck" McKeon (Calif.)
John Mica (Fla.)
Bud Shuster (Pa.)
(ex officio)

Staff

	Room	Telephone
David Heymsfeld, counsel	2251 Rayburn	(202) 225-9161
David Schaffer, minority counsel	2251 Rayburn	(202) 226-3220

Public Works
Subcommittee on Investigations and Oversight

Democrats
Robert Borski (Pa.), Chair
Barbara-Rose Collins (Mich.), Vice chair
Robert Wise (W.Va.)
Greg Laughlin (Texas)
Lucien Blackwell (Pa.)
Leslie Byrne (Va.)
James Barcia (Mich.)
Bob Filner (Calif.)
Eddie-Bernice Johnson (Texas)

Republicans
James Inhofe (Okla.)
John Duncan (Tenn.)
Susan Molinari (N.Y.)
William Zeliff (N.H.)
Wayne Gilchrest (Md.)
Bill Baker (Calif.)

Staff

	Room	Telephone
John Wells, staff director	586 Ford	(202) 225-3274
Charles Ziegler, minority counsel	585 Ford	(202) 225-5504

House Science, Space and Technology Committee

Democrats
George Brown (Calif.), Chair
Marilyn Lloyd (Tenn.)
Dan Glickman (Kan.)
Harold Volkmer (Mo.)
Ralph Hall (Texas)
Dave McCurdy (Okla.)
Tim Valentine (N.C.)
Robert Torricelli (N.J.)

Republicans
Robert Walker (Pa.)
James Sensenbrenner (Wis.)
Sherwood Boehlert (N.Y.)
Tom Lewis (Fla.)
Paul Henry (Mich.)
Harris Fawell (Ill.)
Constance Morella (Md.)
Dana Rohrabacher (Calif.)

Rick Boucher (Va.)
James Traficant (Ohio)
Jimmy Hayes (La.)
John Tanner (Tenn.)
Glen Browder (Ala.)
Pete Geren (Texas)
Jim Bacchus (Fla.)
Tim Roemer (Ind.)
Bud Cramer (Ala.)
Dick Swett (N.H.)
James Barcia (Mich.)
Herbert Klein (N.J.)
Eric Fingerhut (Ohio)
Paul McHale (Pa.)
Jane Harman (Calif.)
Don Johnson (Ga.)
Sam Coppersmith (Ariz.)
Anna Eshoo (Calif.)
Jay Inslee (Wash.)
Eddie-Bernice Johnson (Texas)
David Minge (Minn.)
Lynn Woolsey (Calif.)
Nathan Deal (Ga.)
Robert Scott (Va.)
Xavier Becerra (Calif.)

Steven Schiff (N.M.)
Joe Barton (Texas)
Dick Zimmer (N.J.)
Sam Johnson (Texas)
Ken Calvert (Calif.)
Martin Hoke (Ohio)
Nick Smith (Mich.)
Ed Royce (Calif.)
Rod Grams (Minn.)
John Linder (Ga.)
Peter Blute (Mass.)
Jennifer Dunn (Wash.)
Bill Baker (Calif.)
Roscoe Bartlett (Md.)

Staff	**Room**	**Telephone**
Radford Byerly, chief of staff	2320 Rayburn	(202) 225-6375
David Clement, minority chief of staff	2320 Rayburn	(202) 225-8772

**Science, Space and Technology
Subcommittee On Energy**

Democrats
Marilyn Lloyd (Tenn.), Chair
Robert Scott (Va.)
Bud Cramer (Ala.)
Dick Swett (N.H.)
Herbert Klein (N.J.)
Paul McHale (Pa.)
Sam Coppersmith (Ariz.)
Jay Inslee (Wash.)
Tim Roemer (Ind.)
Dave McCurdy (Okla.)

Republicans
Harris Fawell (Ill.)
Steven Schiff (N.M.)
Bill Baker (Calif.)
Rod Grams (Minn.)
Roscoe Bartlett (Md.)

Staff	**Room**	**Telephone**
Francis Murray, staff director	390 Ford	(202) 225-2884
Mason Wiggins, Republican counsel	388 Ford	(202) 225-6928

Science, Space and Technology
Subcommittee on Investigations and Oversight

Democrats	Republicans
Jimmy Hayes (La.), Chair	Paul Henry (Mich.)
John Tanner (Tenn.)	Constance Morella (Md.)
Marilyn Lloyd (Tenn.)	Joe Barton (Texas)
Don Johnson (Ga.)	
Sam Coppersmith (Ariz.)	

Staff	Room	Telephone
Keith Laughlin, staff director	822 O'Neill	(202) 226-4494
Barry Beringer, Republican counsel	2320 Rayburn	(202) 225-8500

Science, Space and Technology
Subcommittee on Science

Democrats	Republicans
Rick Boucher (Va.), Chair	Sherwood Boehlert (N.Y.)
Ralph Hall (Texas)	Joe Barton (Texas)
Tim Valentine (N.C.)	Sam Johnson (Texas)
James Barcia (Mich.)	Nick Smith (Mich.)
Don Johnson (Ga.)	Peter Blute (Mass.)
Anna Eshoo (Calif.)	
Eddie-Bernice Johnson (Texas)	
David Minge (Minn.)	

Staff	Room	Telephone
Grace Ostenso, staff director	2319 Rayburn	(202) 225-1060
Christopher Roosa, Republican legislative asst.	2319 Rayburn	(202) 225-5848

Science, Space and Technology
Subcommittee on Space

Democrats	Republicans
Ralph Hall (Texas), Chair	James Sensenbrenner (Wis.)
Harold Volkmer (Mo.)	Dana Rohrabacher (Calif.)
Robert Torricelli (N.J.)	Dick Zimmer (N.J.)
James Traficant (Ohio)	Sam Johnson (Texas)
Jim Bacchus (Fla.)	Martin Hoke (Ohio)
Bud Cramer (Ala.)	Ed Royce (Calif.)
James Barcia (Mich.)	Jennifer Dunn (Wash.)
Eric Fingerhut (Ohio)	Steven Schiff (N.M.)
Jimmy Hayes (La.)	Ken Calvert (Calif.)
John Tanner (Tenn.)	
Pete Geren (Texas)	
Tim Roemer (Ind.)	
Jane Harman (Calif.)	
Anna Eshoo (Calif.)	

Staff	Room	Telephone
William Smith, staff director	2324 Rayburn	(202) 225-7858
Shana Dale, Republican counsel	B-374 Rayburn	(202) 225-2656

Science, Space and Technology
Subcommittee on Technology, Environment and Aviation

Democrats	Republicans
Tim Valentine (N.C.), Chair	Tom Lewis (Fla.)
Dan Glickman (Kan.)	Constance Morella (Md.)
Pete Geren (Texas)	Ken Calvert (Calif.)
Dick Swett (N.H.)	Nick Smith (Mich.)
Herbert Klein (N.J.)	Rod Grams (N.M.)
Paul McHale (Pa.)	John Linder (Ga.)
Jane Harman (Calif.)	Peter Blute (Mass.)
Don Johnson (Ga.)	Roscoe Bartlett (Md.)
Sam Coppersmith (Ariz.)	Dana Rohrabacher (Calif.)
Anna Eshoo (Calif.)	Dick Zimmer (N.J.)
Jay Inslee (Wash.)	Martin Hoke (Ohio)
Eddie-Bernice Johnson (Texas)	Ed Royce (Calif.)
David Minge (Minn.)	
Lynn Woolsey (Calif.)	
Nathan Deal (Ga.)	
Xavier Becerra (Calif.)	
Robert Torricelli (N.J.)	
Jim Bacchus (Fla.)	

Staff	Room	Telephone
James Turner, staff director	B-374 Rayburn	(202) 225-8128
Paddy Link, Republican legislative asst.	B-374 Rayburn	(202) 226-1371

House Ways and Means Committee

Democrats	Republicans
Dan Rostenkowski (Ill.), Chair	Bill Archer (Texas)
Sam Gibbons (Fla.)	Philip Crane (Ill.)
J.J. Pickle (Texas)	Bill Thomas (Calif.)
Charles Rangel (N.Y.)	E. Clay Shaw (Fla.)
Fortney Stark (Calif.)	Don Sundquist (Tenn.)
Andrew Jacobs (Ind.)	Nancy Johnson (Conn.)
Harold Ford (Tenn.)	Jim Bunning (Ky.)
Robert Matsui (Calif.)	Fred Grandy (Iowa)
Barbara Kennelly (Conn.)	Amo Houghton (N.Y.)
William Coyne (Pa.)	Wally Herger (Calif.)
Michael Andrews (Texas)	Jim McCrery (La.)
Sander Levin (Mich.)	Mel Hancock (Mo.)
Benjamin Cardin (Md.)	Richard Santorum (Pa.)
Jim McDermott (Wash.)	Dave Camp (Mich.)
Gerald Kleczka (Wis.)	
John Lewis (Ga.)	
L.F. Payne (Va.)	
Richard Neal (Mass.)	
Peter Hoagland (Neb.)	
Michael McNulty (N.Y.)	
Michael Kopetski (Ore.)	
William Jefferson (La.)	
Bill Brewster (Okla.)	

Mel Reynolds (Ill.)

Staff	Room	Telephone
Janice Mays, chief counsel and staff director	102 Longworth	(202) 225-3625
Phillip Moseley, minority chief of staff1	106 Longworth	(202) 225-4021

Mel Reynolds (Ill.)

Staff	Room	Telephone
Janice Mays, chief counsel and staff director	102 Longworth	(202) 225-3625
Phillip Moseley, minority chief of staff1	106 Longworth	(202) 225-4021

U.S. SENATE

Senate Appropriations Committee

Democrats	Republicans
Robert Byrd (W.Va.), Chair	Mark Hatfield (Ore.)
Daniel Inouye (Hawaii)	Ted Stevens (Alaska)
Ernest Hollings (S.C.)	Thad Cochran (Miss.)
Bennett Johnston (La.)	Alfonse D'Amato (N.Y.)
Patrick Leahy (Vt.)	Arlen Specter (Pa.)
Jim Sasser (Tenn.)	Pete Domenici (N.M.)
Dennis DeConcini (Ariz.)	Don Nickles (Okla.)
Dale Bumpers (Ark.)	Phil Gramm (Texas)
Frank Lautenberg (N.J.)	Christopher Bond (Mo.)
Tom Harkin (Iowa)	Slade Gorton (Wash.)
Barbara Mikulski (Md.)	Mitch McConnell (Ky.)
Harry Reid (Nev.)	Connie Mack (Fla.)
Bob Kerrey (Neb.)	Conrad Burns (Mont.)
Herb Kohl (Wis.)	
Patty Murray (Wash.)	
Dianne Feinstein (Calif.)	

Staff	**Room**	**Telephone**
James English, staff director	S-128 Capitol	(202) 224-7200
Keith Kennedy, minority staff director	119 Dirksen	(202) 224-7335

Appropriations

Subcommittee on Commerce, Justice, State and Judiciary

Democrats	Republicans
Ernest Hollings (S.C.), Chair	Pete Domenici (N.M.)
Daniel Inouye (Hawaii)	Ted Stevens (Alaska)
Dale Bumpers (Ark.)	Mark Hatfield (Ore.)
Frank Lautenberg (N.J.)	Phil Gramm (Texas)
Jim Sasser (Tenn.)	Mitch McConnell (Ky.)
Bob Kerrey (Neb.)	

Staff	**Room**	**Telephone**
Scott Gudes, clerk	S-146 Capitol	(202) 224-7277
John Shank, minority clerk	152 Dirksen	(202) 224-7219

Appropriations

Subcommittee on Defense

Democrats	Republicans
Daniel Inouye (Hawaii), Chair	Ted Stevens (Alaska)
Ernest Hollings (S.C.)	Alfonse D'Amato (N.Y.)
Bennett Johnston (La.)	Thad Cochran (Miss.)
Robert Byrd (W.Va.)	Arlen Specter (Pa.)
Patrick Leahy (Vt.)	Pete Domenici (N.M.)
Jim Sasser (Tenn.)	Don Nickles (Okla.)
Dennis DeConcini (Ariz.)	Phil Gramm (Texas)
Dale Bumpers (Ark.)	Christopher Bond (Mo.)

Frank Lautenberg (N.J.)
Tom Harkin (Iowa)

Staff	**Room**	**Telephone**
Richard Collins, clerk	119 Dirksen	(202) 224-7205
Steven Cortese, minority clerk	117 Dirksen	(202) 224-2739

Appropriations
Subcommittee on Foreign Operations

Democrats	**Republicans**
Patrick Leahy (Vt.), Chair	Mitch McConnell (Ky.)
Daniel Inouye (Hawaii)	Alfonse D'Amato (N.Y.)
Dennis DeConcini (Ariz.)	Arlen Specter (Pa.)
Frank Lautenberg (N.J.)	Don Nickles (Okla.)
Tom Harkin (Iowa)	Connie Mack (Fla.)
Barbara Mikulski (Md.)	Phil Gramm (Texas)
Dianne Feinstein (Calif.)	

Staff	**Room**	**Telephone**
Eric Newson, clerk	137 Dirksen	(202) 224-7284
James Bond, minority clerk	135 Dirksen	(202) 224-7274

Appropriations
Subcommittee on Military Construction

Democrats	**Republicans**
Jim Sasser (Tenn.), Chair	Slade Gorton (Wash.)
Daniel Inouye (Hawaii)	Ted Stevens (Alaska)
Harry Reid (Nev.)	Mitch McConnell (Ky.)
Herb Kohl (Wis.)	

Staff	**Room**	**Telephone**
Michael Walker, clerk	131 Dirksen	(202) 224-7276
James Morhard, minority clerk	157 Dirksen	(202) 224-7271

Appropriations
Subcommittee on Transportation

Democrats	**Republicans**
Frank Lautenberg (N.J.), Chair	Alfonse D'Amato (N.Y.)
Robert Byrd (W.Va.)	Pete Domenici (N.M.)
Tom Harkin (Iowa)	Mark Hatfield (Ore.)
Jim Sasser (Tenn.)	Arlen Specter (Pa.)
Barbara Mikulski (Md.)	

Staff	**Room**	**Telephone**
Patrick McCann, clerk	156 Dirksen	(202) 224-0330
Anne Miano, minority clerk	153 Dirksen	(202) 224-7213

Appropriations
Subcommittee on VA, HUD and Independent Agencies (including NASA)

Democrats	**Republicans**
Barbara Mikulski (Md.), Chair	Phil Gramm (Texas)
Patrick Leahy (Vt.)	Alfonse D'Amato (N.Y.)

Bennett Johnston (La.)	Don Nickles (Okla.)
Frank Lautenberg (N.J.)	Christopher Bond (Mo.)
Bob Kerrey (Neb.)	Conrad Burns (Mont.)
Dianne Feinstein (Calif.)	

Staff	**Room**	**Telephone**
Kevin Kelly, clerk	142 Dirksen	(202) 224-7231
Stephen Kohashi, minority clerk	157 Dirksen	(202) 224-7253

Senate Armed Services Committee

Democrats	**Republicans**
Sam Nunn (Ga.), Chair	Strom Thurmond (S.C.)
James Exon (Neb.)	John Warner (Va.)
Carl Levin (Mich.)	William Cohen (Maine)
Edward Kennedy (Mass.)	John McCain (Ariz.)
Jeff Bingaman (N.M.)	Trent Lott (Miss.)
John Glenn (Ohio)	Dan Coats (Ind.)
Richard Shelby (Ala.)	Robert Smith (N.H.)
Robert Byrd (W.Va.)	Dirk Kempthorne (Idaho)
Bob Graham (Fla.)	Lauch Faircloth (N.C.)
Charles Robb (Va.)	
Joseph Lieberman (Conn.)	

Staff	**Room**	**Telephone**
Arnold Punaro, staff director	228 Russell	(202) 224-3871
Anthony Principi, minority staff director	228 Russell	(202) 224-3871

Armed Services
Subcommittee on Coalition Defense and Reinforcing Forces

Democrats	**Republicans**
Carl Levin (Mich.), Chair	John Warner (Va.)
James Exon (Neb.)	William Cohen (Maine)
John Glenn (Ohio)	Dan Coats (Ind.)
Richard Shelby (Ala.)	Robert Smith (N.H.)
Robert Byrd (W.Va.)	Dirk Kempthorne (Idaho)
Bob Graham (Fla.)	

Staff	**Room**	**Telephone**
John Hamre, professional staff member	228 Russell	(202) 224-9338

Armed Services
Subcommittee on Defense Technology, Acquisition and Industrial Base

Democrats	**Republicans**
Jeff Bingaman (N.M.), Chair	Robert Smith (N.H.)
Carl Levin (Mich.)	William Cohen (Maine)
Edward Kennedy (Mass.)	Trent Lott (Miss.)
Robert Byrd (W.Va.)	Dan Coats (Ind.)
Bob Graham (Fla.)	Dirk Kempthorne (Idaho)
Charles Robb (Va.)	Lauch Faircloth (N.C.)
Joseph Lieberman (Conn.)	

Staff	**Room**	**Telephone**
John Douglass, professional staff member	228 Russell	(202) 224-8638

Armed Services
Subcommittee on Force Requirements and Personnel

Democrats	Republicans
Richard Shelby (Ala.), Chair	Dan Coats (Ind.)
Edward Kennedy (Mass.)	John McCain (Ariz.)
Robert Byrd (W.Va.)	Lauch Faircloth (N.C.)
Joseph Lieberman (Conn.)	

Staff	Room	Telephone
Frederick Pang, professional staff member	228 Russell	(202) 224-3128

Armed Services
Subcommittee on Military Readiness and Defense Infrastructure

Democrats	Republicans
John Glenn (Ohio), Chair	John McCain (Ariz.)
Jeff Bingaman (N.M.)	Robert Smith (N.H.)
Richard Shelby (Ala.)	Lauch Faircloth (N.C.)
Charles Robb (Va.)	

Staff	Room	Telephone
David Lyles, professional staff member	228 Russell	(202) 224-9344

Armed Services
Subcommittee on Nuclear Deterrence, Arms Control and Defense

Democrats	Republicans
James Exon (Neb.), Chair	Trent Lott (Miss.)
Carl Levin (Mich.)	John Warner (Va.)
Jeff Bingaman (N.M.)	Dirk Kempthorne (Idaho)
John Glenn (Ohio)	

Staff	Room	Telephone
William Hoehn, professional staff member	228 Russell	(202) 224-9343

Armed Services
Subcommittee on Regional Defense and Contingency Forces

Democrats	Republicans
Edward Kennedy (Mass.), Chair	William Cohen (Maine)
James Exon (Neb.)	John Warner (Va.)
Bob Graham (Fla.)	John McCain (Ariz.)
Charles Robb (Va.)	Trent Lott (Miss.)
Joseph Lieberman (Conn.)	

Staff	Room	Telephone
Creighton Greene, professional staff member	228 Russell	(202) 224-6115

Senate Commerce Committee

Democrats	Republicans
Ernest Hollings (S.C.), Chair	John Danforth (Mo.)
Daniel Inouye (Hawaii)	Bob Packwood (Ore.)
Wendell Ford (Ky.)	Larry Pressler (S.D.)
James Exon (Neb.)	Ted Stevens (Alaska)
Jay Rockefeller (W.Va.)	John McCain (Ariz.)

John Kerry (Mass.)
John Breaux (La.)
Richard Bryan (Nev.)
Charles Robb (Va.)
Byron Dorgan (N.D.)

Staff	**Room**	**Telephone**
Kevin Curtin, chief counsel and staff director	254 Russell	(202) 224-0427
Jonathan Chambers, minority staff director	554 Dirksen	(202) 224-5183

Commerce
Subcommittee on Aviation

Democrats	**Republicans**
Wendell Ford (Ky.), Chair	Larry Pressler (S.D.)
James Exon (Neb.)	John McCain (Ariz.)
Daniel Inouye (Hawaii)	Ted Stevens (Alaska)
Robert Kerrey (Neb.)	Slade Gorton (Wash.)
Richard Bryan (Nev.)	

Staff	**Room**	**Telephone**
Samuel Whitehorn, senior counsel	428 Hart	(202) 224-9350
Alan Maness, minority senior counsel	516 Dirksen	(202) 224-4852

Commerce
Subcommittee on Science, Technology and Space

Democrats	**Republicans**
Jay Rockefeller (W.Va.), Chair	Conrad Burns (Mont.)
Ernest Hollings (S.C.)	Larry Pressler (S.D.)
John Kerry (Mass.)	Trent Lott (Miss.)
Richard Bryan (Nev.)	Judd Gregg (N.H.)
Charles Robb (Va.)	

Staff	**Room**	**Telophone**
Steven Palmer, senior staff	427 Hart	(202) 224-9360
Louis Whitsett, minority counsel	566 Dirksen	(202) 224-8172

Senate Finance Committee

Democrats	**Republicans**
Daniel Patrick Moynihan (N.Y.), Chair	Bob Packwood (Ore.)
Max Baucus (Mont.)	Robert Dole (Kan.)
David Boren (Okla.)	William Roth (Del.)
Bill Bradley (N.J.)	John Danforth (Mo.)
George Mitchell (Maine)	John Chafee (R.I.)
David Pryor (Ark.)	Dave Durenberger (Minn.)
Donald Riegle (Mich.)	Charles Grassley (Iowa)
John Rockefeller (W.Va.)	Orrin Hatch (Utah)
Thomas Daschle (S.D.)	Malcolm Wallop (Wyo.)
John Breaux (La.)	
Kent Conrad (N.D.)	

Staff	**Room**	**Telephone**
Lawrence O'Donnell, staff director	205 Dirksen	(202) 224-4515
Edmund Mihalski, minority chief of staff	205 Dirksen	(202) 224-5315

Section XIV.

Special Interest Groups

SPECIAL INTEREST GROUPS

September 1992

Acronym	Organization	Key Contact	Phone/Fax
AIAA	Aerospace Industries Association of America	Don Fuqua President	202-371-8400 202-371-8470
	Suite 1100 1250 I Street, N.W. Washington, DC 20005		
ALPA	Air Line Pilots Association	Randolph Babbitt President	703-689-2270
	535 Herndon Parkway P.O. Box 1169 Herndon, VA 22070		
ATCA	Air Traffic Control Association	Gabe Hartl President	703-522-5717
	Suite 410 2020 N. 14th Street Arlington, VA 22201		
ATA	Air Transport Association of America	James Landry President	202-626-4000
	1301 Pennsylvania Ave, N.W. Suite 1100 Washington, DC 20004-1707		
AEA	Aircraft Electronics Association	Monte Mitchell Executive Director	816-373-6565
	P.O. Box 1981 Independence, MO 64055-1981		
AOPA	Aircraft Owners & Pilots Association	Phil Boyer President	301-695-2000 301-695-2375
	421 Aviation Way Frederick, MD 21701-4756		

Acronym	Organization	Key Contact	Phone/Fax
AOCI	Airport Operators Council International	George Howard President	202-293-8500

Suite 800
1220 19th Street, N.W.
Washington, DC 20036

| **APA** | Allied Pilots Association | Frederick Vogel President | 214-988-3188 |

P.O. Box 5524
Arlington, TX 76005-5524

| **AAAE** | American Assoc. of Airport Executives | Charles Barclay President | 703-824-0500 |

4212 King Street
Alexandria, VA 22302

| **AHS** | American Helicopter Society | Rhett Flater Executive Director | 703-684-6777 |

217 N. Washington Street
703-739-9279
Alexandria, VA 22314

| **AIA** | American Institute of Aeronautics and Astronautics, Inc. 370 L'Enfant Promenade, S.W. Washington, DC 20024 | Cort Durocher Executive Director | 202-646-7400 202-646-7508 |

| **AFA** | Association of Flight Attendants | Dee Maki President | 202-328-5400 |

1625 Massachusetts Avenue, N.W.
Washington, DC 20036

| **APFA** | Association of Professional Flight Attendants | Michael Kalliher President | 817-540-0108 |

1004 West Euless Blvd.
Euless, TX 76040

| **ADMA** | Aviation Distributors and Manufacturers Association | Patricia Lilly Exec Vice President | 215-564-3484 |

1900 Arch Street
Philadelphia, PA 19103

Acronym	Organization	Key Contact	Phone/Fax
DOD/ FAA SAFRL, Room 4D865 Pentagon Washington, DC 20330-1000	Department of Defense Advisory Committee on Federal Aviation	Frank Colson Executive Director	202-697-6937
EAA Whitman Field Oshkosh, WI 54903-3086	Experimental Aircraft Association	Paul Poberezny President	414-426-4800
FSF 2200 Wilson Boulevard Suite 500 Arlington, VA 22201-3306	Flight Safety Foundation, Inc.	John Enders Vice Chairman	703-522-8300 703-525-6047
GAMA Suite 801 1400 K Street, N.W. Washington, DC 20005	General Aviation Manufacturers Assoc.	Edward Stimpson President	202-393-1500 202-842-4063
HAI 1619 Duke Street Alexandria, VA 22314-3439	Helicopter Association International	Frank Jensen President	703-683-4646 703-683-4745
NAA 1815 N. Fort Myer Dr. Suite 700 Arlington, VA 22209	National Aeronautics Association, Inc.	Walter Miller Executive Director	703-527-0226 703-527-0229
NAAA 1005 E Street, S.E. Washington, DC 20003	National Agricultural Aviation Assoc.	Harold Collins, Jr. Executive Director	202-546-5722 202-546-5726
NACA Suite 710 1730 M Street, N.W. Washington, DC 20036	National Air Carrier Association, Inc.	Edward Driscoll President	202-833-8200

Acronym	Organization	Key Contact	Phone/Fax
NARA	National Aircraft Resale Association 4226 King Street Alexandria, VA 22302	Susan Sheets Executive Director	703-671-8273
NATCA	National Air Traffic Controllers Assoc. Suite 845 444 N. Capitol Street, N.W. Washington, DC 20001	Barry Krasner President	202-347-4572
NATA	National Air Transportation Association 4226 King Street Alexandria, VA 22302	Lawrence Burian President	703-845-9000 703-845-8176
NAATS	National Association of Air Traffic Specialists, Incorporated Suite C 4740 Corridor Place Beltsville, MD 20705-1165	Marsha Brown President	301-595-2012
NAFI	National Assoc. of Flight Instructors P.O. Box 793 Dublin, OH 43017	Jack Eggspuehler President	614-889-6148
NASAO	National Association of State Aviation Officials Metro Plaza One, Suite 505 8401 Colesville Road Silver Spring, MD 20910	Ed Scott Exec. Vice President	301-588-0587 301-588-1288
NBC/ FAE	National Black Coalition of Federal Aviation Employees 2233 West Rosedale Ft. Worth, TX 7611	Evelyn Washington President	817-924-5836
NBAA	National Business Aircraft Association, Inc. 1200 Eighteenth Street, N.W. Washington, DC 20036	John W. Olcott President	202-783-9000

Acronym	Organization	Key Contact	Phone/Fax
NFFE FAA Technical Center Atlanta International Airport Local 1340 Atlantic City, NJ 08405	National Federation of Federal Employees	Gerry Berry President	609-484-4321
NHC/ FAE 708 East Lakefront Circle Smithville, NJ 08201	National Hispanic Coalition of Federal Aviation Employees	René Matos President	609-748-8918
PASS Suite 840 444 North Capitol Street, N.W. Washington, DC 20001	Professional Airways Systems Specialists	Howard Johannessen President	202-347-6065
PAMA Suite 401 500 NW Plaza St. Ann, MO 63074	Professional Aviation Maintenance Association	David Wadsworth Executive Director	314-739-2580
PWC P.O. Box 44085 Oklahoma City, OK 73144	Professional Women Controllers	Rose Marino President	213-297-1379
RAA Suite 700 1101 Connecticut Avenue, N.W. Washington, DC 20036	Regional Airline Association	Walt Coleman President	202-857-1170
SAP 635 Carefree Way Friday Harbor, WA 98250	Society of Airway Pioneers	Leon Daugherty Executive Director	206-378-3881
TWO 920 Westlink Wichita, KS 67212	Technical Women's Organization	Abby Correll President	316-946-4480

Acronym	Organization	Key Contact	Phone/Fax
RTCA	Requirements and Technical Concepts for Aviation, Inc.	David Watrous President	202-833-9339

1140 Connecticut Avenue, N.W.
Suite 1020
Washington, DC 20036
(Formerly Radio Technical Communications Association)

Section XV.

Federal Aviation Administration

FEDERAL AVIATION ADMINISTRATION
Public Inquiry Center—267-3484
Employee Information—267-3229
Personnel Locator-366-4000

Area Code 202 except where noted.
800 Independence Avenue, S.W.
Washington, D.C. 20591
(Working Hours: 8:30 a.m.-5:00 p.m.)

THE ADMINISTRATOR

AOA-1	Administrator, David Hinson	
	Room 1010	267-3111
AOA-I	Secretary, (Vacant)	
	Room 1010	267-3111
AOA-I	Staff Assistant, Rochelle Claypoole,	
	Room 1010	267-3111
AOA-2	Chief of Staff, Margaret Gilligan	
	Room 1010	267-3111
AOA-2	Secretary, Shelley Buckler	
	Room 1010	267-3111
AOA-3	Executive Secretariat Director	
	Denise A. Howe	
	Room 1007	267-3512
AOA-10	Administrative Officer, Suzanne C. Holloway	
	Room 1004B	267-3852
AOA-20	Administrator's Hotline, Joe Stevens	
		267-9532

DEPUTY ADMINISTRATOR

ADA-1	Deputy Administrator, Vacant	
	Room 1010	267-8111
ADA-1	Staff Assistant, Pauline Carter	
	Room 1010	267-8111
ADA-1	Secretary, Cecilia Harley	
	Room 1010	267-8111
ADA-1A	Special Assistant to Deputy Administrator	
	Christine M. George	
	Room 1010	267-8111
ADA-20	Emergency Operations Staff, Jack Mills	
	Room 1015	267-3523
ADA-20A	FAA Records Center Facility,	
	Raymond Derby	700-652-0631
ADA-20B	NEOF 11, James Campbell	
	(Hampton, GA)	700-248-7872
ADA-20C	NEOF 111, Wayne Sallee (Longmont,CO)	
		303-651-4131
ADA-30	Operations Center, Mae L. Avery, Manager	
	Room 1014A	863-5100

| ADA-40 | Telecommunications Center, Lonnie Gordon, Manager |
| | Room 128   267-3500 |

CHIEF COUNSEL, OFFICE OF THE

AGC-1	Chief Counsel, Mark Gerchick
	Room 900E. 267-3222
AGC-1	Confidential Staff Assistant, Laura Salovitch
	Room 900E. 267-3222
AGC-2	Deputy Chief Counsel, John H. Cassady
	Room 900E 267-3773
AGC-2	Secretary, (Vacant)
	Room 900E 267-3773
AGC-3	Special Assistant to the Chief Counsel, Nancy D. Lo Bue
	Room 900E. 267- 3773
AGC-7	International Affairs & Legal Policy Staff
	Michael B.Jennison, Asst. Chief Counsel
	Room 919D 267-3515
AGC-10	Program Management Staff, Denise Castaldo, Manager
	Room 924 267-3216
AGC-10	FAR Editor, Clara Thieling
	Room 913 267-3123
AGC-10	Statistics and Dockets Section, Gwenda Claiborne
	Room 924 267-3129
AGC-10	Rules Docket, Michael D. Triplett
	Room 915G 267-3132
AGC-10	Slot Administration, Scott Donley
	Room 916H 267-3915
AGC-60	Legislative Staff, Albert B. Randall, Assistant Chief Counsel
	Room 921D 267-3217
AGC-100	General Legal Services Division, James W. Whitlow, Asst. Chief Counsel
	Room 922 267-3473
AGC-110	General Law Branch, John M.Walsh, Manager
	Room 922 267-3362
AGC-120	Personnel and Labor Law Branch, Mary N. W. Jones, Manager
	Room 922 267-3473
AGC-130	Airports/Environmental Law Branch, Richard W. Danforth, Manager
	Room 921 267-3199
AGC-200	Regulations and Enforcement Division, Donald P. Byrne, Asst. Chief Counsel
	Room 915A 267-3073
AGC-200	Deputy Assistant Chief Counsel, (Vacant)
	Room 914A 267-3491
AGC-210	Airworthiness Law Branch, Gary A. Michel, Manager
	Room 915C 267-8756
AGC-220	Operations Law Branch, Richard C. Beitel, Manager,
	Room 915C 267-8756
AGC-230	Airspace and Air Traffic Law Branch, David Bennett, Manager
	Room 9170 267-3073
AGC-240	Certification Law Branch, Michael Chase, Manager,
	Room 916D 267-3491

AGC-250   Enforcement Proceedings Branch, Peter Lynch, Manager
           Room 918B   267-3641
AGC-260   Enforcement Policy Branch, Mardi Ruth Thompson, Manager
           Room 916F   267-3137
AGC-270   Enforcement Litigation Branch, Allan Horowitz, Manager
           Room 917B   267-3137
AGC-400   Litigation Division, James S. Dillman, Asst. Chief Counsel
           Room 912B   267-3661
AGC-410   General and Administrative Litigation Branch, David M. Wiegand, Manager
           Room 913B   267-3671
AGC-420   Accident Counsel Branch, Andrew J. Dilh, Manager,
           Room 912A   267-3167
AGC-430   Adjudication Branch, Vicki S. Leemon, Manager
           Room   912-267-3661
AGC-500   Procurement Legal Division, John R. McCaw, Jr. Asst. Chief Counsel
           Room 920F   267-3480
AGC-510   Acquisition Policy and Commercial Law Branch, (Vacant) , Manager
           Room 920C   267-7368
AGC-520   Contracts and Litigation Branch,Sybil Horowitz, Manager
           Room 920B   267-3824
AGC-530   Policy and Regional Procurement Branch, Patricia McNall, Manager
           Room 920G   267-3480

CIVIL RIGHTS, OFFICE OF

ACR-1   Assistant Administrator, Leon C. Watkins
           Room 1030G   267-3254
           Secretary, Mary Mesa
           Room 10306   267-3258
ACR-2   Deputy Assistant Administrator, George F. Gordon,
           Room 1030H   267-3264
ACR-3   Internal Program, Gwen Jones, Manager
           Room 1030F   267-3259
ACR-3   Hispanic Employment Program, Alfred Mendez, Manager
           Room 1030B   267-3262
ACR-3   Federal Women's Program Manager, (Vacant)
           Room 1030E   267-3256
ACR-4   External/Public Program, David Micklin, Lead Specialist
           Room 1030N   267-3270
ACR-4   Historically Black Colleges and Universities, Dr. George Thomas
           Room 1030M   267-3267
ACR-9   Headquarter Civil Rights Officer, Naite (Tina) Stephens
           Room 1030J   267-3253
ACR-10   Management and Program, Lorraine Harris, Manager
           Room 1030G   267-3256
ACR-10   Administrative Officer, Mary A.Winston
           Room 1030G   267-7551

GOVERNMENT AND INDUSTRY AFFAIRS, OFFICE OF

AGI-1   Assistant Administrator, Vacant
           Room 1020C   267-3277

AGI-1	Secretary, Marvel A. Gardner
	Room 1020 267-3277

AGI-1   Secretary, Marvel A. Gardner
       Room 1020   267-3277

AGI-2   Deputy Assistant Administrator, Saralee C. Boteler
       Room 1020E   267-8211

AGI-2   Secretary, Helen M. Gross
       Room 1020   267-8211

AGI-3   Administrative Officer, Robert R. Cripe
       Room 1020B   267-8211

AGI-5   Manager, Government Affairs, (Vacant)
       Room 1020   267-3277

AGI-5   FAA Congressional Reports Officer, Barbara G. Tauben
       Room 1020F   267-3277

AGI-5   Congressional Liaison Officer, (Vacant)
       Room 1020D   267-3277

AGI-5   Congressional Liaison Specialist, R.Willene Minnick
       Room 1018C   267-3277

AGI-5   Congressional Liaison Specialist,Gwen B.Caudle
       Room 1020A   267-3277

AGI-5   Congressional Liaison Specialist, Patricia A. Lyons
       Room 1018D   267-3277

AGI-6   Manager, Industry Affairs, (Vacant)
       Room 1020   267-8211

AGI-6   Industry Liaison Specialist,(Vacant)
       Room 1018A   267-8211

AGI-6   Industry Liaison Specialist, Brian K. Langdon
       Room 1018B   267-8211

FAX   Room 1020   267-8210

INFORMATION TECHNOLOGY, OFFICE OF

AIT-1   Assistant Administrator for Information Technology, Theron A.Gray
       Room 1016D   267-7203

AIT-1   Secretary, Angie B. Williams
       Room 1016D   267-7203

AIT-2   Deputy Assistant Administrator, Dana L. Scott (Acting)
       Room 600W   267-3316

AIT-2   Secretary, Tonie Williams
       Room 600W   267-3316

AIT-3   Special Assistant Margaret Binns (Acting)
       Room 1016D   267-8078

AIT-100   Program and Resource Management Division, Barbara A. Edwards, (Acting) Manager
       Room 600W   267-9979

AIT-100   Administrative Officer, Margo Inskeep
       Room 640   267-8067

AIT-200   IT Policy and Plans Division, Paula Lewis, (Acting) , Manager
       Room 607   267-8063

AIT-300   IT Architecture and Data Management Division, Roger Cooley, (Acting) Manager
       Room 604   267-9980

AIT-400   IT Applications Division, Sharon Darnell, Manager (Acting)
            Room 604   267-9980
AIT-500   IT Contracts Management Division, Ellen Cook, Manager (Acting)
            Room 604   267-3112
AIT-600   IT Systems Division, Nada Harris, Manager (Acting)
            Room 604   267-3112
AIT-700   IT Services Division, Gary Titsworth, Manager (Acting)
            Room 607   267-8063
AIT-900   CORN Special Projects Office, Tim Carrico, Manager (Acting)
            Room 601   267-7806
FAX   Room 600W   267-5080

PUBLIC AFFAIRS, OFFICE OF

APA-1   Assistant Administrator, Vacant
            Room 911A   267-3883
APA-1   Secretary Dorothy Mason
            Room 911   267-3883
APA-2   Deputy Assistant Administrator, Robert Buckhorn
            Room 911C   267-3462
APA-2   Secretary, Kim Glorius
            Room 911   267-3462
APA-3   Program Management, Gerald E. Lavey, Manager
            Room 911B   2673883
APA-3   Audio-Visuals, Thomas E. Gillette
            Room 907C   267-3474
APA-4   Agency Historian, Ned Preston
            Room 907A   267-3478
APA-10   Administrative Officer, Connie Housewright
            Room 911D   267-3458
APA-100   Civil Aviation Information Distribution Division,   Phillip Woodruff, Manager
            Room 907B   267-3471
APA-120   Aviation Education, Valarie Collins
            Room 907F   267-3476
APA-200   Community & Consumer Liaison Division, Frederick H. Pelzman, Manager
            Room 906A   267-3479
APA-220   FOIA Officer, Gladys Stewart
            Room 907E   267-3490
APA-300   Public & Employee Communications Division, Paul Steucke, Manager
            Room 908A   267-8521
APA-310   Media Operations Branch, Fred Farrar, Manager
            Room 908B   267-8521
APA-340   Employee Communiution Branch, Partick Cariseo, Manager
            Room 908C   267-8521

AIRPORTS, ASSISTANT ADMINISTRATOR FOR

ARP-1   Assistant Administrator for Airports, Vacant
            Room 600E.   267-9471
ARP-1   Administrative Staff Specialist, Audrey Linehan
            Room 600E   267-9471

ARP-2	Deputy Associate Administrator, Quentin S. Taylor
	Room 600E 267-8738
ARP-2	Secretary, Bessie M. Waiters
	Room 600E 267-8738
ARP-10	Management Staff, Elizabeth A. Walker, Manager
	Room 621A 267-3040
ARP-11	Program Support Branch, Clarence O. Baron, Manager
	Room 621B 267-3040
ARP-12	Management Analysis and ADP Support Branch, Timothy D. Booth, Manager
	Room 620A 267-3040
FAX	267-5383

AIRPORT PLANNING AND PROGRAMMING, OFFICE OF

APP-1	Director, Paul L Galis
	Room 600E 267-8775
APP-1	Secretary, Jean Mattingly
	Room 600E 267-8775
APP-400	National Planning Division, James V. Mottley, Manager
	Room 615C 267-3451
APP-500	Grants-N-Aids Division, Lowell H. Johnson, Manager
	Room 615D 267-3831
APP-510	Program Guidance Branch, Ellis Ohnstad, Manager
	Room 615F 267-3831
APP-520	Programming Branch, Stan Lou, Manager
	Room 615A 267-8809
APP-530	Passenger Facility Charge Branch, Donna Taylor, Manager
	Room 615 267-3263
APP-600	Community and Environmental Needs Division, Lynne S. Pickard, Manager
	Room 615B 267-3263
FAX	Room 620 267-5683

AIRPORT SAFETY AND STANDARDS, OFFICE OF

AAS-1	Director, Leonard E. Mudd
	Room 600E 267-3053
AAS-1	Secretary, Karen Rosso
	Room 600E 267-3053
AAS-2	Deputy Director, Raymond T. Uhl
	Room 600E 267-3053
AAS-3	Technical Officer (Int'l), Harold Smetana
	Room 600E 267-8778
AAS-100	Design and Operations Criteria Division, James W. Bushee, Manager
	Room 622H 267-3446
AAS-200	Engineering and Specifications Division, Richard Worch, Manager
	Room 616C 267-3826
AAS-300	Airport Safety and Operations Division Manager (Vacant)
	Room 616D 267-3085
AAS-310	Airport Safety and Compliance Branch, Ben Castellano, Manager
	Room 616E 267-3085

AAS-330	Airport Safety Data Branch, M. Darlene Hickox, Manager
	Room 616A 267-8730
FAX	Room 620 267-5383

BUDGET AND ACCOUNTING, ASSISTANT ADMINISTRATOR FOR

ABA-1	Assistant Administrator, Nicholas S.Stoer
	Room 1003 267-3721
ABA-1	Secretary, Isabella Mickle
	Room 1003 267-3721

ACCOUNTING, OFFICE OF

AAA-1	Director, Ernest M. Keeling
	Room 539B 267-8002
AAA-1	Secretary, Patricia A.Lisowicz
	Room 539 267-8002
AAA-2	Deputy Director, George B. Fineberg
	Room 539C 267-8010
AAA-2	Secretary, Vernell C. Staton
	Room 539C 267-8010
AAA-60	Financial Programs Staff, Deborah A. Osipchak, Manager
	Room 538C 267-8040
AAA-100	Payroll and Administrative Systems Division, John W. Brown, Jr., Manager
	Room 539D 267-8938
AAA-110	Payroll Systems Branch, George H. Pierce, Manager
	Room 540A 267-9006
AAA-120	Administrative Services, Linda D. Brown
	Room 540B 267-8939
AAA-200	Accounting Operations Division, Frederick N. Whiteside, Manager
	Room 532A 267-8242
AAA-205	Internal Review Staff, Gilbert G. Donn, Manager
	Room 535E 267-8949
AAA-210	General Accounting Branch, Rufus Smith, Manager
	Room 533A 267-3018
AAA-211	Cost and Property Section, Dawna L. Scott, Supervisor
	Room 533 267-8958
AAA-212	Reports and Analysis Section, Charlotte Harrison, Supervisor
	Room 533 267-8658
AAA-220	Accounts Payable Branch, Stephen I. Newborn, Manager
	Room 535B 267-8968
AAA-221	Travel and Transportation Section, Andree C. Hall, Supervisor
	Room 535 267-8952
AAA-222	Contract and Miscellaneous Section, Joey C. Gray, Supervisor
	Room 535 267-7734
AAA-300	Travel and Relocation Systems Division, Donald E. Hansen, Manager
	Room 538A 267-8141
AAA-400	Accounting Systems Division, Patrick J. Heidenthal
	Room 537A 267-8154
AAA-500	Financial Information Division, (Vacant) Manager
	Room 537C 267-8057

BUDGET, OFFICE OF

ABU-1	Director, Ruth A. Leverenz	
	Room 1001	267-3751
ABU-1	Secretary (Vacant)	
	Room 1001	267-3751
FAX	Room 1001	267-7761
ABU-2	Deputy Director, Larry Covington (Acting)	
	Room 1001	267-3751
ABU-2	Secretary, Charlene Washington	
	Room 1001	267-3751
ABU-3	Special Projects Officer (Vacant)	
	Room 1001	267-9064
ABU-4	Administrative Officer, Mary Ann Hall	
	Room 1034-1038	267-9052
ABU-10	Budget Review Staff, Steve Unthank, Manager	
	Room 1034-1038	267-9054
ABU-20	Systems Staff, Wilbert G. Laird, Manager	
	Room 1034-1038	267-9054
ABU-30	Policy and Evaluation Staff, Paulette Lutjens, Manager	
	Room 1034-1038	267-3744
ABU-200	Operations Division, Robert R. Woodruff, Manager	
	Room 1034-1038	267-7249
ABU-210	Coordination Branch, Rodney Herron, Manager	
	Room 1034-1038	267-2749
ABU-220	Field Operations Branch, Colleen Kennedy-Roberts, Manager	
	Room 1034-1038	267-9057
ABU-230	Regulatory and Grants Branch, Christopher Reese, Manager	
	Room 1034-1038	267-9057
ABU-300	Capital Division, James S. Kemp, Manager	
	Room 1034-1038	267-3740
ABU-310	Program Analysis Branch, Donald Briggs, Manager	
	Room 1034-1038	267-3398
ABU-320	Program Control Branch, Shirley S. Miller, Manager	
	Room 1034-1038	267-3742
FAX	(ABU-200/300)	
	Room 1034	267-5382

CIVIL AVIATION SECURITY, ASSISTANT ADMINISTRATOR FOR

ACS-1	Assistant Administrator, General O. K. Steele	
	Room 300	267-9863
ACS-lA	Special Assistant, Kay Payne	
	Room 300	267-3970
ACS-1	Secretary, Sally Krtanjek	
	Room 300	267-9863
ACS-2	Deputy Assistant Administrator, Jack L. Gregory	
	Room 300 East	267-3969
ACS-2	Technical Assistant, Quinten T. Johnson	
	Room 300	267-3370
ACS-2	Secretary, Pat Miller	
	Room 300	267-3969

ACS-20	Scientific Advisor, Lyle Malotky
	Room 315 267-3967
ACS-30	International Liaison Staff, David Leach
	Room 300 East 267-7200
ACS-40	IRM Staff, Mike Dunlap
	Room 308 267-9673

CIVIL AVIATION SECURITY INTELLIGENCE, OFFICE OF

ACI-1	Director of Civil Aviation Security Intelligence, Patrick T. McDonnell
	Room 319 267-9075
ACI-2	Deputy Director, Michael Morse
	Room 319 267-9075
ACI-100	Intelligence Operations Division, Manager (Vacant)
	Room 319 267-9075
ACI-200	Strategic Intelligence Division, Stephanie Stauffer , Manager
	Room 319 267-3492
ACI-300	Sensitive Activities Division, Paula Todrin, Manager
	Room 319 267-7726
ACI-400	Special Evaluation Division, Leo Bovin, Manager
	(Dulles Airport) 703-661-0310

CIVIL AVIATION SECURITY OPERATIONS, OFFICE OF

ACO-1	Director of Civil Aviation Security Operations, Lynne Osmus
	Room 315 267-8537
ACO-1	Secretary, Doretha Boston
	Room 315 267-8537
ACO-2	Deputy Director, Robert Blunk
	Room 315 267-7262
ACO-2	Secretary, Sherry Purvis
	Room 315 267-7262
ACO-100	Security Operations, Division Manager, Susan Commisso
	Room 316 267-8231
ACO-110	Foreign Operations, Branch Manager, Charley Carrington
	Room 325 267-7724
ACO-120	Domestic Operations, Branch Manager, Rich Stevens
	Room 315 267-7362
ACO-130	Dangerous Goods & Explosives, Branch Manager (Vacant)
	Room 325 267-3951
ACO-140	Federal Air Marshall, Branch Manager, Carolyn J. Reilly
	(Dulles Airport) 703-661-0312
ACO-200	Airport/AD Carrier Liaison, Division Manager, Charlotte Bryan
	Room 316 267-7018
ACO-300	Investigations and Security Division, Dennis Hupp, Manager, TAR
	Room 326 724-0573
ACO-310	Investigations Branch, (Vacant) Manager, TAR
	Room 328 724-0576
ACO-320	Internal Security and AIS Branch, Steven Smith, Manager, TAR
	Room 334 724-0585
ACO-330	Drug Investigation Branch, (Vacant) Manager, TAR
	Room 313 724-0783

CIVIL AVIATION SECURITY POLICY AND PLANNING, OFFICE OF

ACP-1   Director, Office of Civil Aviation Security Policy and Planning,
Bruce R. Butterworth
Room 325   267-8058

ACP-1   Secretary, Yvette Brooks
Room 325   267-8058

ACP-2   Deputy Director, Max Payne
Room 325   267-8058

ACP-100   Policy and Standards Division, Larry Bruno, Manager
Room 325   267- 3413

ACP-110   Airport Policy Branch Manager, Bob Cammaroto
Room 325   267-3947

ACP-120   Internal Branch, David McFadden, Manager
Room 325   267- 8505

ACP-200   Strategic Initiatives Divison, Ken Lauterstein, Manager,
Room 325   267-7201

CIVIL AVIATION SECURITY PROGRAM AND RESOURCE MANAGEMENT, OFFICE OF

ACZ-1   Director of Civil Aviation Security Program and Resource Management
(Vacant)
Room 312   267-8152

ACZ-100   Evaluation and Program Analysis Division, Maria Tavenner, Manager
Room 312   267-9653

ACZ-200   Training and Career Division, David Smith, Manager
Room 325   267-3412

ACZ-300   Budget and Program Management Division, Betty Davis, Manager
Room 319   267-9078

HUMAN RESOURCE AND MANAGEMENT, ASSISTANT ADMINISTRATOR FOR

AHR-1   Assistant Administrator for Human Resource Management
Herbert R. McLure
Room 500E   267-3456

AHR-1   Secretary, Patricia P. Carlton
Room 500E   267-3456

AHR-2   Deputy Assistant Administrator, Dorothy H.Berry
Room 500E   267-3850

AHR-2   Secretary, Gloria J. Terrell
Room 500E   267-3850

AHR-3   Special Assistant, Carol M. Arnold
Room 500E   267-3850

AHR-10   Executive Staff, Ann Peavey Hoffer, Manager
Room 520   267-3855

AHR-20   Senior Executive Resource Staff, Debora L. Clough
Room 522C   267-8886

AHR-100   Human Resource Management Division, Pamela Foss, Manager
Room 516A   267-8012

AHR-102   Human Resource Management Division, Glenda J. Whiting, Assistant Manager
        Room 516B   267-8012

AHR-105   Organizational Development Specialist, Cynthia T. Zook
        Room 516C   267-3867

AHR-120   Training and Career Development Branch, John Ogden, Manager
        Room 516   267-7785

AHR-130   Information Resource Management Branch, (Vacant), Manager
        Room 516D   267-3886

AHR-130   Processing & Records Information
        Room 515   267-3882

               Employment Information
        Room 515   267-3229

AHR-140   Employee Relations and Career Development Branch, Gerrie Cappello
        Room 514   267-8916

AHR-150   Employment and Staffing Branch Manager, Amelia Robbins
        Room 515A   267-3780

HUMAN RESOURCE DEVELOPMENT, OFFICE OF

AHD-1   Director of Human Resource Development, Ann Rosenwald
        Room 522A   267-3911

AHD-1   Secretary, Vivian W. Grissinger
        Room 522   267-3911

AHD-3   Special Projects Officer, Richard M. Rice
        Room 522   267-3911

AHD-100   Organizational Planning and Development Division, William D. Masters, Manager
        Room 527   267-8844

AHD-200   Career Systems Division, Shelley A. Thomas, Manager
        Room 527   267-8550

AHD-300   Human Resource Management Planning and Resource Div., Karen Franco, Manager
        Room 527   267-8098

LABOR AND EMPLOYEE RELATIONS, OFFICE OF

ALR-1   Director of Labor and Employee Relations, Joseph W. Noonan
        Room 519A   267-3979

ALR-1   Secretary, Christine R. Coates
        Room 519   267-3979

ALR-2   Deputy Director Office of Labor and Employee Relations, Gary W. Baldwin
        Room 519B   267-3979

ALR-100   Union / Management Relations Division, (Vacant) Manager
        Room 519C   267-8895

ALR-200   Employee Relations Division, James E. Adler, Manager
        Room 519D   267-3990

PERSONNEL, OFFICE OF

APN-1	Director of Personnel, Kay Frances Dolan	
	Room 521	267-9041
APN-1	Secretary, Betty Sowell	
	Room 521	267-9041
APN-2	Deputy Director of Personnel, Stephen M. Soffe	
	Room 521	267-9041
APN-3	Special Assistant, John S. Walker	
	Room 521	267-9041
APN-100	Human Resource Management Automation Division, Kenneth Macomber, Manager	
	Room 523	267-3024
APN-200	Staffing Policy Division, Wanda Kay Reyna, Manager	
	Room 525	267-8832
APN-300	Position and Pay Policy Division, Cindy Medlock, Manager	
	Room 524	267-8811

TRAINING AND HIGHER EDUCATION, OFFICE OF

AHT-1	Director of Training and Higher Education, Joseph P. Kisicki, NASSIF	
	Room PL-100	366-7503
AHT-1	Secretary, Angie B. Williams, NASSIF	
	Room PL-100	366-7503
AHT-2	Deputy Director of Training and Higher Education, (Vacant), NASSIF Building	
	Room PL-100.	366-7503
AHT-3	Special Assistant, John R Garrett, NASSIF	
	Room PL-100	366-7503
AHT-10	Strategic Planning, Policy, and Budget Staff, (Vacant) Manager, NASSIF	
	Room PL-100	366-7092
AHT-20	Quality Assurance Staff, Richard E. Cullen, Manager, NASSIF	
	Room PL-100	366-7092
AHT-30	Higher Education and Advanced Technology Staff, Michael Kruger, Manager, NASSIF	
	Room PL-100	366-7073
AHT-200	Regulations and Standards Training Program Divsion, (Vacant) Manager, NASSIF	
	Room PL-100	366-7200
AHT 400	Airway Facilities Training Program Division (Vacant) Manager, NASSIF	
	Room PL-100	366-7400
AHT-500	Air Traffic Training Program Division, Vacant, Manager, NASSIF	
	Room PL-100	366-7500

POLICY, PLANNIING, AND INTERNATIONAL AVIATION, ASSISTANT ADMINISTRATOR FOR

API-1	Assistant Administrator for Policy, Planning, and Int'l Aviation, Vacant	
	Room 1005D	267-3033
API-1	Secretary, Carol A. Strong	
	Room 1005D	267-3033

API-2	Deputy Assistant Administrator for Policy Planning, and Int'l Aviation, Dale E. McDaniel
	Room 1005D 267-9105
API-2	Secretary, Edith Egypt
	Room 1005D 267-9105
API-10	Program Management Staff, William A. Tusaie, Manager
	Room 1004 267-9106
API-18	Budget and Financial Management, Juergen G. Tooren
	Room 933 267-9098
API-19	Administrative Systems and Overseas Support, Raymond K. Smith
	Room 1028F 267-9085

AVIATION POLICY, PLANS, AND MANAGEMENT ANALYSIS, OFFICE OF

APO-1	Director, John M. Rodgers
	Room 939B 267-3274
APO-1	Secretary, Mary Glotzback
	Room 939 267-3274
APO-2	Deputy Director, John F. Hennigan
	Room 939A 267-3276
APO-2	Secretary, Martha Finger
	Room 939 267-3275
APO-3	Assistant to the Director, Norm Weil
	Room 939 267-3275
APO-100	Planning Division, John Mathewson, Manager
	Room 933C 267-8444
APO-110	Statistics and Forecast Branch, Gene S. Mercer, Manager
	Room 935F 267-3355
APO-120	Planning Branch, Chuck Dennis, Manager
	Room 934BF 267-3307
APO-130	Information Systems Branch, Carlton Wine, Manager
	Room 934BF 267-3350
APO-200	Policy and Systems Analysis Division, Mike Evans, Manager, TAR
	Room 109 724-0359
APO-210	Policy and Management Studies Branch, Ray Colangelo, (Acting) Manager, TAR
	Room 111 724-0378
APO-220	Economic Analysis Branch, Stefan Hoffer, Manager
	Room 936F 267-3308
APO-230	Management Engineering Branch, Susan Helzer, (Acting), Manager, TAR
	Room 124 376-1232
APO-300	Regulation and Organizational Analysis Division, Paul Larson, Manager
	Room 937 267-3296
APO-310	Operations Regulatory Analysis Branch, Thomas Smith, Manager
	Room 938 267-3284
APO-320	Aircraft Regulatory Analysis Branch, Ward Keech, Manager
	Room 938B 267-3284
APO-330	Organizational Analysis and Programs Branch, Kenneth Harris, Manager
	Room 940A 267-3711

ENVIRONMENT AND ENERGY, OFFICE OF

AEE-1	Director, Louise Maillet	
	Room 432C	267-3576
AEE-1	Secretary, Peggy Saben	
	Room 432	267-3576
AEE-2	Deputy Director, Paul R. Dykeman	
	Room 432B	267-3576
AEE-2	Secretary, Eileen J. Scehnet	
	Room 432	267-3577
AEE-3	Chief International and Scientific Program Officer, Nicholas P. Krull (Acting)	
	Room 431B	267-3577
AEE-4	Program Support Specialist, Clara Ullmann	
	Room 432	267-3419
AEE-20	Hazardous Materials and Special Projects Staff, (Vacant)	
	Room 432	267-3495
AEE-100	Technology Division, Thomas L. Connor (Acting) Manager	
	Room 431	267-8933
AEE-110	Research and Engineering Branch, Haney H. Van Wyen	
	Room 432D	267-3558
AEE-120	Analysis and Evaluation Branch, (Vacant)	
	Room 432F	267-3570
AEE-300	Policy and Regulatory Division, William W. Albee, Manager	
	Room 432E	267-3553

INTERNATIONAL AVIATION, OFFICE OF

AIA-1	Director, Joan W. Bauerlein	
	Room 1027	267-3213
AIA-1	Secretary, Shirley Courtney	
	Room 1027	267-3213
AIA-2	Deputy Director, John R. Hancock	
	Room 1027	267-8112
AIA-2	Secretary, Valinda Lewis	
	Room 1027	267-8112
AIA-3	Special Assistant to the Director, Frank McCabe	
	Room 1027	267-8172
AIA-6	UK Mission to FAA, Victor Brennan	
	Room 1028C	267-8226
AIA-7	Australian Aviation Representive to FAA, Kevin Moore	
	Room 1028C	267-8122
AIA-8	French Aviation Representative to FAA, Edmond Boullay,	
	(French Embassy)	944-6054
AIA-10	Program Management Staff, Michael O'Neill, Manager	
	Room 1024C	267-8937
AIA-100	International Operations Organizations Division, Anthony Fazio, Manager	
	Room 1025C	267-3230
AIA-110	International Organizations Branch, Dave DeCaRoome, Manager	
	Room 1027	267-8172
AIA-120	International Operations Branch, (Vacant) Manager	
	Room 1026E	267-3210
FAX	Room 1026	267-5082

FAX	Room 1023	267-5618
AIA-200	International Assistance Division, Herbert Bachner, Manager	
	Room 1025B	267-3173
AIA-210	International Assistance Management Branch, Marci Kenney,Manager	
	Room 10216	267-3175
AIA-220	International External Development Branch, Peter M. Keefe, Manager	
	Room 1023	267-3190
FAX	Room 1025	267-5619
FAX	Room 1023	267-5306

EXECUTIVE DIRECTOR FOR SYSTEM OPERATIONS

AXO-1	Executive Director, Joseph M. Del Balzo	
	Room 1021	267-7111
AXO-1	Support Services Specialist, Carol Owensby	
	Room 1021	267-7111
AXO-1	Secretary, Ruth Wood	
	Room 1021	267-3368
AXO-2	Executive Officer, Jacqueline Rehmann	
	Room 1021	267-3368
AXO-3	Technical Assistant, Luis F. Castro	
	Room 1021	267-3368
AXO-4	Technical Assistant, Richard Weiss	
	Room 1021A	267-3368
AXO-5	Special Assistant, Wanda Munoz	
	Room 1021A	267-7715
AXO-10	Management Analyst, Madeline Taylor	
	Room 1021A	267-3499

AIR TRAFFIC, ASSOCIATE ADMINISTRATOR FOR

AAT-1	Associate Administrator for Air Traffic, William H.Pollard	
	Room 1002B	267-3666
AAT-1	Secretary, Carol Montgomery	
	Room 1002	267-3666
FAX	Room 1002	267-5456
AAT-2	Deputy Associate Administrator for Air Traffic	
	Room 1002D	267-8558
AAT-2	Secretary, Jane Kenney	
	Room 1002	267-8558
AAT-3	Special Assistant	
	Room 1002A	267-3666
AAT-4	USAF Director of Operations, Liaison Officer, (Vacant)	
	Room 1033A	267-3197
AAT-5	Assistant to AFCC Liaison Officer, COL Robert B. Nicholson	
	Room 613A	267-3031
AAT-6	USN Liaison Officer, CMDR Jesse E. Salters	
	Room 1033A	267-9431
AAT-8	USMC Liaison Officer, COL Lawrence Reed	
	Room 1033A	267-9428

AIR TRAFFIC PLANS AND REQUIREMENTS SERVICE

ATR-1	Director, Walter H. Mitchell	
	Room 400E	267-3136
ATR-1	Secretary, Patricia A. Lewis	
	Room 400E	267-3136
FAX		267-5304
ATR-3	Special Assistant, Mary Ellen Kraus	
	Room 400E	267-3136
ATR-4	Administrative Officer, (Vacant)	
	Room 400E	267-7190
ATR-100	System Plans and Programs Division, William Behan, Manager	
	Room 421A	267-9162
FAX		267-5992
ATR-101	Assistant Manager, Willis C. Nelson	
	Room 421D	267-3769
ATR-110	En Route Branch, Geoffrey L. Shearer, Manager	
	Room 423A	267-8693
ATR-120	Terminal Enroute Branch Larry Gray Manager	
	Room 422B	267-9172
ATR-130	Flight Services and Weather Branch, , Manager	
	Room 422A	267-8517
ATR-140	Privatization Branch, (Vacant) Manager	
	Room 421	267-9090
ATR-200	Automation Software Division, Rodman D. Bourne, Manager	
	Room 635	267-9433
	FAX	267-5566
ATR-201	Assistant Manager, Gordon W. Heritag	
	Room 635	267-9433
ATR-210	Automation Planning Branch, Jim Hevelone, Manager	
	Room 635	267-9434
ATR-220	FSS Automation Branch, William H. Brodie, Manager	
	Room 635	267-9379
ATR-230	Automation Policy Branch, Jerry McCaleb, Manager	
	Room 635	267-9435
ATR-300	Advanced Systems and Facilities Division, Dennis Koehler, (Acting) Manager	
	Room 423B	267-7195
FAX		267-9199
ATR-301	Assistant Division Manager, Dennis Koehler	
	Room 423B	267-7195
ATR-310	ACF/MCF Implementation and Air Traffic Plans Branch, JohnStaples, Manager	
	Room 427B	267-9090
ATR-320	Air Traffic Advanced Automation Sys. Requirements Branch,	
	John H. Timmerman, Manager	
	Room 423C	267-8760
ATR-330	NASP and Future Systems Branch Joseph Pitts, Manager,	
	Room 412B	267-9182
FAX	(ATR-400)	609-484-4221
AOS-1	Director, Operational Support Service, Charles L. Stith	
AOS-100	Manager, National System Requirements Analysis Division, Jim Hevelone	
AOS-200	Manager, National Engineering Field Support Division, Charles Cage, (FAAAC)	

AOS-300   Manager, National Automation Engineering Field Support Division, Ed Schuman
AOS-400   National Automation Field Support Division,Richard B. Shinpaugh, Manager ,
            FAA Technical Center
                        609-484-5665
AOS-402   Plans and Programs Staff, James R. Clinton, Manager, FAA Technical Center
                        609-484-5693
AOS-410   Terminal Field Support Branch, Robert L. Welch, Manager,
            FAA Technical Center
                        609-484-6401
AOS-420   EnRoute Field Support Branch, (Vacant) Manager, FAA Technical Center
                        609-484-6550
AOS-430   Flight Service Field Support Branch, Norman Hopkins, Manager,
            FAA Technical Center
                        609-484-5145

AIR TRAFFIC PROGRAM MANAGEMENT, OFFICE OF

ATZ-1   Director of Air Traffic Program Management, Neil R. Planzer
            Room 1033E   267-3022
ATZ-1   Secretary, Lee Norvell
            Room 1033B   267-3022
FAX   Room 1033   267-5211
ATZ-2   Deputy Director, Nancy Kalinowski
            Room 1033D   267-3022
ATZ-10   Executive Staff, Lynda F. Deaton, Manager
            Room 1005C   267-9220
ATZ 100   Training Requirements Program, Ned Reese, Manager, NASSIF
            Room 5103   366-1300
FAX   DOT 5103   366-7620
ATZ-110   Advanced Systems, Sherelle Carper, Manager, NASSIF
            Room 5103   366-6840
ATZ-120   Field Development, (Vacant) Manager, NASSIF
            Room 5103   366-1297
ATZ-200   Employee Systems Program, Sharon B. Flemming, Manager
            Room 428D   267-9214
FAX   Room 429   267-5592
ATZ-210   Support Systems, Peter N. Kovalick, Manager
            Room 428B   267-8941
ATZ-220   Career Systems, Don Kimball, Manager
            Room 428C   267-9214
ATZ 300   Resource Management Program, Mary Barnett, Manager
            Room 932   267-7161
            FAX Room 932   267-5455
ATZ-310   Fiscal Resource, Everett Brown, Manager
            Room 932   267-9146
ATZ-320   Staffing Systems, Mike Scott, Manager
            Room 932   267-7161

AIR TRAFFIC RULES AND PROCEDURES SERVICE

ATP-1   Director, L. Lane Speck
            Room 400E   267-9205

ATP-1	Secretary, Peggy Glorius
	Room 400E 267-9205
FAX	267-5304
ATP-100	Procedures Division, Paul H. Strybing, Manager
	Room 417A 267-3725
FAX	267-5120
ATP-101	Assistant Division Manager, Ted Davies
	Room 417B 267-3725
ATP-110	FSS Procedures Branch, Kenneth L. Johnson, Manager
	Room 420 267-8802
ATP-120	Terminal Procedures Branch, Lee McGlamery, Manager
	Room 420D 267-8460
ATP-130	En Route Procedures Branch, Charles F. Criswell, Manager
	Room 417C 267-8630
ATP-140	International Procedures Branch, William F. Price
	Room 417E 267-9317
ATP-200	Airspace Rules and Aero Information Division, Harold W. Becker, Manager
	Room 414B 267-3731
FAX	267-5809
ATP-201	Assistant Division Manager, (Vacant)
	Room 414C 267-3731
ATP-210	Publication Branch, Fred Gibbs, Manager
	Room 411A 267-8790
ATP-220	Cartographic Standards, Dick Powell, Manager
	Room 411 267-8790
ATP-230	Air Traffic Rules Branch, William C. Davis, Manager
	Room 415 267-8783
ATP-240	Airspace and Obstruction Evaluation Branch
	(Vacant) Manager
	Room 415B 267-3075

AIR TRAFFIC SYSTEM EFFECTIVENESS, OFFICE OF

ATH-1	Director, John D. Canoles
	Room 428A 267-8781
ATH-1	Secretary, Karen C. Palmersheim
	Room 428 267-8781
	FAX 267-5754
ATH-3	Special Assistant, Thomas Lintner
	Room 427A 267-7019
ATH-100	Evaluations Division, Timothy E. Halpin, Manager
	Room 424C 267-9342
	FAX Room 708 267-8129
ATH-110	Evaluation Standards and Coordination Branch, Ron Cooper, Manager
	Room 412B 267-3260
ATH-120	National Air Traffic Sys. Effectiveness/Evaluations Branch, (IAD) Gary Romano
	703-661-0132
ATH-130	National Air Traffic Sys. Effectiveness/Evaluations
	Branch, (DFW) Richard P Burgess, Manager
	817-574-5588

ATH-140	National Air Traffic Sys. Effectiveness/Evaluations Branch, Bill Ellis, Manager
	206-764-3412
ATH-150	National Air Traffic Sys. Effectiveness/Evaluations Branch, (ATL)
	Walter R. Coker, Manager
	404-994-5670
ATH-200	Quality Assurance Division, James Buckles, Manager
	Room 424B 267-9126
ATH-210	Investigation Branch, Samuel F. Woods, Manager
	Room 416E 267-7901
ATH-220	Analysis System Branch, Lawrence Silvious, Manager
	Room 416F 267-7120
ATH-230	Tracking/Coordination Branch, Debra Timmerman, Manager
	Room 416C 267-7902

AIR TRAFFIC SYSTEM MANAGEMENT, OFFICE OF

ATM-1	Director, David J. Hurley
	Room 400E 267-9155
ATM-1	Secretary, Mary Ann Bourgeois
	Room 400E 267-9155
FAX	267-5304
ATM-2	Deputy Director, Richard Cox
	Room 425A 267-9358
ATM-3	Special Assistant, Frank Hatfield
	Room 400E 267-9155
ATM-5	Budget Analyst, Jane M. Miller
	Room 420D 267-9455
ATM-6	Personnel Specialist, Sandra McCluney
	Room 420D 267-7046
ATM-100	Civil Operations, William Marx
	Room 426E 267-8343
	FAX 267-5305
ATM-200	Air Traffic Control System Command Center, Charles H. Hall, Manager
	Room 626 267-8670
	FAX 267-5307
ATM-210	Assistant Program Manager for Administration, Ronald LaMarche (Acting)
	Room 626 267-9387
ATM-211	Assistant Manager for Plans, Programs, and Procedures (Vacant)
	Room 626 267-7957
ATM-212	Assistant Manager for Quality Assurance and Training, Carl E. Fullner
	Room 626 267-7958
ATM-220	Assistant Program Manager for Operations, Donald G. Eddy
	Room 626 267-7960
ATM-300	NAS Analysis, Patricia A. Beam, Manager
	Room 424A 267-9401
FAX	267-5305
ATM-310	ATOMS/Operations, Leonard Williams, Manager
	Room 424A 267-9801
ATM-320	NAS Analysis and Development Branch
	Dr. Alan L. Breitler, Manager
	Room 424A 267-9801

ATM-400	Military Operations, William O. Byberg, Manager
	Room 426C 267-9362
FAX	267-5305
ATM-410	Operations and Procedures Branch, Dan Walczak (Acting), Manager
	Room 426 267-9322
ATM-410	Tactical Air Command (Air Combat Command) Liaison,(Virginia)
	Ralph Priestley, AFB 804-441-3880
ATM-410	AWACS Liaison (Oklahoma), Charles Wood AFB
	405-231-5115
ATM-410	Air Defense Liaison Officer (Washington), Edward Enkerud, AFB
	206-383-7913
ATM-4 10	FAA Air Defense Liaison Officer (Colorado), Authur A. Hanlon, AFB
	303-231-5688
ATM-410	Air Training Command, HQ Liaison (Texas), Robert Butcher, (ATC)
	512- 229-4162
ATM-410	UNCINCLANT/CINCLANTFLT HQ Liaison (Virginia), Thomas Zurhorst, AFB
	310-363-6256
ATM-410	Air Defense Liaison Officer (California), Roger B. Osgood, AFB
	714-276-6429
ATM-410	FAA Air Liaison Officer (Florida), Wiley J. Helms AFB
	703-603-5491
ATM-410	FAA Air Defense Liaison (New York), Kurt P. Frenzel, AFB
	303-391-6724
ATM-410	CBI East (Florida), Jerry Ball
	703-822-0636
ATM-410	CBI West (California), Herbert Hackett
	714-796-8022
ATM-420	Airspace and CARF Branch, Lorraine Nealis, Manager
	Room 612A 267-9361
ATM-500	Airspace Management Systems, Timothy G. Fleming, Manager
	Room 426B 267-7247
FAX	267-5305
ATM-510	Future Systems, Dave Ross, Manager
	Room 634 267-7461
ATM-520	Current Systems (Vacant) Manager
	Room 626 267-7010
ATM-600	National Flight Data Center, William Wallis
	Room 634 267-9311
FAX	267-5322
ATM-610	Assistant Manager for Operations, Alberta H. Santora
	Room 634 267-9311
ATM-611	U.S. NOTAM Office, Lyle Miller,(Acting) Manager
	Room 630 267-3390
ATM-612	Airports, Facilities and Communications, Brenda Hawkins, Manager
	Room 631 267-9277
ATM-613	Airspace and Flight Procedures, Carol Santelia, Manager
	Room 638 267-9288
ATM-620	Assistant Manager for Plans and Programs, Fitzhugh Stephens
	Room 628 267-9268
ATM-700	Environmental Issues, Charles R. Reavis,Manager
	Room 426 267-9367

AIRWAY FACILITIES, ASSOCIATE ADMINISTRATOR FOR

AAF-1   Associate Administrator, Arnold Aquilano
         Room 700E.　　267-8181

AAF-1   Secretary ,Carol A. Brooks
         Room 700E　　267-8181

AAF-2   Deputy Associate Administrator, Edward M. Kelly
         Room 700E　　267-3366

AAF-2   Secretary, Ponjola V. Short
         Room 700E　　267-3366

AAF-3   Special Assistant, Calvin Fischer
         Room 700E　　267-7304

AAF-10   Manager, Resource Staff, Michael Upton
         Room 720　　267-8118

AAF-11   Planning Branch, Valdis I. Krumins, Manager
         Room 719　　267-9757

AAF-12   F&E Branch, William Kamsier
         Room 720　　267-3083

AAF-13   Operations Branch, John Ward, Manager
         Room 720　　267-8312

AAF-20   Manager Evaluation Staff, William Dixon
         Room 725　　267-3054

AAF-30   Manager Management Staff, John Reilly
         Room 721　　267-7133

AAF-40   Manager Information Resources Management Staff, William Ford
         Room 731　　267-8970

AAF 50   Manager Building Management Staff, Joanne Eakin
         Room 712　　267-8855

NAS TRANSITION AND IMPLEMENTATION SERVICE

ANS-1   Director, Steven I. Rothschild
         Room 700W　　267-9835

ANS IA   Secretary, (Vacant)
         Room 700W　　267-9835

ANS-2   Deputy Director, James R. Stagner
         Room 700W　　267-7973

ANS-2A   Secretary, (Vacant)
         Room 700W　　267-7973

ANS-3   Special Assistant,Elizabeth Rand
         Room 700W　　267-7442

ANS-10   Program and Resource Management Staff, (Vacant), Manager
         Room 701W　　267-8617

ANS-100   F & E Resource and Planning Division, Thomas Proeschel, Manager
         Room 704W　　267-7022

ANS-110   NAS Planning and Support Branch, Don Lombard, Manager
         Room 704W　　267-7371

ANS-120   Human Resource Planning Branch Axel R Granholm , Manager
         Room 704W　　267-7312

ANS-130   Resource Management Branch, Dennis Holton, Manager
         Room 704W　　267-7369

ANS-200	Facility Programs and Transition Division, John Carlson, Manager
	Room 704W 267-8679
ANS 220	Air Route Traffic Control Branch, Donald Swinney, Manager
	Room 704W 267-8570
ANS-230	GNAS Facilities Program Branch, Phillip Shelstad
	Room 704W 267-7807
ANS 240	ATCT/TRACON Program Branch, Wilmer Hunter, Manager
	Room 704W 267-7834
ANS-300	Special Programs Integration Division, Dean Christy, Manager
	Room 704W 267-7331
ANS-400	NAILS Program Division Mary Karen Cronin, Manager
	Room 707W 267-7795
ANS-410	NAILS Policy and Planning Branch, Paul Yantchook, Manager
	Room 707W 267-7926
ANS-420	NAILS Implementation Branch, Bill Bernhardt, Manager
	Room 707W 267-7796

SYSTEMS MAINTENANCE SERVICE

ASM-1	Director, Stanley Rivers
	Room 700W 267-3034
ASM-1	Secretary, Sharon Jeffries
	Room 700W 267-3034
FAX	Room 700W 267-5935
ASM-2	Deputy Director, Joaquin Archilla
	Room 700W 267-7409
ASM-2	Secretary, Patricia Worthington
	Room 700W 267-7409
ASM-4	Technical Assistant, Betty Jones
	Room 700W 267-3056
ASM-10	Resource Management Staff, Marcia V. Corey, Manager
	Room 739 267-3077
FAX	Room 739 267-5141
ASM-100	Maintenance Engineering Division, George Terrell, Manager
	Room 735 267-3551
FAX	Room 735 267-5061
ASM-101	Assistant Division Manager, David Yaeger, Manager
	Room 735 267-3551
ASM-110	Performance Analysis Branch, Thomas Gassert, Manager
	Room 735 267-3051
ASM-111	National Maintenance Coordination Branch
	Frank DeMarco, Manager
	Room 735 267-7359
ASM-111	National Maintenance Coordination Ctr.(NMCC)
	Room 626 267-6622
ASM-120	Technical Standards Branch, (Vacant) , Manager
	Room 736 267-8496
ASM-140	Facilities Engineering Branch, Wendie F. Chapman, Manager
	Room 735 267-8274
ASM-200	Maintenance Operations Division, Carey L. Weigel, Manager
	Room 734 267-3597

FAX   Room 734   267-5632

ASM-201   Assistant Division Manager, (Vacant)
Room 734   267-3597

ASM-240   Policy and Requirements Branch, Alice Wong, Manager
Room 734   267-8353

ASM-250   Technical Training and Certification Branch, Harry Grindstaff
Room 734   267-3628

ASM-260   Workforce Standards and Analysis Branch, Jose E. Justiniano
Room 734   267-8284

ASM-270   Maintenance Automation Branch, James C. Nocker, Manager
Room 734   267-8280

ASM-300   Telecommunications Management and Operations Division,
Michael Gariazo, Manager
Room 731   267-8225

FAX   Room 731   267-5543

ASM-301   Assistant Division Manager, Michael Gariazzo
Room 731   267-8225

ASM-301A   Technical Assistant, Douglas Kay
Room 731   267-8225

ASM-310   Program Support and International Telecommunications Branch (Vacant)
Room 731   267-7509

ASM-320   Telecommunications Network Planning & Engineering Branch, (Vacant)
Manager, (Nassif Bldg.)
Room 5410   366-6125

ASM-330   Telecommunications Acquisition and Implementation Branch
Room 731   267-3055

ASM-340   Telecommunications Network Management Branch, David Lantzy, Manager
Room 731   267-8610

ASM-400   National Automation Engineering Field Support Division,
Edward Schuman, Manager, FAA Technical Center
609-484-6611

FAX   FAA TECH   609-484-5126

ASM-401   Assistant Division Manager, Brian Riehle, FAA Technical Center
609-484-6611

ASM-410   Systems Engineering Branch, Luan Jones, Manager, FAA Technical Center
609-484-5737

ASM-420   ATC Surveillance Engineering and Data Analysis Branch, Paul Tucci, Manager
FAA Technical Center
609-484-5856

ASM-430   Enroute Systems Engineering Branch, Kenneth Fore, Manager
FAA Technical Center
609-484-5735

ASM-440   Terminal Systems Engineering Branch, Warren Mills Manager
FAA Technical Center
609-484-6274

ASM-450   Data Systems Communications Engineering Branch, Frank Happel,Manager
FAA Technical Center
609-484-5249

ASM-500   Spectrum Engineering Division, Manager, Gerald J. Markey
Room 714A   267-9710

FAX   Room 714   267-5901
ASM-501   Assistant Division Manager, George Sakai
                 Room 714   267- 9710
ASM-510   Frequency Management Branch, Edward Keefe, Manager
                 Room 714   267- 8534
ASM-520   Frequency Engineering Branch Manager, Victor Foose
                 Room 714   267 -9712
ASM-600   National Engineering Field Support Division, Charles Gage, Manager
                 Aeronautical Center   405-680-3647
FAX   AAC   405-680-4674
ASM-602   Program Support Staff, James Pritchard, Manager, Aeronautical Center
                                        405-680-3481
ASM-603   Technical Support Staff, Joe L. Downs, Manager
                                        405-680-3647
ASM-610   Systems Engineering Branch, (Vacant) , Manager, Aeronautical Center
                                        405-608-3724
ASM-620   Environmental Engineering M Branch, David Fleming, Manager,
                 Aeronautical Center   405-680-4635
ASM-630   Radar Engineering Branch, David E. Campbell
                 Aeronautical Center   405-608-4271
ASM-640   NAV/COM Engineering Branch, Jim E. Evanoff, Manager,
                 Aeronautical Center   405-608-3611
ASM-700   NAS Support Division, Richard Clevenger, Manager
                 Room 712   267-3153
FAX   267-5753
ASM-701   Assistant Division Manager, Walter Anderson, Manager
                 Room 433C   267-8269
ASM-720   Materiel Management Branch, Dorothy Combs, Manager
                 Room 708   267-8837
ASM-730   NAS Contract Support Branch, Janice Williamson, Manager
                 Room 708   267-3036
ASM-740   Real Property Branch, Murry Camp, Manager
                 Room 708   267-8246

AVIATION STANDARDS, ASSOCIATE ADMINISTRATOR FOR

AVS-1   Associate Administrator for Aviation Standards, Garland P. Castleberry
                 Room 1040A-C   267-3133
AVS-1   Secretary, Linda Blythe
                 Room 1040A-C   267-3133
FAX   Room 1040A-C   267-5621
AVS-2   Deputy Associate Administrator for Aviation Standards, Brooks C. Goldman
                 Room 1040A-C   267-7555
AVS-2   Secretary, (Vacant)
                 Room 1040A-C   267-7555
AVS-3   Special Assistant to the Associate Administrator, Judy G. Nauman
                 Room 1040A-C   267-3133
AVS-10   Programs and Plans Staff, Robert J. Dame
                 Room 312   267-9650
AVS-30   Financial Management Staff, Joseph Flaim
                 Room 312   267-9659

AVS-40   Information Resource Management Staff, Thomas Brown
           Room 308   267-9659
FAX   (AVS-10/30/40)
           Room 312   267-5191
AVS-60   Washington Flight Program Staff, Paul H. Wiater
           Hanger 6   703-557-0380
FAX   HANGER 6   703-557-7024

ACCIDENT INVESTIGATION, OFFICE OF

AAl-1   Director, William R. Hendricks
           Room 332A   267-9612
AAl-1   Secretary, Mary Alice Bartock
           Room 332   267-9612
AAl-2   Deputy Director,David F. Thomas
           Room 332B   267-9612
AAl-3   Program Analysis Officer, Steve R. Fewell
           Room 332   267-9612
AAl-3A   Administrative Officer, Glenda Kottka
           Room 329   267-7124
AAl-100   Accident Investigation Division, John D. Rawson
           Room 331A   267-8190
AAl-101   Assistant Division Manager, Harold Donner
           Room 329   267-8190
AAl-110   Accident Coordination Unit, Lyle Streeter
           Room 329   267-8190
AAl-120   Accident Duty Room, Sam Mangano
           Room 329   267-3120
AAl-200   Recommendation and Quality Assurance Division, Frank S.Del Gandio
           Room 329   267-9622
AAl-210   Recommendation Branch, Judith Leach, Manager
           Room 329   267-9629
AAl-220   Quality Assurance Branch, Brian Poole
           Room 329   267-9628

AVIATION MEDICINE, OFFICE OF

AAM-1   Federal Air Surgeon, Jon L Jordan, MD
           Room 322   267-3535
AAM-1   Secretary, Louise C. Dille
           Room 322   267-3535
AAM-2   Deputy Federal Air Surgeon, (Vacant)
           Room 322   267-3537
AAM-2   Secretary, Deborah A. Ford
           Room 322   267-3537
AAM-3   Director, Civil Aeromedical Institute, William E. Collins, PhD,
           Aeronautical Center   405-680-4806*
AAM-6   Program Management Staff, Virginia Hicks, Aeronautical Center
                                  405-680-4811*
AAM-100   Plans, Evaluation and Management Support Division, Curtis B. Maloy, Manager
           Room 2404   366-1045

AAM-110	Plans and Evaluation Branch, R. Mark Adams
	Room 2404 366-1048

AAM-110   Plans and Evaluation Branch, R. Mark Adams
          Room 2404   366-1048
AAM-120   Management Support Branch, Martin (Doc) Miller
          Room 2404   366-0938
AAM-200   Medical Specialties Division William H Hark, MD
          Room 2336,   366-6350
AAM-201   Chief Psychiatrist, Barton Pakull, MD, Manager
          Room 327.   267-3382
AAM-210   Aeromedical Standards Branch, Dennis McEachen, Manager
          Room 2336   366-6354
AAM-220   Drug Abatement Branch, William R. McAndrew, Manager
          Room 2336   366-6710
AAM-230   Employee Health Branch, Donald M. Watkin, MD, Manager
          Room 327   267-3767
AAM-240   Biomedical and Behavioral Sciences Branch, William T.Shepherd, PhD, Manager
          Room 2404   267-6910
AAM-300   Aeromedical Certification Division, Audie W. Davis, MD, Manager
          Aeronautical Center   405-680-482*
AAM-400   Aeromedical Education Division, (Vacant) Manager, Aeronautical Center
                              405-680-4803*
AAM-500   Human Resources Research Division, David A.Schroeder, PhD, Manager
                              405-680-4846*
AAM-600   Aeromedical Research Division, Jerry R. Hordinsky, MD, Manager,
          Aeronautical Center   405-680-4808*
AAM-700   Occupational Health Division,William E. Davis, Manager,
          Aeronautical Center   405-680-3711*

(*These organizations are located at the Mike Monroney Aeronautical Center, CAMIBuilding Oklahoma City, OK)

AVIATION SYSTEM STANDARDS, OFFICE OF

P.O. Box 25082, Oklahoma City, OK 73125
Commercial Telephone (Dial 405-680 and extension)

AVN-1   Director,William H.Williams,Jr.,
          ARB-214   405-680-3305
AVN-1   Office Administrator, Irma Barrigan
          ARB-214   405-680-3305
FAX   OKLAHOMA CITY 405-255-3436
AVN-2   Deputy Director, Louis D. Ludwig
          ARB-214   405-680-3306
AVN-1AF   Air Force Flight Inspection Center, COL Dean Alexander, Manager
          ARB-207D   405-680-4402
AVN-4   Special Assistant, Phyllis A. Howard
          ARB-214   405-680-3156
AVN-7   Fleet Modernization, James G. Duea
          HG8W-215   405-680-3995
AVN-9   FAA Senior Flight Safety Officer, Fred P. Laird
          ARB-207E   405-680-4272
AVN-10   Management Programs Staff, Carolyn E. Hohmann, Manager
          ARB-207A   405-680-4366

AVN-11	Program Analysis Branch, Cheryl I, Jackson, Manager	
	ARB-207B	405-680-3497
AVN-12	Management and Evaluation Branch, (Vacant), Manager	
	ARB-207C	405-680-3736
AVN-13	SystemsTechnology Branch, Ted McCarley, Manager	
	ARB-111 D	405-680-4464
AVN-20	Standardization and Evaluation Staff	
	Laurence P. Musser, Manager	
	HG9E-207	405-680-3733
AVN-30	Financial Management Staff, Linda J. Biship, Manager	
	ARB-115E	405-680-3641
AVN-31	Operations Branch, Jaurita Voyles, Manager	
	ARB-115B	405-680-3641
AVN-32	Program Office Branch, Michael J. Lovvorn, Manager	
	ARB-I I 5B .	405-680-3641
AVN-40	Registry Modernization Staff, Earl F. Mahoney, Manager	
	ARB-313L	405-680-7357
AVN-100	Regulatory Support Division, Dustin L. Sloan, Manager	
	ARB-212C2	405-680-4431
AVN-110	Engineering and Manufacturing Branch, James H. Vaughan, Manager	
	ARB-207	405-680-4374
AVN-120	Operational System Branch, (Vacant), Manager,	
	ARB-212C	405-680-4391
AVN-130	Operations Support Branch, Wallace L. Emory, Manager	
	ARB-212E	405-680-4151
AVN-140	Maintenance Support Branch (Vacant)	
	ARB-212	405-680-6495
AVN-200	Flight Procedures and Inspection Division, Jim C. Savage, Manager	
	ARB-201	405-680-3766
AVN-201	Assistant Manager, (Vacant)	
	ARB-201	405-680-3766
AVN-220	Flight Procedures Branch, Lyle G.Wink, Manager	
	ARB B-6	405-680-3382
AVN-230	Flight Inspection Operations Branch, Steve Siebo, Manager	
	ARB-201	405-680-3505
AVN-240	Data Analysis Branch, Charles Hale, Manager	
	ARB B-6	405-680-3411
AVN-270	National Flight Procedures Development Branch, James Y. Kakazu, Manager	
	HG8W-308	405-680-3027
AVN-300	Aircraft Maintenance and Engineering Division, Wayne C.Jacox, Manager	
	HQ9E-347	405-680-4126
AVN-301	Assistant Manager, Robert L. Mullican	
	HQ9E-347	405-680-4126
AVN-310	Fleet Support Branch, John B. Nix, Manager	
	HQ9E-118	405-680-3101
AVN-320	Quality Assurance Branch, George W. Dove, Manager	
	HQ8W-206	405-680-3321
AVN-330	Aircraft and Avionics Maintenance Branch, Jimmy D. King, Manager	
	HG9E-343	405-680-4728

AVN-340	Engineering Branch, (Vacant), Manager	
	HG9E-343.	405-680-4728
AVN-350	ACY Aircraft Services Branch, J. Robert Cone, Manager	
	209A Tech CTR, Atlantic City	
		609-484-4558
AVN-400	Airmen and Aircraft Registry Division, Geoffrey W. McLellen, Manager	
	ARB-313H	405-680-4331
AVN-450	Aircraft Registration Branch, James E. Henderson	
	ARB-308	405-680-3131
AVN-460	Airmen Certification Branch, Mary N. Hill, Manager	
	ARB-301	405-680-3205
AVN-500	Aircraft Program Division, Donald D. Snow	
	ARB-210	405-680-4208
AVN-501	Assistant Manager, William E. Traylor	
	ARB-201	405-680-4208
AVN-510	Fleet Management Branch, Janice L. Sledge	
	ARB-201	405-680-4614
AVN-520	Financial Management Branch, (Vacant), Manager	
	ARB-201	405-680-3641
AVN-530	Plans and Projects Branch, Marvin A. Conway, Manager	
	ARB-201	405-680-4488
AVN-540	Standards Development Branch, Donald P. Pate, Manager	
	ARB-B3.	405-680-4164
AVN-550	Flight Inspection Policy and Standards Branch, Marc E. Lewand	
	ARB-115	405-680-4526
	FAA Technical Center, Atlantic City Airport, Atlantic City, NJ 08405	

REGULATION AND CERTIFICATION, ASSOCIATE ADMINISTRATOR FOR

AVR-1	Associate Administrator for Regulation and Certification, Anthony J. Broderick	
	Room 1000W	267-3131
AVR-1	Secretary, Norma Jean Vanagas	
	Room 1000W	267-3131
FAX		267-9769
AVR-1A	Special Assistant, Denise Miller	
	Room 1000W	267-7250
AVR-2	Deputy Associate Administrator, Daniel C. Beaudette	
	Room 1000W	267-7804
AVR-2	Secretary, Debrah Walker	
	Room 1000W	267-7804
AVR-3	Special Assistant, Colonel Peter A.Kacerguis, USA	
	Room 1000W	267-9707
AVR-10	Manager, Executive Staff, Gerald G. Froelke	
	Room 312E	267-7099
AVR-10	Secretary, Suzanne Ponchock	
	Room 312	267-7099

AIRCRAFT CERTIFICATION SERVICE

AIR-1   Director, M. Craig Beard
         Room 300 W   267-8235

AIR-1   Secretary, Margie C. Ross
         Room 300W   267 -8235

AIR-2   Deputy Director, Thomas E. McSweeny
         Room 300W   267-7270

AIR-2   Secretary, Katherine A. Ball
         Room 300W   267-7270

AIR-3   Assistant Director, William J. Sullivan
         Room 300W   267-9554

AIR-4   International Airworthiness Programs Officer
         Adolofo O. Astorga
         Room 335   267-9559

AIR-100   Aircraft Engineering Division, John K. McGrath, Manager
         Room 335   267-9580

AIR-101   Assistant Division Manager, Daniel P. Salvano
         Room 335   267-3785

AIR-107   Continued Airworthiness Staff, George H. Marania, Manager
         Room 335   267-9569

AIR-110   Policy and Procedures Branch, David W. Ostrowski, Manager
         Room 335   267-9588

AIR-120   Technical Analysis Branch, Henri P. Branting (Acting)
         Room 335   267-9578

AIR-121   Airframe & Powerplant Section, Henri Branting
         Room 335   267-9578

AIR-122   Systems & Flight Section, Richard Kirsch
         Room 335   267-3343

AIR-200   Aircraft Manufacturing Division, Ronald T. Wojnar, Manager
         Room 333   267-8361

AIR-201   Assistant Division Manager, Dana D.Lakeman
         Room 333   267-8361

AIR-220   Production Certification Branch, Renton S. Bean
         Room 333   267-7142

AIR-230   Airworthiness Certification Branch, Frank P. Paskiewicz
         Room 329   267-7990

AIR-300   System Surveillance and Analysis Division, Michael Gallagher, Dulles Airport
         Gateway Building   703-661-0111

AIR-500   Planning and Program Management Division, Elizabeth Yoest
         Room 340   267-9372

AIR-510   Program Planning and Analysis Branch, Philip A. Canal
         Room 340   267-8623

AIR-520   Automated Systems Branch, Eugene N. Newman
         Room 324   267-7024

AIR-530   Administrative Management Branch, Alice M. Payne, Manager
         Room 324   267-7025

FLIGHT STANDARDS SERVICE

AFS-1   Director, Thomas C. Accardi
         Room 302C   267-8237

AFS-1	Secretary, Susan C. Fenn	
	Room 302	267-8237
FAX		267-5230
AFS-2	Deputy Director, William J. White	
	Room 302D	267-8237
AFS-2	Secretary, Gwendolyn A. Hargrove	
	Room 302	267-3651
AFS-4	Military Liaison, LTC Bob Ullom, Dulles Airport, Gateway Building, Suite 131	
		703-661-0357
AFS-5	International Liaison Staff, James B. Kenney, Manager	
	Room 302	267-3716
AFS-10	Executive Officer, Barbara A. Aleshire	
	Room 301A	267-3651
AFS-11	Evaluation and Analysis Branch, Richard Lea	
	Room 303D	267-3433
AFS-12	Planning and Program Management Branch, Josheph K.Tintera	
	Room 302E	267-3787
AFS-13	Administrative Management Branch (Vacant)	
	Room 301A	267-8441
AFS-14	Project SAFE Staff, Lionel R. Driscoll	
	Room 325	267-3708
AFS-200	Air Transportation Division, David R. Harrington, Manager	
	Room 303	267-8166
AFS-201	Assistant Manager, David S. Potter	
	Room 303	267-8166
AFS-210	Air Carrier Training Branch (Vacant)	
	Room 303D	267-3718
AFS-214	Program/Policy Section, Thomas Toula	
	Room 305B	267-3718
AFS-215	Training Systems Development, Application, and Policy Section (Vacant)	
		703-661-0275
AFS-216	Data Management and Analysis Section, Thomas M. Longridge	
		703-661-0279
AFS-220	Air Carrier Branch, David L. Catey	
	Room 304A	267-8094
AFS-240	Regulations Branch, Gary E. Davis	
	Room 304B	267-8096
AFS-250	Commuter and Air Taxi Branch, David W. Kress, Manager	
	Room 304C	267-8086
AFS-260	Program Management Branch, Thomas M. Penland	
	Room 306A	267-3460
FAX		267-5229
AFS-300	Aircraft Maintenance Division, Frederick J. Leonelli, Manager	
	Room 340	267-3546
AFS-301	Assistant Manager, Dennis H. Piotrowski	
	Room 340D	267-3546
AFS-320	Repair Station Branch, Pericles C. Hantis	
	Room 340C	267-8283
AFS-330	Air Carrier Branch, John R. Tutora	
	Room 339A	267-3440

AFS-340	General Aviation and Commercial Branch, Larry G. Kephart
	Room 338 267-8203
AFS-350	Avionics Branch (Vacant)
	Room 337 267-8177
AFS-360	Regulations Branch, James E. Siegman
	Room 339 267-3794
FAX	267-5115
AFS-400	Technical Programs Division, Louis C. Casimano (Acting)
	Room 334 267-8452
AFS-401	Assistant Division Manager, H. Guice Tinsley (Acting)
	Room 334 267-8452
AFS-410	All Weather Operations Branch, James H. Enias
	Room 321 267-7211
AFS-420	Flight Procedures Standards Branch, Paul J. Best
	Room 305C 267-8277
AFS-430	Special Programs Branch, Howard Hess (Acting)
	Room 334 267-7301
AFS-440	Flight Technical Programs (Vacant)
	Room 334 267-8452
AFS-450	Technical Analysis and Support Branch, William F. Petruzel
	Room 321 267-8452
FAX	267-5086
AFS-500	2 Field Programs Division, William C. Withycombe, Manager
	703-661-0333
AFS-501	2 Assistant Division Manager, Quentin J. Smith, Jr.
	703-661-0333
AFS-510	2 Certificate Management Branch, Marion B. Dittman
	703-661-0351
AFS-520	3 Inspector Resource Management Branch, Richard C. Berg
	703-661-0200
AFS-530	Automation Branch, George I. Petros
	703-661-0219
AFS-540	Current Operations Branch, Debra J. Entricken
	703-661-0345
AFS-550	Technical Standards Branch, Marlene G. Livack (Acting)
	703-661-0211
AFS-800	General Aviation and Commercial Division (Vacant), Manager
	Room 307A 267-8212
AFS-801	Assistant Manager, Robert A. Wright
	Room 307B 267-8212
AFS-810	Accident Prevention Program Branch, Roger M. Baker, Jr., Manager, DOT
	Room 2322 366-6320
AFS-820	Operations Branch, Robert M. Barton, Manager
	Room 307 267-8194
AFS-840	Certification Branch, James I. Riddle, Manager
	Room 306 267-8196
AFS-850	Regulations Branch, Ronald W. Myres, Manager
	Room 308A 267-8150
FAX	267-5094

RULEMAKING, OFFICE OF

ARM-1	Director, Chris A. Christie	
	Room 314E	267-9677
ARM-1	Secretary, Charlene T. Brown	
	Room 314E	267-9677
ARM-2	Deputy Director, Joseph A. Hawkins	
	Room 314E	267-9680
ARM-2	Secretary (Vacant)	
	Room 314E	267-9680
ARM-20	Program Analysis Officer Manuel F Vega	
	Room 314E	267-5575
ARM-100	Airmen and Airspace Rules Division, Ida M. Klepper, Manager	
	Room 314E	267-7625
ARM-200	Aircraft and Airport Rules Division, Brenda D. Courtney	
	Room 314E	267-9680

SYSTEM CAPACITY AND REQUIREMENTS, OFFICE OF

ASC-1	Director, Edward T. Harris	
	Room 723	267-7370
ASC-1	Secretary, Penelope Ronnie	
	Room 723	267-7370
FAX		267-5767
ASC-2	Deputy Director, James R. Smith	
	Room 723	267-8789
ASC-2	Secretary, Meredith Spencer	
	Room 723	267-8789
ASC-10	Program Management Staff, Leonard B. Bell, Manager	
	Room 723	267-3310
ASC-100	Airport Capacity Planning and Development Division, James McMahon, (Aciting) Manager	
	Room 723	267-7425
FAX		267-5767
ASC-200	Airspace Capacity Planning and Development Division, Richard L. Danz, Manager	
	Room 723	267-8733
FAX	Room 620	267-5767
ASC-300	Aviation Weather Division, Richard J. Heuwinkel, Manager	
	Room 723	267-7443

EXECUTIVE DIRECTOR FOR SYSTEM DEVELOPMENT

AXD-1	Executive DIrector, John A. Burt	
	Room 1019	267-7222
AXD-1	Secretary, Judith Bain	
	Room 1019	267-7222
AXD-1	Secretary/Correspondence Control, Patricia R. Schauer	
	Room 1019 DOT	267-7223
AXD-3	Special Assistant, Jane P. Caldwell	
	Room 1019D	267-9866
AXD-4	Chief Scientific and Technical Advisor for Human Factors, H. Clayton Foushee	
	Room 329	267-7125

AXD-4	Deputy Chief Scientific and Technical Advisor for Human Factors, Phyllis Kayton
	Room 329 267-7125
AXD-5	Special Assistant, Dennis L. Kline
	Room 1019C 267-3912

NAS DEVELOPMENT, ASSOCIATE ADMINISTRATOR FOR

AND-1	Associate Administrator, John E. Turner
	Room 800W 267-3555
AND-1	Secretary, Evelyn Chaney
	Room 800W 267-3555
AND-2	Secretary ,(Vacant)
	Room 800W 267-3200
AND-3	Special Assistant, Herman Tharrington
	Room 800W 267-8218
AND-4	Special Assistant, Ken Ward
	Room 801A 267-9080
AND-7	Military Assistant, Bob Jones, Lt. Col.
	Room 801A 267-8711
AND-8	Special Assistant, Robert Bernard
	Room 801A 267-8710
AND-10	Financial Management Staff, Julie Gatling, Manager
	Room 838 267-9062
AND-20	Resource Management Staff, Brenda L. Brooks, Manager
	Room 838 267-9714
AND-30	MLS Program Office, Donald Stadtler, Manager
	Room 507 267-5857

PROGRAM MANAGER FOR ADVANCED AUTOMATION

AAP-1	Program Manager, Michael E. Perie, MSQ
	Room 625 376-6500
AAP-1	Secretary, Rose E. Mara, MSQ
	Room 625 376-6500
AAP-2	Deputy Program Manager, (Vacant), MSQ
	Room 625 376-6506
AAP-2	Secretary, (Vacant), MSQ
	Room 625 376-6506
AAP-3	Special Assistant, Paul M. Rich, MSQ
	Room 625 376-6539
AAP-4	Technical Assistant, William H. Koch, MSQ
	Room 625 376-6540
AAP-10	Program Management Staff, Glenn W. Halbert, Manager, MSQ
	Room 625 376-6509
AAP-200	Advanced Automation System Division, (Vacant) Manager, Philip J. Gill, MSQ
	Room 625 376-6512
AAP-201	Assistant Division, Philip J. Gill, Manager, MSQ
	Room 625 376-6512
AAP-210	Systems Development Branch, G. Michael Bateman, Manager, MSQ
	Room 625 376-6516
AAP-220	Hardware Development Branch, Gary R. Rowland, Manager, MSQ
	Room 625 376-6516

AAP-230   Software Development Branch, Brian C. Andrews, Manager, MSQ
            Room 625   376-6519
AAP-240   Implementation Branch, Richard J. Marek, Manager, MSQ
            Room 625   376-6587
AAP-250   COTR and Segment Management Branch, (Vacant) Manager, MSQ
            Room 625   376-6578
AAP-400   Voice Switching and Control System Division, H. Lee Tucker, MSQ, Manager
            Room 625   376-6528
AAP-401   Assistant Division, Richard J. Cassidy, Manager, MSQ
            Room 625   376-6528
AAP-410   VSCS Development Branch, James R. Monnie, Manager, MSQ
            Room 625   376-6531
AAP-420   VSCS Implementation Branch, Joseph Morelli, Manager, MSQ
            Room 625   376-6534
AAP-430   VSCS COTR Branch (Vacant), Manager, MSQ
            Room 625   376-6282

PROGRAM DIRECTOR FOR AUTOMATION

ANA-1   Director, Ken Byram
            Room 801A   267-7113
ANA-1   Secretary, Patti Mueller
            Room 801A   267-7113
ANA-100   Automation Engineering Division, William L. Umbaugh, Manager
            Room 512B   267-3290
ANA-120   Maintenance Automation Associate Program Manager for Engineering,
            Allen F. Beard, WTC
            Room 431   646-2093
ANA-130   En Route/TMS Associate Program Manager for Engineering, Delois K. Smith
            Room 512C   267-8347
ANA-140   Terminal Automation Associate Program Manager for Engineering, Robert J. Reed
            Room 511B   267-8364
ANA-150   Oceanic Auto Associate Program Manager for Engineering, Juanita D. Callahan
            Room 510A   267-8363
FAX   Room 512   267-8365
ANA-200   Program Manager for Maintenance Automation, Rondel Lipps, Manager, WTC
            Room 429   646-2021
ANA-300   Program Manager for En Route Automation/TMS, Harry B. Kane, Manager
            Room 505   267-8336
ANA-400   Program Manager for Terminal Automation, ARTS II, Robert Voss
            Room 511B   267-8349
ANA-500   Program Manager for Terminal Automation, ARTS III, Mitchell Narins, Manager
            Room 511A   267-8339

PROGRAM DIRECTOR FOR COMMUNICATIONS

ANC-1   Director, Arthur R. Feinburg, WTC
            Room 368   646-4963
ANC-1   Secretary, Chiquita Forrest, WTC
            Room 366   646-4963
ANC-100   Communication Engineering Division, John Bisaga, Manager, WTC
            Room 368   646-4962

ANC-120	Voice Switching and Recording Program Branch, Stephen Dash, Manager, WTC
	Room 365 646-4961
ANC-130	Air/Ground Communication and Control Branch, Nathaniel Johnson, Manager, WTC
	Room 378 646-4957
ANC-140	Interfacility Communications Program Branch, Kenneth Clark, Manager, WTC
	Room 376 646-5416
ANC-200	Program Manager for Voice Switching and Recording Program, Joann Kansier, Manager, WTC
	Room 370 646-6960
ANC-300	Program Manager for Air/Ground Communication and Control, Lockett Yee, Manager, WTC
	Room 374 646-4960
ANC-400	Program Manager for Interfacility Communications, Joan Gariazzo, Manager, WTC
	Room 372 646-2390
ANC-500	Program Manager for Aircraft Acquisition James Van Namee, Manager, WTC
	Room 364 646-2171

PROGRAM DIRECTOR FOR NAVIGATION AND LANDING

ANN-1	Director, Rodman Gill, CGHQ
	Room B432 267-6595
ANN-1	Secretary, Kathryn Roberts, CGHQ
	Room B432 267-6531
ANN-100	Navigation and Landing Engineering Division, Reuben Powell, Manager, CGHQ
	Room B432 267-6511
FAX	CGHQ
	Room B432 267-4587
ANN-120	APME, ILS Program, Kenneth L. Harris, CGHQ
	Room B423 267-6563
ANN-130	APME Navigation Program ,John M. Williams, CGHQ
	Room B620 267-6552
ANN-140	APME Visual Aids Program, William Chin, Manager, CGHQ
	Room B424 267-6676
ANN-150	APME/MLS/LORAN C/GPS Program, Rial Sloan, Manager, CGHQ
	Room B448E 267-6660
FAX	CGHQ
	Room B448 267-6512
ANN-200	Program Manager for Landing, Gary Skillicorn, Manager, CGHQ
	Room B442C 267-6675
ANN-300	Program Manager for Navigation, Charles Ochoa, Manager, CGHQ
	Room B442B 267-6601
FAX	(ANN-200/300) CGHQ
	Room B442 267-5229

PROGRAM DIRECTOR FOR SURVEILLANCE

ANR-1	Director Terry Hannah, UNS
	Room 812 606-4531

ANR-1	Secretary, Patricia Finley, UNS
	Room 812 606-4531

ANR-1   Secretary, Patricia Finley, UNS
                  Room 812   606-4531

ANR-100   Surveillance Engineering Division, Carmine Primeggia, Manager, UNS
                  Room 812   606-4532

ANR-110   Radar Engineering Branch, Donald E. Johnson, Manager, UNS
                  Room 818   606-4574

ANR-120   Terminal Radar Branch, John S. Horrocks, Manager, UNS
                  Room 810   606-4613

ANR-130   Mode S Branch Secondary Radar, Bryan Johnson, Manager, UNS
                  Room 803   606-4644

ANR-140   ARSR-4 Branch, Dennis Kolb, Manager, UNS
                  Room 826   606-4648

ANR-150   Weather Radar Branch, Ray Weimer, Manager, UNS
                  Room 821   606-4683

ANR-200   Program Manager for Terminal Radar, Gerold Taylor, Manager, UNS
                  Room 810   606-4622

ANR-300   Program Manager for Secondary Radar, Pike Reynolds, Manager
                  Room 836   267-8258

ANR-400   Program Manager for En Route Radar, Richard Lay, Manager, UNS
                  Room 826   606-4661

ANR-500   Program Manager for Weather Radar, Donald Turnbull, Manager, UNS
                  Room 822   606-4693

PROGRAM DIRECTOR FOR WEATHER AND FLIGHT SERVICE SYSTEMS

ANW-1   Director, Alvin L. Thomas
                  Room 502   267-7198

ANW-1   Secretary, Rose Schmokel
                  Room 502   267-7198

ANW-100   Weather and Flight Service Station Engineering Division, Henry Wade, Manager
                  Room 505   267-8691

ANW-120   Flight Service Station Branch, Rudolph Watkins
                  Room 712   267-8656

ANW-130   Weather Processors Branch, Kevin Young
                  Room 505   267-8547

ANW-140   Weather Sensors Branch, George McConnell, Manager
                  Room 713   267-8671

ANW-200   Program Manager for Flight Service Stations, (Vacant), Manager
                  Room 712   267-7799

ANW-300   Program Manager for Weather Processors, (Vacant), Manager
                  Room 506   267-5857

ANW-400   Program Manager for Weather Sensors, Steven Hodges, Manager
                  Room 712   267-7849

AQUISITION SUPPORT, OFFICE OF

ASU-1   Director, Carolyn Blum
                  Room 400W   267-8513

ASU-1   Secretary, Elizabeth A. Bugay
                  Room 400W   267-8513

ASU-2   Deputy Director, Dennis DeGaetano
                  Room 400W   267-8515

ASU-2	Secretary, Peggy A. Carter
	Room 400W 267-8515
ASU-4	Special Assistant, Clayton Deaton
	Room 400W 267-8864
ASU-10	Manager of Administrative Systems, Ann-Marie S. Sadler
	Room 400W 267-8448
ASU-11	Administrative Management Branch, Philip Fitzhugh.
	Room 2330, DOT 366-6981
ASU-12	Financial Management Branch, Maggie Baker
	Room 401B 267-7202
ASU-20	Technical Support Staff, Thomas L. Marker
	Room 808 267-8889
ASU-100	Management, Plans and Evaluation Division, Richard Heironimus, Manager
	Room 2330, DOT 366-6951
ASU-110	Evaluation Support Branch, (Vacant)
	Room 2330, DOT 366-6950
ASU-120	Plans Branch, Timothy O'Hara, Manager,
	Room 2330, DOT 366-6950
ASU-130	Procurement Management Branch, Dorothy Sirk, Manager
	Room 2330, DOT 366-6950
ASU-300	Contracts Division, Gilbert B. Devey, Jr., Manager
	Room 402A 267-3580
FAX	202-267-5149
ASU-301	Contracts Division, Deborah W. Wilson
	Assistant Manager
	Room 402B 267-3582
ASU-305	Pricing Staff, Harold O'Connell, Manager
	Room 122, TAR 376-3716
ASU-310	NAVAIDS/MLS Branch, Abe Tenenbaum, Manager
	Room 408 267-3655
ASU-320	Surveillance Branch, William T. Hohe, Manager
	Room 814, UNS 606-4220
ASU-330	Communications/Aircraft and Weather Branch, Brian Isham, Manager
	Room 408 267-3631
ASU-340	Systems Operations Engineering Branch, Algie Guy, Manager
	Room 408 267-3648
ASU-350	Automation/Advanced Automation Branch, David Bailey, Manager
	Room 406 267-3660
ASU-360	Resource Management Branch, Lauraline Gregory, Manager
	Room 408 267-3598
ASU-400	Industrial Division, Alphonso Barr, Manager
	Room 708 267-8268
ASU-401	Assistant Division, Terrence Greenwood, Manager
	Room 711 267-8268
ASU-410	Industrial Evaluation Branch, Edward Huntzinger, Manager
	Room 708 267-8270
ASU-420	Quality Assurance Branch, Kenneth Laenger, Manager
	Room 708 267-7037
ASU-430	Quality Standards Branch, Nathan Lichter, Manager
	Room 708 267-8270

EXECUTIVE DIRECTOR FOR AQUISITION AND SAFETY OVERSIGHT

AXQ-1 Executive Director, Monte R. Belger
     Room 1040   267-7720
AXQ-1 Secretary, Jan Barrett
     Room 1040   267-7720
AXQ-3 Special Assistant for TQM, H. Theodore Criswell
     Room 929   267-7925
AXQ-4 Special Assistant for MBE, Inez Williams
     Room 929   267-8881
AXQ-5 Special Assistant (Admin), Bonnie Richards-Ryan
     Room 929   267-8445
AXQ-6 Special Assistant for Aquisition Improvement, Alice Harball
     Room 929   267-3459

AVIATION SAFETY, ASSOCIATE ADMINISTRATOR FOR

ASF-1 Associate Administrator, Darlene Freeman
     Room 1000E   267-9613
ASF-1 Secretary Norma E. Hussey
     Room 1000E   267-9613
ASF-2 Deputy Associate Administrator, Charles Huettner
     Room 1000E   267-3611
ASF-2 Secretary, Dee McHugh
     Room 1000E   267-3611
ASF-10 Executive Officer, Steve Springmann
     Room 1000E   267-7227

INTEGRATED SAFETY ANALYSIS, OFFICE OF

ASV-1 Director Charles Fluet (Acting)
     Room 2221A, DOT 366-6003
ASV-1 Secretary, Christy Holcombe
     Room 2221A, DOT 366-6003
ASV-1 Administrative Officer, Joyce Gantt
     Room 2221A, DOT 366-6173
ASV-100 National Aviation Safety Data Division
     Robert Toenniessen, Manager
     Room 221A, DOT 366-6003
ASV-200 Safety Indicators Division, Carolyn Edwards, Manager
     Room 2221A, DOT 366-6004

SAFETY INFORMATION AND PROMOTION, OFFICE OF

AOV-1 Director (Vacant)
     Room 2227, DOT 366-6300
AOV-1 Secretary, Marcia Berritt
     Room 2227, DOT 366-6300
AOV-1 Special Assistant, Dale Ruoff
     Room 2207, DOT 366-6429
AOV-100 National Airspace System Division, Richard Bair, Manager
     Room 2227, DOT 366-6300

AOV-200   Flight Operations and Maintenance Division, Peter McHugh, Manager
            Room 2227, DOT   366-6006
AOV-300   Aircraft and Technology Division, James P. McVicker
            Room 2227, DOT   366-6006

ACQUISITION POLICY AND OVERSIGHT, OFFICE OF

ACQ-1   Director, David J. Morrissey
            Room 929   267-3320
ACQ-1   Secretary, Brenda Sullivan
            Room 929   267-3320

INDEPENDENT OPERATIONAL TEST AND EVALUATION OVERSIGHT, OFFICE OF

ATQ-1   Director, A. Martin, Phillips
            Room 929   267-8926
ATQ-1   Secretary, Barbara Jones
            Tech Center   609-484-5031

FAA FIELD OFFICES

•ALASKA
Anchorage   Anchorage Area Corrdinator
Frank Babiak, Jr.   8-907-243-2348
1:00 p.m.-9:30 p.m.
4500 West 50th Street
Coordinator
Anchorage, AK 99502

Fairbanks   Fairbanks Area Coordinator
Paul Gallagher   8-907-456-4600
1:00 p.m.-9:30 p.m.
5640 Airport Way   Coordinator
Fairbanks, AK 99701

Juneau   Juneau Area Coordinator
Harold Guthrie   8-907-789-2193
11:00 a.m.-8:00 p.m.
P.O. Box 2157
Coordinator
Juneau, AK 99803

King Salmon   King Salmon Area Coordinator
Carl E. Fundeen   8-907-246-3312
1:00 p.m.-9:30 p.m.
P.O. Box 67
Coordinator
King Salmon, AK 99613

•FLORIDA
Miami   FAA Representative
Thomas P. Martin   8-407-861-2551
8:00 a.m.-4:30p.m.
Box 59-2815
Representative
Miami, FL 33159

Palm Coast   FAA Center for Management
J. William Newman   8-202-205-6381
7:30 a.m.-5:30 p.m.
Development Manager
4500 Palm Coast Parkway East
Palm Coast, FL 32037

•NEW JERSEY
Atlantic City   FAA Technical Center
Edward T. Harris   8-609-484-6641
8:00 a.m.-4:30 p.m.
Atlantic City Airport
Director
Atlantic CIty, NJ 08405

•OKLAHOMA
Oklahoma City
Mike Monroney   Aeronautical
Homer C. McClure   8-405-680-4521
9:00 a.m.-5:30 p.m.
Center
Director
P.O. Box 25082
Oklahoma City, OK 73125

•WEST VIRGINIA
Martinsburg   FAA Headquarters Record Center
Jerome Cohen   8-304-267-2981
8:00 a.m.-4:30 p.m.
West King Street & Maple Avenue
Facility Manager
Martinsburg, WV 25401

FAA OVERSEAS OFFICES

* When APO is used, address to be 3 lines only; DO NOT use name of country
** Mail addresses to Department of State should be marked as to classification.

•BELGIUM, BRUSSELS
3:00 a.m.-12:00 noon
Europe, Africa and Middle East Office
Patrick N. Poe, Director
U.S. Postal System Air Mail Address: AA c/o American Embassy, APO New York
09667-1011 Location: 15 Rue De La Loi 1040 Brussels, Belgium
International Air Mail Address: FAA c/o American Embassy, 27 Boulevard
DuRegent, 1000 Brussels, Belgium
Tel AUTOVON: 793-2110, COMM: County Code: 32, City Code: 2 513.38.30;
Night answering: COMM or AUTOVON: ext. 2236

•BRAZIL, RIO DE JANEIRO
FAA Representative, Raymond Ybarra
IFO/RIO, c/o U.S. Consulate General, APO Miami 34030
TEL: 292-7117, ext. 298

•CANADA, MONTREAL
9:00 a.m.-5:30 p.m.
U.S. Member, Robert E. Cook
ICAO Air Navigation Commission, Room 753, 1000 Sherbrooke Street, West
Montreal, Quebec, Canada
Tel: 285-8304, FTS: 953-3211

•ENGLAND, LONDON
3:00 a.m.-12:00 noon
FAA Representative, Emily Trapnell
U.S. Postal System Air Mail Address: FAA c/o American Embassy, Box 40, FPO,
New York 09510 Location and International Air Mail Address: FAA c/o American
Embassy, Grosvenor Square, London W1AE
Tel: Country Code: 44, City Code: 1 499-9000, ext. 2497

•FRANCE, PARIS
4:00 a.m.-1:00 p.m.
FAA Representative, Robert T. Francis
U.S. Postal System Air Mail Address: FAA, c/o American Embassy, APO New York
09777 Location and International Air Mail Address: FAA c/o American Embassy, 2
Avenue Gabriel, 75382 Paris Cedex 08, France
Tel: Country Code: 33, City Code: 42.96.1202 or 261-8075, ext. 2901 or 2937

•GERMANY, BERLIN
2:00 a.m.-11:00 a.m.
U.S. Administrator for Aeronautics, Robert F. Smith
U.S. Postal System Air Mail Address: U.S. Administrator for Aeronautics, EU-9,
U.S. Mission, Berlin, Air P.O. Box 5500, APO New York 09742
Tel: Country Code: 49, City Code: 30, Berlin, Civil 819-5351 or 690-9666 Berlin,
Military 5351

•GERMANY, FRANKFURT
2:30 a.m.-11:00 a.m.
Flight Inspection Field Office, Manager, Gary Wirt
U.S. Postal System Air Mail Address: c/o American Consulate General, APO New
York 09757 FAA, c/o American Consulate General, Siesmayer Strasse 21,
Frankfurt/Main, Germany
Tel: Frankfurt Civilian 69.705.211, Autovon: 330-1110, ext. 6069 International
Field Office-51,

Manager, James Coffey
U.S. Postal System Air Mail Address: c/o American Consulate General, APO New
York 09213. FAA, c/o American Consulate General Siesmayer Strasse 21,
Frankfurt/Main, Germany
Tel: Country Code: 49, City Code: 69, Frankfurt Civilian 69.705-110, Military
330-1110, ext 6089

•GUAM
6:00 p.m.-2:30 a.m.
Guam Office, WP, Roy Pickett
Resident Director/Sector Manager, Western Pacific Region, Route 008, Finegayan,
Guam 96912
Tel: 355-5026

•ITALY, ROME
12:00-8:30 a.m.
FAA, Representative, (Vacant)
U.S.Postal System Air Mail Address: FAA c/o American Embassy, APO New York
09794-0007 Location and International Air Mail Address: FAA, American Embassy
Annex, Via V. Veneto 119, Rome, Italy
Tel: Country Code: 39, City Code: 6.4674.2241 or 4674.2298

•JAPAN, TOKYO
7:00p.m.-3:30 a.m.
FAA Representative, WP, Dennis Warth
East Asian IFO, c/o American Embassy, APO San Francisco 96503
Tel: Tokyo 583-7141, ext 7625
8:00 p.m.-4:30 a.m.
Flight Inspection Field Office, Manager, Charles P. Jewell
Building 1376, APO San Francisco 96328
Tel: Tokyo 225-5682 or 57302

•JORDAN, AMMAN
(Saturday thru Thursday)
12 Midnight-6:30 a.m.
FAA Representative, Donald V. Jones
c/o American Embassy, FAA; APO New York 09892
Tel: Embassy 962-6-443 716

•OMAN, MUSCAT
(Saturday thru Thursday)
10:00 p.m.-5:00 a.m.
Chief, Richard M. Gavin
Civil Aviation Assistance Group, c/o American Embassy Muscat, Department of
State Mail Room, Washington, D.C. 20520
Tel: (Embassy) Country Code 968-519390

•SAMOA
2:30 p.m.-11:00 p.m.
Samoa Office, WP, Dean Dowell, Resident Director
Western Pacific Region, P.O. Box 8, Pago Pago, American Samoa 96799
Tel: Samoa 639-9485

•SAUDI ARABIA, JEDDAH

Senior Advisor, Chester Chang
Civil Aviation Assistance Group, FAA, c/o American Consulate General, APO, New York 09697
Tel: Country Code: 966, City Code: 2, 685-5257

•SENEGAL, DAKAR

FAA Representative, Richard L.Hurst
U.S. Postal System Air Mail Address: Federal Aviation Administration, Dakar, Department of State, Washington, D.C. 20520
Tel: Country Code: 221-22.67.53

•SPAIN, MADRID

(Monday thru Friday)
3:30 a.m.-11:30 a.m.
Chief, Rudolph J. Escobedo
Civil Aviation Assistance Group, Box 38, American Embassy, APO, New York 09285
Tel: City Code (341)413-4139

FAA REGIONAL OFFICES

•ALASKAN REGION

Anchorage   Alaskan Region
(Vacant)   907-271-5645
1:00 p.m.-9:30 p.m.
701 C Street
Director
P.O. Box 14
Anchorage, AK 99513

•WESTERN PACIFIC REGION

Los Angeles   Western-Pacific Region
(Vacant)   310-297-1427
10:30 a.m.-7:00
P.O. Box 92007
Director
WorldWay Postal Center
Los Angeles, CA 90009

•SOUTHERN REGION

Atlanta   Southern Region
Vacant   404-763-7222
8:00 a.m.-4:30 p.m.
P.O. Box 20636
Director
Atlanta, GA 30320

•GREAT LAKES REGION
Chicago Great Lakes Region
Timothy P. Forte 312-694-7294
8:30 a.m.-5:00 p.m.
2300 East Devon Street
Director
Des Plaines, IL 60018

•NEW ENGLAND REGION
Boston New England Region
Arlene B. Feldman 617-273-7244
8:00 a.m.-4:30 p.m.
12 New England Executive Park
Director
Box 510
Burlington, MA 01803

•CENTRAL REGION
Kansas City Central Region
Paul K. Bohr 816-374-5626
8:30 a.m.-5:00 p.m.
601 E. 12th Street
Director
Kansas City, MO 64106

•EASTERN REGION
New York City Eastern Region
Daniel J Peterson 718-553-1005
8:00 a.m.-4:30 p.m.
Federal Building
Director
JFK International Airport
Jamaica, NY 11430

•SOUTHWEST REGION
Fort Worth Southwest Region
Donald P. Watson 817-624-5001
9:00 a.m.-5:30 p.m.
P.O. Box 1689
Director
Fort Worth, TX 76101

•NORTHWEST MOUNTAIN REGION

Seattle Northwest Mountain Region
Frederick M. Isaac 206-431-2001
10:30 a.m.-7:00 p.m.
1601 Lind Avenue, S.W.
Director
Renton, WA 98055

OFFICE AND SERVICE REPRESENTATIVES

Directives Management Distribution Officers Representatives	Forms Officers	Records Officers	Reports Control Officers	Incentive Awards Coordinators	
AAA-100	AAA-100	AAA-100	AAA-100	AAA-100	AAA-100
AAF-30	AAF-30	AAF-30	AAF-30	AAF-30	AAF-30
AAP-10	AAP-10	AAP-10	AAP-10	AAP-10	AAP-10
AAM-120	AAM-120	AAM-120	AAM-120	AAM-120	AAM-120
ATZ-10	ATZ-10	ATZ-10	ATZ-10	ATZ-10	ATZ-10
ARP-11	ARP-12	ARP-11	ARP-12	ARP-11	ARP-11
ABU-20	ABU-20	ABU-20	ABU-20	ABU-20	ABU-20
ACR-1	ACR-1	ACR-1	ACR-1	ACR-1	ACR-1
ACS-3	ACS-3	ACS-3	ACS-3	ACS-3	ACS-3
AFS-13	AFS-13	AFS-13	AFS-13	AFS-13	AFS-13
AGC-10	AGC-10	AGC-10	AGC-10	AGC-10	AGC-10
AHR-10	AHR-10	AHR-10	AHR-10	AHR-10	AHR-10
AIR-3	AIR-3	AIR-3	AIR-3	AIR-3	AIR-3
ALR-1	ALR-1	ALR-1	ALR-1	ALR-1	ALR-1
AND-20	AND-20	AND-20	AND-20	AND-20	AND-20
APA-10	APA-10	APA-10	APA-10	APA-10	APA-10
API-19	API-19	API-19	API-19	API-19	API-19
ASC-10	ASC-10	ASC-10	ASC-10	ASC-10	ASC-10
ASF-10	ASF-10	ASF-10	ASF-10	ASF-10	ASF-10
ASM-10	ASM-10	ASM-10	ASM-10	ASM-10	ASM-10